Using Surveys to Value Public Goods: The Contingent Valuation Method

Robert Cameron Mitchell ▪ Richard T. Carson

Resources for the Future
Washington, D.C.

Second Printing 1990

Printed in the United States of America

Published by Resources for the Future
1616 P Street, N.W., Washington, D.C. 20036
Books from Resources for the Future are distributed worldwide by The Johns Hopkins University Press

Library of Congress Cataloging-in-Publication Data

Mitchell, Robert Cameron.
Using surveys to value public goods: the contingent valuation method

Bibliography: p.
Includes index.
1. Public goods—Valuation. 2. Public goods—
Cost effectiveness. I. Carson, Richard T.
II. Title.
HB846.5.M58 1989 363 87-28633
ISBN 0-915 707-32-2

The book was edited by Samuel Allen and designed by Peggy Friedlander.

♾ The paper in this book meets the guidelines for permanence and durability of the Committee on Production Guidelines for Book Longevity of the Council on Library Resources.

To Clifford S. Russell

Contents

List of Tables

List of Figures

Foreword

Public policy decision making often involves balancing the costs of a policy with the benefits. When a policy affects goods and services traded in normal markets, costs and benefits result from consumer responses to changes in prices faced and incomes received. A large body of empirical evidence exists that links price and income change to consumer behavior. This evidence may be employed in a reasonably straightforward fashion to calculate a policy's costs and benefits. On the other hand, when a policy affects the availability or character of public goods—goods such as national parks, wilderness areas, drinking water, and numerous other environmental and natural resources—one does not observe price and income changes, and thus must infer the changes in consumer behavior by using more roundabout methods.

Using Surveys to Value Public Goods: The Contingent Valuation Method provides decision makers, policy analysts, and social scientists with a detailed discussion of a new technique for the valuation of goods not traded in private markets. Termed contingent valuation, the technique draws upon economic theory and the methods of survey research to elicit directly from consumers the values they place upon public goods. This technique for public goods valuation complements earlier advances made at Resources for the Future in the late fifties and early sixties. These earlier approaches inferred the value individuals place on public goods from observable behavior indirectly related to the good. For example, the seminal work of Marion Clawson in 1959 estimated the value of outdoor recreation experiences from the observed distances individuals traveled to particular recreational sites.

The advantage of contingent valuation over the previous indirect approaches concerns the nature of the values that may be addressed. The indirect methods, exemplified by the Clawson travel-distance approach, are best applied to estimate values concerning the use of a public good, while the contingent valuation method may be employed to estimate values not intimately linked to use—for example, the desire of individuals to pass pristine natural environments on to future generations.

The contingent valuation method is still in its infancy and research regarding appropriate application procedures and tests of reliability are under way. *Using Surveys to Value Public Goods* provides the reader with a summary of this current research and a critical evaluation of the method as a tool for public goods valuation. Through the authors' experiences in developing and applying the method, the reader gains an understanding of the method's strengths and limitations and the proper interpretation of contingent valuation results.

Raymond J. Kopp
Senior Fellow and Director
Quality of the Environment Division
Resources for the Future

October 1988

Preface

The original impetus for this book was a report we wrote in 1981 presenting the findings of our first attempt to use the contingent valuation (CV) method to measure the amounts people would be willing to pay (WTP amounts) for a public good. That project, which was funded by the U.S. Environmental Protection Agency as part of its effort to develop benefit measurement methodologies, was an exploratory study of national freshwater pollution control benefits. Although our report was intended to be primarily an analysis of water quality benefits, we found it necessary to devote almost half of it to basic methodological issues. That experience convinced us that we should undertake a study of the contingent valuation method's theory and practice in a more systematic and thorough fashion. This book is the result.

We have written this book primarily for two audiences. The first is comprised of economists and policy analysts who have an interest in measuring the benefits of nonmarketed goods, but no particular experience with the contingent valuation method. The book is intended to provide them with an introduction to the CV method and the theoretical and methodological rationale for its validity as a technique for measuring benefits. The second audience consists of those who conduct (or would conduct) CV studies and those who use (or would use) their findings. Many of these practitioners are resource economists with relatively little background in survey research; some are sociologists or other types of social scientists who may be unfamiliar with economic theory or benefit-cost analysis. This volume is intended to provide these people with a basic background on the method and an analysis of the methodological issues which must be addressed before the results of a CV study can be regarded as valid.

The issues we consider here have broader significance as well. The results of the various methodological experiments conducted by CV researchers and our efforts to develop a model of respondent error are relevant to the current efforts of survey researchers to develop a theory of response behavior. Our discussions of strategic behavior, the nature of benefits, and the debate over why people consistently set a much higher price for giving up an amenity than they are willing to pay to obtain it raise questions about the validity of a number of well-accepted conclusions drawn from economic theory. The possibility that surveys can simulate "informed" referenda has implications for democratic theory and normative economics. Finally, our analysis of the factors that determine preference patterns and their aggregation is relevant to the efforts of sociological theorists to develop a new theory of action (see, for example, Coleman, 1986).

In the course of our work we have incurred numerous debts which we are pleased to acknowledge, with the usual disclaimer that none of the following agencies and individuals is in any way responsible for the views we express here.

Financial support for our work came from three sources. Cooperative agreements with the U.S. Environmental Protection Agency to conduct additional research on freshwater benefits and to measure drinking water risk-reduction benefits enabled us to develop our ideas further while working on these projects. A grant to Resources for the Future (RFF) from the Electric Power Research Institute for a study of how contingent valuation surveys might be used to value visibility benefits provided the stimulus and support necessary to expand our analysis to address experimental design issues and the relationship between contingent valuation and the other benefit measurement techniques. Finally, we are indebted to Resources for the Future for providing the institutional context and, in the project's final stages, the financial support which made the publication of this book possible.

As we worked on the book we were stimulated by discussions with a large number of colleagues at other institutions. These include Icek Ajzen, William Balson, Richard Bishop, David Brookshire, Albert H. Cantril, Jr., Alan Carlin, Peter Caulkins, Kenneth Craik, Ronald Cummings, Robert Davis, William Desvousges, B. L. Driver, Larry Espel, Marvin Feldman, Baruch Fischhoff, Ann Fisher, A. Myrick Freeman III, Theodore Graham-Tomasi, Robin Gregory, John Hoehn, Paul Kleindorfer, Howard Kunreuther, Edna Loehman, John Loomis, William O'Neill, Joe Oppenheimer, George Parsons, Kirk Pate, George Peterson, Stanley Presser, Paul Ruud, Robert Rowe, William Schulze, Paul Slovic, John Stoll, Mark Thayer, George Tolley, William Wade, Thomas Wegge, Ronald Wyzga, and Elizabeth Wilman.

Our colleagues at Resources for the Future have been supportive and helpful throughout the years we have worked on this book. We are especially grateful to John Ahearne, Alan Kneese, Alan Krupnick, Raymond

Kopp, Paul Portney, Brian Norton, Walter Spofford, Thomas Tietenberg, and William J. Vaughan. At the University of California-Berkeley and the University of California-San Diego, where Richard T. Carson was in residence during the book's preparation, he benefited from discussions with Peter Berck, Leo Breiman, Gary Casterline, W. Michael Hanemann, William Foster, Mark Machina, and Selma Monsky. At various points in the manuscript's evolution the following colleagues were kind enough to offer comments after reading some or all of it: Peter Bohm, Michael Hanemann, Thomas Heberlein, Alan Randall, Clifford Russell, V. Kerry Smith, and several anonymous reviewers. Robert Cameron Mitchell especially wishes to thank his wife, Susan J. Pharr, for her editorial comments and moral support.

Valuable editorial, graphic, and research assistance was provided at RFF, Berkeley, and San Diego by Diane Burton, James Conway, James Jasper, Christine Joiner, Katherine Wagner, Kim Weeks, Richard Weeks, and Verna Wefald. Kerry Martin prepared much of the material in appendix A. Betty Cawthorne performed her secretarial tasks with her usual panache despite the frustrations of having to do battle with two different word-processing systems over the extended period of the book's authorship. With meticulous attention to vital matters both great and small, our editor, Samuel Allen, exercised his craft with great and much appreciated skill.

We would like to note that just as the book has evolved since it was originally conceived, so has our colleagueship. The sequence of our names on the cover reflects the book's history; the final product is truly a joint effort for which we share equal credit and equal blame.

Finally, we owe a special debt to Clifford S. Russell, to whom we are pleased to dedicate this book. Now the head of Vanderbilt University's Institute for Policy Studies, Cliff was director of RFF's Quality of the Environment Division from the time we began the book until 1985. He was responsible for introducing us to the contingent valuation method, and, as we struggled with the methodological ramifications of that and other benefit measurement efforts, his unflagging interest in the project and his intellectual support made it possible for us to conceive and write this book.

1

Valuing Public Goods Using the Contingent Valuation Method

Our national commitment to a cleaner and safer environment has persisted in the face of oil embargos, stagflation, concerns about economic competitiveness, and competing budgetary claims. But as we progress toward the goal of a cleaner environment, each successive improvement becomes more costly to accomplish than its predecessor. Given finite public resources and restless taxpayers, this inevitably raises some difficult policy questions. How clean should we make the air? Should we attempt to make the lower Mississippi River as pure as the lakes in Wisconsin? Just how high a level of impurity should we tolerate in our drinking water? Is a further expansion of a state's park system justified in the face of the legitimate needs of industrial developers? Do we buy another B1 bomber? How much more do Medicare recipients value access to their traditional doctor than being enrolled in a health maintenance organization? Economists believe that questions like these can be addressed with empirical research in the form of benefit-cost analysis. By balancing the costs of public goods against their benefits, decision makers can arrive at more informed choices, or so the logic goes. In recent years the demand for such an accounting has found increasing favor among federal and state policy makers.

Unfortunately, few endeavors are more difficult than assigning a dollar value to something as elusive as increases in air visibility, or keeping the option of paddling a canoe in a wilderness preserve. Economists have long measured the value of goods that are routinely bought and sold in the marketplace. But ordinary markets do not exist for "public" goods such as national defense, the Apollo program to send man to the moon, and many environmental amenities.[1] Sometimes, as in the case of recreation sites,

[1] Pure public goods are characterized by the conditions of non-excludability of and non-rivalry congestion between individuals who wish to use the good (Cornes and Sandler,

this is because under public policy there is no charge for the good or service, or there is an arbitrarily determined charge (which does not reflect the full cost of providing the service or its true market value). In other cases, such as air and water quality improvements, which are public goods in the truest sense of the word, a charge would not be feasible because once the amenities are provided, people cannot be excluded from enjoying them.

For decades economists have grappled with the challenge of valuing public goods. The contingent valuation (CV) method is one of a number of ingenious ways they have developed to accomplish this demanding and important task. For reasons to be presented throughout this book, we argue that as things now stand, contingent valuation represents the most promising approach yet developed for determining the public's willingness to pay for public goods. Generally speaking, it appears as accurate as other available methods, it requires the researcher to make fewer assumptions, and it is capable of measuring types of benefits that other methods can measure only with difficulty, if at all. Our message is one of optimism tempered with realism. Like all sophisticated methodologies, contingent valuation presents challenges, and an important focus of the book is on the pitfalls of using the method. Contingent valuation's use of surveys to obtain consumer responses to hypothetical situations makes it vulnerable to various types of error, which we consider in detail so the researcher can take steps to avoid them and so the policy maker can evaluate and use CV findings with confidence.

The Contingent Valuation Method

The CV method uses survey questions to elicit people's preferences for public goods by finding out what they would be willing to pay for specified improvements in them. The method is thus aimed at eliciting their willingness to pay (WTP) in dollar amounts.[2] It circumvents the absence of markets for public goods by presenting consumers with hypothetical markets

---

1986). They may be seen as a special type of externality. In the real world, few public goods meet these strict conditions; we discuss the implications of deviations from this definition for contingent valuation in chapter 3. Whether something is a "good" or a "bad" depends on one's perspective. For example, an increase in environmental quality may be a "good" to consumers, and a "bad" to producers who happen to bear the immediate cost of pollution control.

[2] Respondents to CV surveys may also be asked what level of compensation they would be willing to accept (WTA) for a loss, but in what follows we refer to the WTP format unless otherwise noted. For reasons elaborated in later chapters, the WTA format should usually be avoided in CV studies because it does not elicit valid data under many circumstances.

in which they have the opportunity to buy the good in question. The hypothetical market may be modeled after either a private goods market or a political market. Because the elicited WTP values are contingent upon the particular hypothetical market described to the respondent, this approach came to be called the contingent valuation method (Brookshire and Eubanks, 1978; Brookshire and Randall, 1978; Schulze and d'Arge, 1978).[3] Respondents are presented with material, often in the course of a personal interview conducted face to face, which consists of three parts:

1. *A detailed description of the good(s) being valued and the hypothetical circumstance under which it is made available to the respondent.* The researcher constructs a model market in considerable detail, which is communicated to the respondent in the form of a scenario that is read by the interviewer during the course of the interview. The market is designed to be as plausible as possible. It describes the good to be valued, the baseline level of provision, the structure under which the good is to be provided, the range of available substitutes, and the method of payment. In order to trace out a demand curve for the good, respondents are usually asked to value several levels of provision.
2. *Questions which elicit the respondents' willingness to pay for the good(s) being valued.* These questions are designed to facilitate the valuation process without themselves biasing the respondent's WTP amounts.
3. *Questions about respondents' characteristics (for example, age, income), their preferences relevant to the good(s) being valued, and their use of the good(s).* This information, some of which is usually elicited preceding and some following reading of the scenario, is used in regression equations to estimate a valuation function for the good. Successful estimations using variables which theory identifies as predictive of people's willingness to pay are partial evidence for reliability and validity.

If the study is well designed and carefully pretested, the respondents' answers to the valuation questions should represent valid WTP responses. The next step is to use these amounts to develop a benefit estimate. If the sample is meticulously selected by means of random sampling procedures, if the response rate is high enough, and if the appropriate adjustments are made to compensate for participants who fail to respond (nonrespondents) and for those who give "poor"-quality data, the results can be generalized with a known margin of error to the population from which

[3] At different times and in various places the contingent valuation method has been called the survey method, the interview method, the direct interview method, the direct questioning method, the hypothetical demand curve estimation method, the difference-mapping method, and the preference elicitation method.

the respondents were sampled. The capability of generalization is a powerful feature of the sample survey method. In this way the responses given by, say, eight hundred people can be used to represent the responses that everyone in a river basin, a state or region of the country, or even the entire country would give if they were all queried.

Illustrative Scenarios

The particular form a CV study takes varies according to the nature of the good being valued, the methodological and theoretical constraints imposed by CV practice, the population being surveyed, and the researcher's imagination and ingenuity. The following brief descriptions of how four CV researchers approached the task of designing their scenarios illustrate this diversity. They offer a glimpse of how researchers have attempted to use the method to measure four different benefits: improvement of visibility impairments affecting a recreational area, improvement of national water quality, reductions of risk from motor vehicle accidents, and the level of services (reductions in services) provided by a local senior citizens' program.

Aesthetic Benefits

Figure 1-1 contains the scenario which was used in in-person interviews for a study (Brookshire, Ives, and Schulze, 1976) which sought to value

> This research is designed to more closely examine some of the *trade-offs* between *industrial development, recreation, and the environment* in the Lake Powell area. In connection with these objectives, I would like to ask you a few *questions* to see how you feel about environmental quality and its future in this area. . . .
>
> There are plans to construct a large electric generating plant north of Lake Powell. This plant is expected to be at least as large as the Navajo Plant on the south side of the lake.
>
> 5. Have you noticed the Navajo Plant or its smokestacks? ____ yes ____ no
>
> Depending on exactly where and how a new plant is constructed, it could have a significant effect on the quality of the environment. If the plant is built near the lake, it could be *visible* for many miles up and down the lake. If *air pollution* is not strictly controlled, visibility in the area may be significantly affected.

Figure 1-1. Wording of an early contingent vaulation scenario used in the Lake Powell visibility study

Figure 1-1 (continued)

These photographs (show) are designed to show how a new powerplant on the north side of the lake might appear. Situation A shows a possible plant site but assumes that the powerplant would be built at some distant location, not visible from the lake area. In situation B the powerplant is easily seen from the lake, but emits very little smoke; visibility is virtually unaffected. Situation C is intended to show the situation with the greatest impact on the environment of recreationists in the area. It is easily seen from the lake, and the smoke substantially reduces visibility.

Vacationers, of course, *spend* considerable amounts of *money* and time and effort to equip themselves with vehicles, boats, camping and fishing gear, and for traveling to the destination of their choice. It is reasonable to assume that the amount of money you are *willing to spend* for a recreational experience depends, among other things, on the *quality of the experience you expect*. An improved experience would be expected to be of greater value to you than a degraded one. Since it does *cost money to improve the environment*, we would like to get an estimate of how much a better environment is worth to you.

First, let's assume that visitors to GCNRA [Glen Canyon National Recreation Area] are to finance environmental improvements by paying an *entrance fee* to be admitted into the recreation area. This will be the *only way* to finance such improvements in the area. Let's also assume that *all visitors* to the area will *pay the same* daily fee as you, and all the money collected will be used to finance the environmental improvements shown in the photos.

6. Would you be willing to pay a $1.00 per day family fee to prevent Situation C from occurring, thus preserving Situation A? $2.00 per day? (increment by $1.00 per day until a negative response is obtained, then decrease the bid by 25¢ per day until a positive response is obtained, and record the amount).

   ________ $/day

7. Would you be willing to pay a $1.00 per day fee to prevent Situation B from occurring, thus preserving Situation A? (repeat bidding procedure).

8. (Answer only if a zero bid was recorded for question 6 or 7 above. Did you bid zero because you believe that:

   ________ the damage is not significant

   ________ it is unfair or immoral to expect the victim of the damage to have to pay the costs of preventing the damage

   ________ Other (specify)

Source: Brookshire, Ives, and Schulze (1976:334); italics in the original.

the aesthetic impact of a proposed power plant on the north shore of the Lake Powell recreational area in Utah. The amenity the respondents were asked to value was the aesthetic benefit of having their view of Lake Powell unobstructed by the visible presence of a power plant and the smoke from the plant. The researchers used sets of specially created photographs to depict the baseline view and two levels of impairment. Situation A represented the view without the power plant. The photographs for situation B showed only some smoke from the hypothetical power plant, which itself was not in view. The photographs for C showed the plant, in addition to a greater amount of smoke. The respondents were asked to say how much they were willing to pay to preserve situation A, first from the degradation represented in B, then from situation C. The elicitation method used in this study is known as the bidding game, an iterative technique illustrated in question 6 in figure 1-1. The payment vehicle was a hypothetical entrance fee to the recreation area.

National Freshwater Quality

When Congress passed the Clean Water Act of 1972, it mandated the attainment of a national goal to make every freshwater body in the country clean enough to be used for boating, fishing, and swimming. A national water quality study conducted by the present authors used the CV method to measure this program's overall benefits (Mitchell and Carson, 1981, 1984; Carson and Mitchell, 1986) (hereafter referred to as the water quality study; for the survey instrument used in this study, see appendix B). A preliminary study was conducted in 1980 to test a draft instrument and the feasibility of using the CV method in a national survey. In 1984 a revised instrument was administered by professional interviewers to a national sample.

The scenario for the water quality study was much longer than the one used for the Lake Powell study, requiring interviews which lasted an average of 45 minutes. A central design issue was how to describe the amenity to the respondents. The description had to be understandable, so that the respondents could clearly grasp what they were valuing, and policy-relevant, so that the results could be of use to policy makers. The baseline from which the respondents were asked to measure the changes to be valued was described as a *minimum* level of water quality in the nation's lakes, rivers, and streams—that is, a level below "boatable" quality.[4] A total of three improvements in minimum national water quality were valued: from below boatable to the boatable level, from boatable to

[4] The boatable level of quality was described as the current minimum level, at which no water body is below boatable in quality, although many are higher in quality.

"fishable," and from fishable to "swimmable" water quality.[5] These levels were described in words and shown on a water quality ladder which was used as a key visual aid throughout the scenario. The payment vehicle, annual taxes and higher product prices, was chosen to correspond with the way citizens actually pay for improvements in water quality.

A good example of the kind of methodological problem which a CV study has to overcome was our difficulty in finding an unbiased way to elicit the WTP amounts from the respondents. We initially tried to use a bidding game method similar to the one used in the Lake Powell study, but our preliminary tests found that the initial amount suggested by the interviewer often influenced respondents' answers because they tended to base their answers on this amount instead of making an independent determination of what the water quality improvements were worth to them.[6] Our solution to this "starting point bias" was to use a payment card which showed a large number of dollar amounts ranging from $0 to high figures. In order to provide a meaningful context for the valuation exercise, five of these dollar figures (the "benchmarks") were identified as the amounts households in the respondents' income groups were currently paying in taxes and in higher prices for nonenvironmental public goods such as defense and police and fire protection.

The water quality study also illustrates the use of the split-sample (split-ballot) technique to examine the influence of scenario components on the respondents' WTP amounts. This ability to conduct experiments within surveys is regularly exploited by CV researchers for that purpose. If the sample is big enough, survey samples can be divided by a random procedure into equivalent subsamples which value the good under different conditions. In our preliminary 1980 study and a large 1983 pretest, we administered different versions of the payment cards to subsamples to see if the amounts we told respondents they were paying for the other public goods biased the WTP amounts they gave for water quality in some fashion.[7]

Transportation Safety

Because of the growing need to make choices about which risk reduction programs are the most worthwhile, there is considerable interest in valuing

[5] The questionnaire also measured the respondents' willingness to pay for partial improvements. Three opportunities were provided during the course of the interview for respondents to revise their amounts, two of them after respondents were provided with new information such as the amount households at their income level currently pay for water quality programs.

[6] The bidding game method is not recommended now, for this reason (Cummings and coauthors, 1986; see also chapter 11 of this book).

[7] These experiments indicated that our respondents' answers were not affected by the benchmarks used (Mitchell and Carson, 1981, 1984). See appendix B for the payment cards used.

safety improvements. The real-market-based revealed preference techniques (see Blomquist, 1982) attempt to identify and observe choices in situations in which people actually trade off income against physical risk. The limited applicability of this approach to many types of risks led Jones-Lee and his colleagues (Jones-Lee, Hammerton, and Philips, 1985) to attempt to use the CV method to measure the value of risk reductions.[8] Their survey consisted of lengthy personal interviews with a sample of people in England, Scotland, and Wales. The centerpiece of the study was a series of questions designed to provide estimates of the sample's willingness to pay for certain types of transportation risk reductions.[9]

The primary challenge faced by these researchers was to accurately convey risk levels and risk reductions to the respondents. The solution found by Jones-Lee and his colleagues for this "far from straightforward" problem was to describe the risks in verbal terms as probabilities of "x in 100,000," and in visual terms as a piece of graph paper with 100,000 squares, x of which were blacked out. Altogether, they obtained WTP amounts for three different risk change situations. Two of the situations involved hypothetical private markets; the other described a hypothetical public goods market. The following is one of their private good scenarios:

> Imagine that you have to make a long coach (bus) trip in a foreign country. You have been given £200 for your travelling expenses, and given the name of a coach service which will take you for exactly £200. The risk of being killed on the journey with this coach firm is 8 in 100,000. You can choose to travel with a safer coach service if you want to, but the fare will be higher, and you will have to pay the extra cost yourself.
>
> (a) How much extra, if anything, would you be prepared to pay to use a coach service with a risk of being killed of 4 in 100,000—that is half the risk of the one at £200?
>
> (b) How much extra, if anything, would you be prepared to pay to use a coach service with a risk of being killed of 1 in 100,000—one eighth the risk of the one at £200? (Jones-Lee, Hammerton, and Philips, 1985:56)

The researchers used still another type of elicitation method in their study. All respondents were asked in an open-ended fashion how much extra they would pay for the risk improvement. If a respondent hesitated (which occurred in about 20 to 30 percent of the interviews), the interviewer then followed a procedure whereby she read out amounts of money

[8] Their study is the first large-scale use of the CV method for this purpose. Other recent CV risk studies based on relatively large samples are Smith, Desvousges, and Freeman (1985) and Mitchell and Carson (1986b).

[9] The survey also probed people's understanding of risk and probability concepts and the degree to which their responses conformed with the standard axioms of rational choice under uncertainty, among other topics.

beginning with "nothing," until an amount was reached beyond which the respondent would not pay. At that point further probes were used to establish the WTP amount. The study also used split-sample techniques to conduct several methodological experiments.

Senior Companion Program for the Elderly

Our last example of a CV scenario comes from a study of the benefits of a social program for the elderly. The senior companion program is a federally funded endeavor which makes it possible for low-income older persons to perform paid volunteer services, as "senior companions," for isolated, needy elderly who live in the community or institutions. Some of the services performed by the companions, such as light household chores, providing transportation, and companionship,[10] could be purchased in the market. Other aspects of these services, such as the volunteer character of the companionship with its potential for friendship, are not traded in markets. In order to capture the full range of the user benefits, Garbacz and Thayer (1983) conducted a CV study in which they interviewed (in person) a sample of program recipients in Phelps County, Missouri. The CV scenario posed a hypothetical situation in which the respondent's companion service would be reduced as a result of federal funding cutbacks. Split samples were used to ascertain the effect of posing the issue in a willingness-to-pay framework versus a willingness-to-accept (WTA) framework. Thus the respondents were asked either (1) how much they would be willing to pay (in reduced social security allotments) to prevent reductions of 25 percent and 75 percent in the services they receive, or (2) how much of an increase in their social security allotment they would require before they would accept these reductions.

The Development of the Contingent Valuation Method

The contingent valuation method first came into use in the early 1960s when economist Robert K. Davis (1963a, 1963b, 1964) used questionnaires to estimate the benefits of outdoor recreation in a Maine backwoods area. Earlier, the well-known resource economist Ciriacy-Wantrup (1947) had suggested the use of the "direct interview method" to measure the values associated with natural resources, and had advocated it in his influential book, *Resource Conservation: Economics and Policies* (1952). But it was Davis, at the time unaware of Ciriacy-Wantrup's suggestion, who played the key role. He had developed interests in social psychology

[10] Garbacz and Thayer (1983) regard companionship as a nonmarketed good. This is not always the case, as wealthy invalids often hire the services of practical nurses who provide companionship in addition to their strictly professional services.

and farmers' attitudes toward wildlife before entering Harvard for doctoral studies in economics. At Harvard, Samuel Stouffer, one of the leading academic survey research practitioners, was teaching a course in survey methods in the Social Relations Department. After sitting in on Stouffer's course, Davis reasoned that it should be possible to "approximate a market" in a survey by describing alternative kinds of areas and facilities to be made available to the public, and simulating market bidding behavior.[11] This method, he later wrote, "would put the interviewer in the position of a seller who elicits the highest possible bid from the user for the services being offered" (Davis, 1963a:245). Thus was born the bidding game, in which the interviewer systematically raises or lowers bids from an arbitrarily chosen starting point until the respondent switches his reaction from inclusion to exclusion (or vice versa) and thus reveals his maximum willingness to pay.

Davis personally interviewed a sample of 121 hunters and recreationists in the Maine woods for his doctoral dissertation. His study sought to measure the benefits of a particular recreational area. According to his account, the respondents took the exercise seriously and expressed credible values: "During the bidding game respondents' comments indicated they were turning over in their minds the alternatives available in much the same way a thrifty shopper considers the price and desirability of different cuts of kinds of meat. Moreover, certain income related responses to other preference questions suggest an economic consistency in the responses" (Davis 1964:397). In order to test the "rational structure" of the responses, Davis estimated an equation which explained a large percentage (R^2 = .59) of the variance in the WTP amounts as a function of income, length of stay in the area in days, and acquaintance with the locale in years. In another test, he compared the CV results with those measured by an alternative method which infers values from the cost of traveling to the recreational site—a study which he also conducted. He found close agreement between the two measures (Knetsch and Davis, 1966:140–142). Davis acknowledged "some rough spots" in his new method which needed to be ironed out, but felt that it showed sufficient promise to merit a "major research effort" to improve it further (Knetsch and Davis, 1966:142).

Influenced by Davis, Ronald Ridker (1967) used the CV method in several studies of air pollution benefits. Although the primary thrust of Ridker's work was to value household soiling and material damages using the hedonic pricing approach (Ridker, 1967; Ridker and Henning, 1967), it was his recognition that people might value air pollution because of its "psychic costs" that led him to include a couple of WTP questions in two

[11] Davis, personal communication, June 16, 1986.

different surveys he conducted in Philadelphia and Syracuse in 1965. These questions, which asked people how much they would pay to avoid "dirt and soot" from air pollution, lacked many of the essentials required for a hypothetical market. In reflecting on his experience with the survey questions, including the WTP questions, Ridker made an observation that prefigured later developments in CV research:

> It now seems evident that a much narrower, deeper, and psychologically sophisticated questionnaire is required to measure and untangle the various determinants of cleaning costs. Such a questionnaire would require substantially more time and expenditure—perhaps three to four times greater—than went into this essentially exploratory study. Even then the probability of success may not be very high, for such a project would raise problems in survey design that have not yet been solved. (Ridker, 1967:84)

Over the next few years several other economists followed Davis's lead and used the CV approach to value various recreational amenities. In 1969, Hammack and Brown (1974) sent a mail questionnaire to a large sample of western hunters which asked them how much they would be willing to pay for and willing to accept to give up their rights to hunt waterfowl.[12] Next, in 1970, Cicchetti and Smith (1973, 1976a, 1976b) asked individuals who were hiking in a wilderness area how much they would be willing to pay to reduce congestion in the area from other hikers. Around 1972, Darling (1973) used the CV method in personal interviews to value the amenities of three urban parks in California. The first attempt to value a pollution control amenity since Ridker's took place in the summer of 1972 when Alan Randall and his colleagues used CV to study air visibility benefits in the Four Corners area in the southwest (Eastman, Randall, and Hoffer, 1974, 1978; Randall, Ives, and Eastman, 1974). At about the same time, Acton (1973) applied the method to valuing programs which reduced the risk of dying from a heart attack.[13] In the last of the major early studies, Hanemann asked a sample of people in 1973 how much they were willing to pay for improved water quality at beaches in the Boston area (Hanemann, 1978; Binkley and Hanemann, 1978).

In an attempt to legitimize the CV method, several of the early CV studies compared their findings with those obtained for the same amenities by other techniques which had already reached widespread acceptance for

[12] Miles (1967), LaPage (1968), Mathews and Brown (1970), Beardsley (1971), Berry (1974), and Meyer (1974a, 1974b) used less formal CV techniques to value recreation. The Department of the Interior's Fish and Wildlife Survey of 1975 contained a question which asked respondents how much costs would have to increase before outdoor recreation would be given up.

[13] The use of surveys to value reductions in risks gained a boost from articles by two well-known economists (Schelling, 1968; Mishan, 1971), who noted that the other techniques

valuing recreational use benefits. Davis used a travel cost model for this purpose, Darling used a property value model, and Hanemann compared his findings with those obtained by a generalized travel cost model.

The most influential of the early studies was that conducted by Randall, Ives, and Eastman (1974). Their effort was notable for, among other things, its theoretical rigor, its valuation of a good which could not be valued by alternative methods (such as travel cost and hedonic pricing), its use of photographs to depict the visibility levels being valued, and its experimental design whereby certain aspects of the bidding game (such as the payment vehicle) were varied systematically to see if they affected the WTP amount in some systematic fashion. Perhaps even more significant, the timely publication of their article on the study in the first volume of the new *Journal of Environmental Economics and Management* brought the method to the attention of a broader audience.

Since the early 1970s the contingent valuation technique has been used by economists to measure the benefits of a wide variety of goods, including recreation (Walsh, Miller, and Gilliam, 1983), hunting (Cocheba and Langford, 1978), water quality (Gramlich, 1977), decreased mortality risk from a nuclear power plant accident (Mulligan, 1978), and toxic waste dumps (Smith, Desvousges, and Freeman, 1985). Other valuation focuses have been on aesthetic benefits from forgoing construction of a geothermal power plant (Thayer, 1981), aesthetic and health benefits of air quality (Brookshire, d'Arge, and Schulze, 1979), benefits of the public collection and dissemination of grocery store price information (Devine and Marion, 1979), and benefits of government support for the arts (Throsby, 1984). (See appendix A for a more complete list of CV studies, the public goods they have valued, and some of their characteristics.)

Until recent years most CV studies were exploratory. Researchers concentrated on refining the CV method by identifying and testing for the possible biases that arise in its use, and on establishing its credibility by making comparisons between the benefits measured in CV studies and those measured for the same goods by one of the established techniques, such as the travel cost method. Much of the pioneering methodological work was conducted by Randall and his colleagues (Randall, Ives, and Eastman, 1974; Randall and coauthors, 1978; Randall, Hoehn, and Tolley, 1981), and by Cummings, d'Arge, Brookshire, Rowe, Schulze, and Thayer at the universities of New Mexico and Wyoming (see Schulze,

---

for measuring benefits in these situations were either unavailable or necessitated a large number of implausible assumptions. They concluded that there was everything to gain and nothing to lose from trying the survey approach to asking people for their willingness to pay.

d'Arge, and Brookshire, 1981, for a summary).[14] Parallel theoretical work by many of the same researchers has established that CV data are generated in forms consistent with the theory of welfare change measurement (Randall, Ives, and Eastman, 1974; Freeman, 1979b; Brookshire, Randall, and Stoll, 1980; Just, Hueth, and Schmitz, 1982; Hanemann, 1986a; Hoehn and Randall, 1987).

In 1979 the Water Resources Council (1979) published its newly revised "Principles and Standards for Water and Related Land Resources Planning" in the *Federal Register.* This important document set forth the guidelines for federal participation in project evaluation, which specified those methods that were acceptable for use in determining project benefits. The inclusion of contingent valuation as one of the three recommended methods (the other two were the travel cost and the unit day value methods) was a sign of contingent valuation's growing respectability.[15] More recently the U.S. Army Corps of Engineers has begun to use the contingent valuation method to measure project benefits. As of July 1986 the various engineer corps districts and the corps' Institute of Water Resources had conducted fifteen to twenty CV studies of varying degrees of sophistication,[16] and had published a handbook on the method for corps personnel (Moser and Dunning, 1986). Contingent valuation has also been recognized as an approved method for measuring benefits and damages under the Comprehensive Environmental Response, Compensation, and Liability Act of 1980 (Superfund), according to the final rule promulgated by the Department of the Interior (1986).

Funding from the U.S. Environmental Protection Agency has played a particularly important role in contingent valuation's development. Agency economists recognized early that the prevailing methods of measuring benefits, such as the travel cost method, would be of limited use in valuing the benefits of pollution control regulations they would be responsible for implementing (see chapter 3). In the mid 1970s the agency began to fund a program of research with the avowed methodological purpose of determining the promise and problems of the CV method.[17] At first, almost all of the CV studies funded by this program were designed to test various

[14] Randall's early work was also carried out at the University of New Mexico.

[15] The Principles and Standards document, which was modified and expanded slightly in 1983, also enshrined the then-prevailing CV practice as the officially prescribed way to conduct CV studies of water project benefits. As late as 1986, contract administrators were known to have required researchers to use the by-then outdated bidding game elicitation technique because the Principles and Standards document had declared it to be the "preferred" elicitation format for CV studies.

[16] C. Mark Dunning, personal communication, July 1986.

[17] The Electric Power Research Institute also sponsored some of the early air visibility studies that used the CV method.

aspects of the method and to establish its theoretical underpinnings. As the method became better understood, and the agency's mandate to subject proposed regulations to benefit-cost analysis was given sharper focus by the Reagan administration under Executive Order 12291 (Smith, 1984), EPA's interest shifted to ascertaining just how effectively contingent valuation can be used for policy purposes. As part of this effort the EPA commissioned a state-of-the-art assessment of the CV method in 1983 as a way to step back and reflect on past achievements, remaining problems, and future possibilities. A notable feature of this assessment was the involvement of a review panel of eminent economists and psychologists, including Nobel laureate Kenneth Arrow. During a conference at Palo Alto, California, in July 1984, these and other scholars who were actively involved in contingent valuation research offered their views about the method's promise as a means for evaluating environmental goods. The leading authors of the state-of-the-art assessment (Cummings and coauthors, 1986) concluded that although the method shows promise, some real challenges remain. They gave the highest priority for future research to the development of an overall framework, based on a theory of individual behavior in contingent market settings, which could serve as a basis for hypothesis testing.

This Book

When we began work on our first contingent valuation study in 1979, our backgrounds in sociology (Mitchell) and political science (Carson) and our previous experience with public opinion (Mitchell) and marketing research (Carson) inclined us to be skeptical about whether CV surveys could obtain values for public goods which would be uncontaminated by "instrument effects," or biases introduced by the wording of the questionnaire. As we gained experience with the method we became more confident that it could obtain useful results if it were solidly grounded both theoretically and methodologically. Its relationship to economic theory is now well established (see chapter 2). In the course of our early work we found ourselves grappling with various basic methodological issues, and eventually decided to pursue a more systematic study of the method's reliability and validity. This book is the result.

In the first five chapters we introduce the method. These chapters describe how contingent valuation works, its theoretical basis in welfare economics, the nature of the benefits it can be used to measure, how it compares to other methods for measuring benefits, and the technique of gathering data on which it is based—survey research.

In chapters 6 through 8 we address the skeptics. Of all the social sciences, economics has been least comfortable with the survey research method. Indeed, according to McCloskey (1983), economists are "extremely hostile" to surveys. On the face of it, this is a surprising phenomenon because almost one out of three articles in the leading economics journals now use survey data in some form (Presser, 1984). A closer inspection reveals, however, that economics differs from its sister social sciences in that economists rely primarily on survey data collected by others, and especially on data produced by the Census Bureau. Most economists eschew measures of subjective phenomena, such as attitudes and behavioral intentions,[18] in favor of objective entities such as labor force participation, earnings, credit, and industrial production in which they have more trust (Presser, 1984).

Regarding the CV method in particular, economists have raised several types of objections. Some believe that survey respondents will engage in strategic behavior by giving answers deliberately calculated to influence policy makers. Others hold that those surveyed will not be motivated to search their preferences with sufficient care to give meaningful answers. Still others feel that answers to hypothetical questions cannot predict behavior. The chapters addressing these objections draw on a wide range of materials from several disciplines to consider the fundamental questions of whether respondents in CV surveys will answer honestly and whether they can answer meaningfully. Our analysis and evidence lead us to conclude that there is no valid basis for dismissing the method out of hand on these grounds.

In chapters 9 through 12 and in appendix C we consider the dangers of believing that there are quick and easy ways to do CV surveys or to use the results of CV studies. As the method increasingly comes into vogue, there is a natural tendency on the part of some economists and policymakers to hope that contingent valuation will be straightforward and cheap to implement, or to treat the findings of one CV study as being as good as those of another under the assumption that they are of uniform quality. We argue here that it is still quite difficult to obtain data from CV surveys which are sufficiently reliable and valid to use for policy purposes, and that the dangers of attempting to draw conclusions from poorly conceived or poorly executed studies are great. Contingent valuation is based on a technique (survey research) with which economists have little expertise, yet it presses this technique close to the limits of its capability. At the same time, the economic theory upon which contingent valuation is based is not well known among sociologists or survey research methodologists, nor is it accepted by them. There is as yet no consensus among CV practitioners

[18] An outstanding exception is the program conducted by the Survey Research Center at the University of Michigan on consumer intentions and the determinants of consumer behavior. See, for instance, Katona (1975).

about such fundamental questions as what type of market structure they should emulate and what types of errors they should be trying to avoid. These chapters offer a systematic framework for understanding and overcoming the barriers to obtaining reliable and valid CV estimates.

The concluding chapter presents a brief review of the major themes of this book and a list of questions that must be asked by any decision maker wishing to use the findings of a CV study. In addition, we offer our views about research priorities for the CV method and about some possible new applications.

In this enterprise we have drawn on the findings of scholars, in our own and other disciplines, which seemed relevant to the methodological issues involved in conducting CV surveys. Market researchers have used surveys to test concepts of new, unmarketed goods in order to learn if people would purchase such goods if they are produced. Cognitive psychologists have studied the ways people make decisions from the information available to them. Experimental social scientists from several disciplines have explored the amount of strategic behavior people actually engage in. Social psychologists have been interested in the degree to which attitudes predict behavior, political scientists have examined referenda voting behavior, and statisticians are interested in devising techniques for compensating for less than perfectly executed surveys. Last, but not least, survey researchers have long worried about the validity and reliability of survey data, and, increasingly, have conducted methodological studies aimed at improving our understanding of why respondents respond the way they do in given situations.

The book is based on these literatures, our own research, and on the ever-growing body of CV studies conducted by others. It is our view that good contingent valuation practice, at least at this stage in the method's development, cannot be achieved by following rigid rules. Those who seek a cookbook on how to do CV studies will be disappointed in this book. Although we offer some guidance for practitioners, and in chapter 13 outline the questions that should be asked of any CV study by someone who wishes to use its findings, our guidance is directed more toward fundamentals than formulas. It addresses the questions of whether the method can produce reliable and valid values, and if so, under what conditions. Many of our conclusions are best regarded as hypotheses, to be refined and tested in future research.

2

Theoretical Basis of the Contingent Valuation Method

The ultimate aim of a contingent valuation survey is typically to obtain an accurate estimate of the benefits (and sometimes the costs) of a change in the level of provision of some public good,[1] which can then be used in a benefit-cost analysis. In order to do this, the survey must simultaneously meet the methodological imperatives of survey research and the requirements of economic theory. To meet the methodological imperatives requires that the scenario be understandable and meaningful to the respondents and free of incentives which might bias the results. To meet the requirements of economic theory—the subject of this chapter and the next—a survey must obtain the correct benefit measures for the good in the context of an appropriate hypothetical market setting. In this chapter we review the theoretical underpinnings of the contingent valuation method.

We begin by introducing the key elements of the conventional theory of consumer behavior as a way of acquainting those readers unfamiliar with welfare economics with the theoretical framework underlying contingent valuation. We then discuss benefit-cost measures from a more technical perspective, and follow that discussion with an analysis of one of the major controversies that arises in planning a contingent valuation survey: whether to use a willingness-to-pay (compensation) measure or a willingness-to-accept (compensation) measure in designing certain types of valuation questions. We explain why, in CV surveys, values measured by a WTA format are much bigger than those measured by a WTP format, and propose a new property rights framework which suggests that the

[1] The terms level, quantity, and quality of a public good are used here interchangeably, depending on the nature of the good being discussed.

WTP measure is the correct benefits measure in a wider range of circumstances than previously assumed.

We then go on to show how the contingent valuation method can be used to emulate both a private goods market and a political (referendum) market. This section also discusses how benefit-cost measures are interpreted and used in each of these types of markets. We then consider the theoretical issues involved in aggregating and disaggregating benefits across benefit subcomponents (including geographic areas), across different sequences of policy changes, and across individuals. Also discussed in that section is how contingent valuation responses can be used to estimate the distribution of policy benefits. In conclusion, we summarize the implications of the previous analysis for the CV method by outlining the requirements which economic theory imposes on the design of a contingent valuation scenario.

The Basis of Welfare Economics

Economics can be divided into two branches, positive and normative (Stiglitz, 1986). Positive economics seeks to describe how the world works, while the normative branch, often referred to as welfare economics, seeks to make judgments about the desirability of having government undertake particular policies, or, put in another way, how the world could work. Much of the history of welfare economics has been dominated by the notion of a "social welfare function," and the "optimal" output of an economy has been seen as determined by the point of tangency between the social welfare function and the production possibility frontier (a positive economic concept depicting how production of one good could be traded off, in a technical sense, for production of another good). That point of optimal output is illustrated in figure 2-1, which depicts a production possibility frontier (PPF) for a simple two-good economy capable of producing a private good x and a public good q, and a social welfare function (SWF); the optimal outputs are denoted as x^* and q^*.

The earliest interpretation of the social welfare function was strictly Benthamite utilitarian in approach—the greatest good for the greatest number of people—so that the social welfare function was defined simply as the sum of the utility of the members of that society for the production of different combinations of goods. Utility was assumed to be measurable in a cardinal sense, and comparable across individuals. By the late 1930s the notion of cardinal utility across individuals had been almost completely rejected by economists in favor of an ordinal definition of utility, with no comparability across individuals, that severely undermined the theoretical basis of the social welfare function. Bergson (1938) and Samuelson (1947) attempted to rebuild the social welfare function in a rigorous fashion on

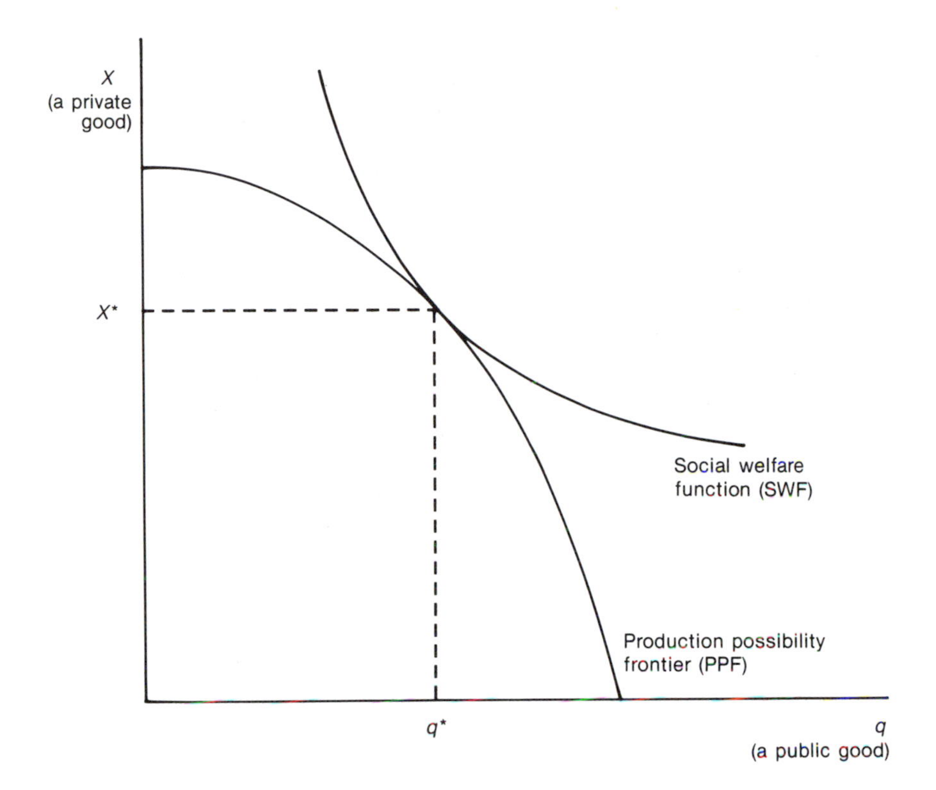

Figure 2-1. Point of optimal production

the new ordinal utility foundation, but their efforts were dealt a fatal blow by the work of Arrow (1951) and his followers.[2] Arrow showed that there was no nondictatorial way to aggregate preference into a social welfare function that did not violate a few simple and quite desirable axioms of behavior and choice. The social welfare function remains frequently used for illustrative purpose in economics texts, but plays no role in applied welfare economics.

In their search for a new welfare criterion, economists turned to the weaker, but perhaps ethically more neutral, Pareto criterion, which stated that policy changes which make at least one person better off without making anyone worse off are Pareto-improving and should be undertaken.[3] Pareto improvements can occur from points in the interior of the production possibility frontier until the production possibility frontier is

[2] See Sen (1986) or Mueller (1979) for a discussion.

[3] To some degree this is a libertarian notion, in that it gives each individual in the society a veto power over possible changes.

reached. Any point on the production possibility frontier is known as a Pareto-optimal position.[4] The Pareto criterion can easily be seen to be a much weaker guideline than a social welfare function by contrasting it with a social welfare function which specifies a unique optimal point for the economy on the production possibility frontier.

Benefit-cost analysis, the applied side of modern welfare economics, operationalizes a variant of the Pareto criterion by trying to find ways to place a dollar value on the gains and losses to those affected by a change in the level of provision of a public good.[5] This allows the calculation of net gain or loss from a policy change, and determination of whether the change is potentially Pareto-improving. Before examining in more detail the Pareto criterion and the particular way it is implemented in benefit-cost analysis, it will be useful to explore two key assumptions of the positive economics upon which welfare economics theory is based.[6]

The first and perhaps most basic assumption of positive economic theory is that economic agents (individuals, households, consumers, or firms), when confronted with a possible choice between two (or more) bundles of goods, have preferences for one bundle over another. The second major assumption is that through its actions and choices an economic agent attempts to maximize its overall level of satisfaction or utility. Both assumptions are controversial and have important implications for the CV method.[7] When coupled with additional conditions involving transitivity and nonsatiation, these assumptions lead to some remarkably powerful predictions about how economic agents will behave in different situations.

When agents are given initial endowments of resources and allowed to trade, the resulting actions and choices demonstrate a clear theory of value—a theory which received its most complete expression in Debreu's 1959 book of that title. Agents trade back and forth until, conditional on the overall initial endowment of resources, there are no possible trades left which will increase the utility of any two agents. The value of a good is the most an agent is willing to give up in exchange for the good out of the

[4] One reason economists are so attached to the notion of Pareto optimality is that a perfectly competitive economic system, in the absence of any externalities, attains an equilibrium position which is always Pareto-optimal (Varian, 1984).

[5] There are a large number of standard texts on welfare economics and benefit-cost analysis, among them Mishan (1976), Atkinson and Stiglitz (1980), Broadway and Bruce (1984), Freidman (1984), and Just, Hueth, and Schmitz (1982).

[6] Among the standard references to the relevant economic theory consulted are Deaton and Muellbauer (1980) and Varian (1984).

[7] The assumption that agents have preferences for and can make choices between dissimilar goods is rejected by some psychologists. The second assumption posits a calculating, rational agent, a notion which has been challenged in particular by Herbert Simon (1982), who has argued that while agents try to improve their utility, they do so by "satisficing" rather than attempting to completely maximize their utility.

resource it controls, or, in terms of another agent's resources, the least, the controlling agent is willing to accept in return for giving up the good.

Applying such a system of value to the provision of public goods involves clear-cut normative assumptions, as the initial (or existing) endowment of resources (ability, income, or wealth) plays a key role in determining value. In an economic context, nothing has value in and of itself; value is possible only in relation to the overall economic system.[8] Even then, value is defined only as the most someone is willing to pay for a good, or the least the owner of a good is willing to take in exchange for it.

Turning to welfare economics, two basic characteristics of benefit-cost analysis follow naturally from this positive economic foundation. The first is the acceptance of consumer sovereignty, a principle embodying the belief that the consumer is a better judge of what gives him utility than anyone else; the second is a tendency of benefit-cost analysis to emphasize economic efficiency rather than distributional issues.[9] Efficiency is a natural focus for economists because its measurement follows directly from the underlying positive economic theory. Distributional effects, in contrast, have been more difficult to analyze because data on them are of poor quality, and many benefit measurement techniques are not well-suited to shedding light on distributional outcomes. The contingent valuation method is consistent with the consumer sovereignty assumption, and is unique among benefit measurement techniques in its ability to obtain detailed distributional information.

The criterion used by welfare economics to judge a given policy is whether that policy is Pareto-improving.[10] Because in practice there are very few, if any, policy changes which make no one worse off, the only way such a criterion can be implemented is to allow those who gain from a policy change to compensate the losers. According to the compensation test, the Pareto criterion is met if, after the gainers have compensated the losers, one agent is better off and no one is worse off. In practice, however, compensation is rarely paid, and the compensation test is not of great practical use.

John Hicks (1939) and Nicholas Kaldor (1939) proposed a welfare criterion which has been alternatively called the potential Pareto-improvement criterion or the potential compensation test. The potential Pareto criterion has been controversial (Just, Hueth, and Schmitz, 1982)

[8] See Kneese and Schulze (1985) for a comparison of the ethical implications of economic theory with those of other philosophical systems.

[9] By economic efficiency we mean that resources are put to their most productive (highest-valued) use.

[10] A Pareto-improving policy change is one which moves an economy to a position which is Pareto-superior (preferred) from a position which is Pareto-inferior (less preferred). Such a change is sometimes referred to as an increase in relative efficiency.

because, without the actual payment of compensation, it is possible to make a very small group of people much better off while making the vast majority worse off. While some economists argue that actually making the compensation transfer is not as difficult as it appears and hence only the compensation test and not the potential compensation test should be considered, the potential compensation test has found remarkably wide acceptance and use among applied economists.

Use of the potential Pareto-improvement criterion has been justified on several grounds. The most common of these is the argument that projects should be decided on a basis of strict economic efficiency, since political authorities can, if necessary, use lump-sum transfers to address any distributional consequences. Closely related to this is the argument that the potential Pareto criterion is only one piece of information available to policymakers, who are free to reject policy changes with adverse distributional consequences if they wish. Another common justification (Friedman, 1984, for example) is that while any single policy change may have adverse consequences for some group, the government undertakes a large number of projects to improve the welfare of its citizens; if each of these projects meets the potential Pareto-improvement criterion, it is likely that everyone, or at least almost everyone, will be better off if they are all implemented.

In accepting the Pareto criterion or the potential Pareto criterion in benefit-cost analysis, one should be aware of what decision criteria have been implicitly rejected.[11] First, paternalism—the notion that the government, scientists, or any other "elite" group knows best what should be done to increase an individual's utility—has been rejected. Also rejected have been the notions that animals or other nonhuman species have rights, and (because initial endowments have been accepted as given) the notion that an individual has a right to some minimum standard of living. Egalitarianism, along Rawlsian or other lines, is likewise precluded. Most voting schemes are also at least potentially at odds with the Pareto criterion (Zeckhauser, 1973; Mueller, 1979; Sen, 1986). One advantage of contingent

[11] While the debate among philosophers on ethical systems is well established (Kneese and Schulze, 1985), those who appear most troubled by the economist's notion of value are not sociologists, political scientists, or market researchers (who largely adhere to, or at least understand, the notion of consumer sovereignty), but environmentalists, lawyers, and psychologists engaged in valuing scenic beauty and recreational experiences and risks to life. For a discussion of the problems that environmentalists, lawyers, and rule makers have with the way economists view the world, see Kelman (1981). For a discussion of the problems which some resource psychologists have with the economist's concept of value, see Brown (1984), Randall and Peterson (1984), Peterson, Driver, and Brown (1986), and Harris, Tinsley, and Donnelly (1986). For a philosophically oriented examination of the implications of consumer sovereignty, see Penz (1986). Baumol (1986) presents a review of work on alternatives to the Pareto criterion based on different concepts of economic fairness.

valuation is that it is capable of providing the information necessary to assess benefits by a variety of criteria, including voting and the potential Pareto-improvement criterion.

Choice of Benefit Measure

The traditional measure of consumer benefits, proposed by Dupuit in the nineteenth century and championed by Marshall, is consumer surplus, which is defined as the area under the ordinary (Marshallian) demand curve and above the price line.[12] In figure 2-2, the ordinary demand curve is the line marked *D* and price is assumed to be zero, as is typically the case with pure public goods. The change in consumer surplus resulting from an increase in provision of the public good from Q_0 to Q_1 is the area $a + b$.[13] Unfortunately, the concept of consumer surplus has been shown to have a number of problems as a measure of the benefits resulting from price or quantity changes (Samuelson, 1947; Silverberg, 1978). These problems are largely a result of the fact that the ordinary (Marshallian) demand curve does not hold the level of utility or satisfaction constant, but rather holds income constant. Hicks (1941, 1943, 1956) suggested two measures of the gain or loss which hold utility constant at the initial level (compensating variation and surplus), and two measures which hold utility constant at some specified alternative level (equivalence variation and surplus). The Hicksian consumer surplus measures may be thought of as Marshallian consumer surplus measures calculated from demand curves where total utility is held constant at different specified levels.[14] Depending on the consumer's property right position vis-à-vis the good in question, each of the four measures may involve either payment or compensation in order to maintain utility at the specified level. The combinations of properties yield the eight welfare measures shown in table 2-1. The Hicksian variation measures are to be used when the consumer is free to vary the quantity of the good considered, and the surplus measures when the consumer is constrained to buy only fixed quantities of the particular good (Randall and Stoll, 1980).[15]

[12] Just, Hueth, and Schmitz (1982) provide a detailed discussion of the material presented in this section.

[13] It will always be assumed here that consumers prefer more of the public good being valued than less. A similar presentation can be developed for a public good which is undesirable, such as pollution.

[14] In a few special cases, as when the agent has a constant marginal utility for money, the Hicksian measures are all equal to the Marshallian consumer surplus.

[15] We have a Hicksian surplus measure only in the partial equilibrium sense, as the agent is not constrained to purchase a fixed quantity in markets for other goods.

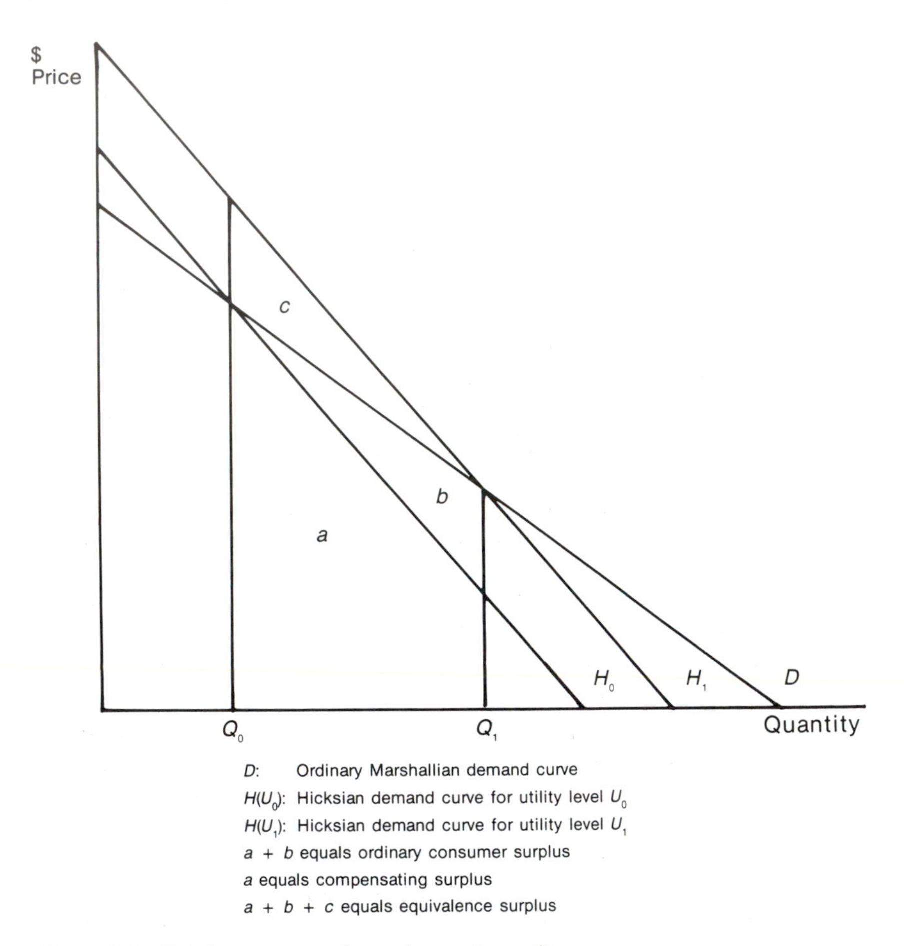

Figure 2-2. Surplus measures for a change in quality

Surplus Measures

In the diagrammatic illustration of the Hicksian surplus measures in figure 2-2, it can be seen that (holding price equal to zero) for a quantity increase from Q_0 to Q_1, ordinary (Marshallian) consumer surplus is the quantity associated with the area under the demand curve labeled D and between Q_0 and Q_1 (that is, $a + b$); compensating surplus is the area under the Hicksian compensated demand curve H_0 (that is, a); and equivalence surplus is the area under the Hicksian compensated demand curve H_1 (that is, $a + b + c$). Note that for a quantity increase, compensating surplus is less than or equal to ordinary consumer surplus, which in turn is less than or equal to equivalence surplus. For a quantity decrease these relationships are reversed.

Table 2-1. Hicksian Welfare Measures for Contingent Valuation Surveys

	WTP	WTA
Quantity increase	CS	ES
Price decrease	CS;CV	ES;EV
Quantity decrease	ES	CS
Price increase	ES;EV	CS;CV

Definitions:
WTP—willingness to pay
WTA—willingness to accept
CS—compensating surplus
CV—compensating variation
ES—equivalence surplus
EV—equivalence variation

WTP is the amount of money an agent would be willing to give up to obtain a change and still be as well off as with his previous entitlement.
WTA is the amount of money which would have to be given to an agent, with a specified entitlement, to forgo a change and still be as well off as if the change had occurred.
Compensating measures assume that the agent is entitled to his current level of utility, or, alternatively, his status quo endowment of property rights.
Equivalence measures assume that the agent is entitled to some alternative level of utility, or, alternatively, to a set of property rights different from those currently held.
Surplus measures constrain the quantity of the good being considered at the quantity which would be purchased at the new (old) price in the *absence* of compensation for the compensating surplus (equivalence surplus).
Variation measures do not constrain the quantity of the good the agent would purchase.

Since policy interest usually lies in the potential benefits as measured from the consumer's current or initial level of utility, the choice of Hicksian measures can often be further narrowed down to the two compensating surplus measures. For a quantity increase such as raising the level of air visibility in a city, the compensating surplus measure can be interpreted as the consumer's maximum willingness to pay in order to gain the quantity increase and still maintain his initial level of utility. In the case of a quantity decrease from the currently available level (for example, reduction in visibility below the current level), the compensating surplus measure may be viewed as the minimum compensation the consumer is willing to accept in return for receiving the decreased quantity.

Some benefit-cost theorists, however, such as McKenzie (1983) and Morey (1984), recommend use of the Hicksian equivalence measures in every situation because of the potential for finding a non-unique solution to the problem of what policy to undertake when using the Hicksian compensation measures to compare more than two policy options.[16] For

[16] The Hicksian equivalence measures can be shown to form a unique money metric, which is usually but not always true of the Hicksian compensating measures. The conditions

a promised quantity increase, equivalence surplus is defined as the minimum amount of compensation the consumer is willing to accept in order to do without the quantity increase and still enjoy the level of utility which would have resulted if the quantity of the good in question had been increased. For a promised quantity decrease, equivalence surplus is defined as the amount the consumer is willing to pay to avoid the decrease.

Welfare Measures as the Difference between Two Expenditure Functions

The measures obtainable by the contingent valuation method can be represented in terms of the difference between two expenditure functions. Such a representation is particularly illuminating because it reveals that in a CV survey a respondent is being asked to determine what change in his income, coupled with the change in the level of the public good, leaves his utility level unchanged.

According to modern consumer theory (Deaton and Muellbauer, 1980; Varian, 1984), the expenditure function is one of the four equivalent ways to represent the constrained utility maximization problem.[17] It is written as

$$e(p, q, U) = Y, \tag{2-1}$$

where p is a vector of prices, q is a vector of fixed public goods, U is a level of utility, and Y is the minimum amount of income needed to maintain utility level u given the price and public goods vectors.

Letting p_0, q_0, U_0, Y_0 represent some initial level of those respective arguments and p_1, q_1, U_1, Y_1 represent some subsequent levels, we can represent the compensation surplus (CS) by

$$\begin{aligned} CS &= [e(p_0, q_0, U_0) = Y_0] - [e(p_0, q_1, U_0) = Y_1] \\ CS &= Y_0 - Y_1 \end{aligned} \tag{2-2}$$

---

under which there are problems with the Hicksian compensating measures seem fairly remote in practice, however, and the equivalence measures have much less intuitive appeal. Moreover, CV is probably not a precise enough measuring instrument to pick up such fine differences in the ranking of projects if they do exist.

[17] The others are the direct utility function $U(x, q)$, the indirect utility function $V(p, q, Y)$, and the distance function $d(x, q, U)$. The four can be shown to be either the dual or inverse functions of each other (Blackorby, Primont, and Russell, 1978). A slightly different form of the expenditure function $\mu(q \mid q^*, Y^*)$, which gives how much income the agent must have with level q of the public good to have the same level of utility as he experiences with level q^* of the public good and income Y^*, is known as the income compensation function and has been used by Willig (1976) and Randall and Stoll (1980). Because we are dealing with public goods where market prices infrequently exist, most of the economic theorems in this area are proved by using compensated and uncompensated inverse demand functions which depict price as a function of quantity. See Deaton and Muellbauer (1980), Anderson (1980), or Hanemann (1986a) for discussions.

If CS is positive, then q_1 is preferred to q_0 and the consumer would be willing to pay up to the point where his utility level was the same as it was in the initial situation.

Extension to the Hicksian equivalence welfare measures is straightforward. Equivalence surplus (ES) may be written as

$$\text{ES} = [e(p_0, q_0, U_1) = Y_0^*] - [e(p_0, q_1, U_1) = Y_1^*] \qquad (2\text{-}3)$$
$$\text{ES} = Y_0^* - Y_1^*,$$

where Y_0^* and Y_1^* are not in general the same Y_1 and Y_0 as in equation (2-2). If q_1 is preferred to q_0, we have a WTA amount, while if q_0 is preferred to q_1 we have a WTP amount.[18] Similar results can be obtained from altering the price vector p_0, and/or combinations of the price and quantity vectors.[19] It is possible to vary the q_1 in equations (2-2) or (2-3) so that valuation estimates for a number of possible quantities can be obtained. In some cases a valuation function can be estimated which gives CS or ES as a function of an arbitrary q_i—an issue taken up in the next subsection. If the quantity in equation (2-2) is negative, then the amount is the minimum compensation the consumer would accept.

Estimating a Fixed Quantity Change or a Valuation Function

The policy choices confronting a decision maker can be characterized in terms of the relationship between the current level of provision and a proposed change in that level. Changes can vary from a large, discrete (and possibly discontinuous) change from the existing level to small, incremental or marginal changes, often used by economists in stylized examples.[20] We will examine each of these extreme cases in turn from the perspective of the contingent valuation method.

For a discrete quantity change which can be specified in advance, contingent valuation is capable of obtaining the appropriate Hicksian measure (that is, equation 2-2 or 2-3) without having to estimate directly any form of either the Marshallian demand curve or one of the Hicksian compensated demand curves.[21] For this type of policy change, contingent valua-

[18] WTP measures are sometimes said to be income-constrained, while WTA measures are not income-constrained. This is strictly true only if borrowing is not possible.

[19] In some instances, such as an increase or decrease in a park entrance fee, the welfare measure of interest can be best reflected as a change between the initial and subsequent price vectors.

[20] Large discrete policy changes pose difficulties for standard economic theory because the appropriate utility functions are not differentiable with respect to q, and hence a unique optimal level of provision may not be identifiable. The standard linear approximation techniques used by economists are not generally well-suited to problems of this kind.

[21] This is not true for some of the discrete-choice formats used in contingent valuation surveys; see chapter 4 for a discussion.

tion enjoys at least one strong advantage over any of the commonly used indirect benefit measurement techniques (for example, hedonic pricing) which *rely* on the estimation of some type of demand curve.[22] To be able to consider the benefits (or costs) or arbitrary marginal changes in the level of the public good q, the CV researcher must also estimate some equation which gives WTP (or WTA) as a function of q. Specifically, one wants to estimate the parameters of the applicable inverse Hicksian compensated demand function

$$\Pi(p, q, T, U_i), \tag{2-4}$$

where p is a vector of prices for the private goods X, q, is the level of the public good, T is a vector of taste variables, and U_i is the level of utility being held constant. The Hicksian compensating surplus measure can be defined as

$$\int_{q_s}^{q_t} \Pi(p\ , q, T, U_0)\, dq, \tag{2-5}$$

while the Hicksian equivalence surplus measures can be defined as

$$\int_{q_s}^{q_t} \Pi(p\ , q, T, U_1)\, dq, \tag{2-6}$$

where q_t is preferred to q_s.[23] The inverse compensated demand function can be shown to be the derivative of the expenditure function (or the

[22] There are three sources of possible error when using one of the indirect valuation methods. (1) Estimating the ordinary demand curve which determines the Marshallian consumer surplus; this type of error may be quite large and usually occurs when the functional form of the ordinary demand curve is unknown (Bockstael and McConnell, 1980a, 1980b; Vaughan and Russell, 1982). (2) Going from the Marshallian consumer surplus to the appropriate Hicksian consumer surplus, discussed below. (3) Benefit components which do not share any weak complementarity with marketed goods may enter the consumer's utility function and hence will be uncounted (Freeman, 1979b). Some of the available indirect techniques are considered in more detail in chapter 3.

[23] A variant of the ordinary (Marshallian) inverse demand function, which involves an income adjustment as well as a price effect, can be represented by

$$\Pi(p, q, T, Y),$$

where now Y rather than U_i is held constant. A Marshallian-type consumer surplus for a quantity change from q_s to q_t is given by

$$\int_{q_s}^{q_t} \Pi = (p, q, T, Y)\, dq,$$

the area under the inverse ordinary demand curve from q_s to q_t. See Hanemann (1986a) for a discussion of the different inverse demand functions and the welfare measures associated with them.

income compensation function), so that equations (2-2) and (2-3) are equivalent to equations (2-5) and (2-6), respectively.

With regard to small incremental changes, estimation of the unknown parameters Θ, in the valuation function $\int\Pi(p, q, T, U_i; \Theta)$, is in general not an easy task since the appropriate taste variables are generally unknown, and even if known there are frequently only poor-quality proxies available for the variables suspected of comprising T.[24] The functional form for Π is also unknown, and there is even less guidance from economic theory on its estimation than for a private good. To the extent that these difficulties can be overcome, using data from CV studies to estimate a valuation function is highly desirable from a policy standpoint. Compared with knowing whether a small number of possible changes are potentially Pareto-improving, a valuation function is usable across a wide range of changes. Thus, even if a CV study is designed with a particular quantity change in mind, it is prudent for the researcher to ask the respondents to value at least one additional quantity change above and below the change of interest. The additional data can be used to determine how sensitive the benefit estimate is to that particular quantity, in addition to making it possible to estimate a valuation function if that becomes a useful thing to do.[25]

Uncertainty and the Ex Ante versus Ex Post Perspectives

Thus far we have assumed certainty. Under this condition, where agents know how much utility they would get from the availability of a public good, the Hicksian welfare measures are appropriate. With the introduction of various types of uncertainty this is no longer necessarily true, and new welfare measures must be defined. This is because the utility an agent expects to receive from a good before knowing what the state of the world will be (the *ex ante* perspective) may differ dramatically from the utility that agent expects to receive after he knows what the state of the world will be (the *ex post* perspective).[26] For most welfare economic purposes, the ex ante perspective is generally considered to be the most appropriate one in situations where uncertainty about outcomes is involved (Graham,

[24] The estimation task is easier if the actual values of $\int\Pi(p, q, T, U_i; \Theta)\, dq$ are available, but it is still possible to estimate some of the Θ if only discrete indicators of $\int\Pi(p, q, T, U_i; \Theta)\, dq$ are available. See Hanemann (1983a, 1984a) for a discussion of how to define the Hicksian consumer surpluses in these cases.

[25] The early Randall, Ives, and Eastman (1974) CV study which set out to measure a valuation function by measuring willingness to pay for three different q has become the norm for many CV studies. Measuring WTP for more q than this is desirable, but runs the risk of overburdening the respondents.

[26] As an example, consider the decision of whether to build a house on a flood plain. Assuming a probability of a flood once every hundred years, an ex ante decision will be quite different from an ex post decision based on knowledge that it will flood next year.

1981; Chavas, Bishop, and Segerson, 1986). Many different rationales have been offered for this view. The principle of consumer sovereignty suggests that policymakers should attempt to implement the current desires of the public—clearly an objective with an ex ante perspective.[27]

Smith (forthcoming) has advanced the concept of a planned expenditure function which allows one to define equations directly analogous to (2-2) and (2-3), or, equivalently, (2-5) and (2-6).[28] The quantities so defined are sometimes referred to as option prices, and are generally considered to be the correct measure of welfare changes in the presence of uncertainty (Graham, 1981; Smith, forthcoming). The implications of the uncertainty issue for the measurement of benefits are considered in the next chapter.

Willingness-to-Pay versus Willingness-to-Accept Measures

Whether an elicitation question in a contingent valuation survey is phrased in terms of willingness to pay or willingness to accept depends on which Hicksian consumer surplus measure the researcher wants to obtain. The choice between the WTP or WTA formulation is a question of property rights: does the agent have the right to sell the good in question or, if he wants to enjoy it, does he have to buy it?[29] Since we are dealing with public goods where rights are held collectively, this question is often not an easy one to answer.

The traditional way of measuring benefits is based on the estimation of some form of the ordinary demand curve which allows the researcher to obtain the Marshallian consumer surplus. The Marshallian consumer surplus avoids the problem of having to make a decision about the appropriate property right, but it suffers from the serious flaw of not being a true measure of an agent's welfare change. As a result, the CV researcher is faced with the task of deciding which Hicksian consumer surplus measure to use for a given welfare change. Should it be one based on WTP, or one based on WTA? The contingent valuation method provides the only way of directly measuring both WTP and WTA. Over the past decade CV

[27] The most obvious problems with the ex ante perspective occur when policymakers believe the public's ex ante and ex post values will differ significantly, or when a group of respondents believes that the government will bail them out of undesirable ex post outcomes (Kleindorfer and Kunreuther, 1986; Kunreuther, 1976).

[28] The planned expenditure function holds an individual agent's ex ante expected utility constant rather than holding expected utility constant in each possible future state of the world.

[29] There is a long and extensive literature on property rights in both law and economics (see Calabresi and Melamed, 1972; Coase, 1960; Furubotn and Pejorvich, 1972; Hirsch, 1979; Knetsch 1983, 1985; Polinsky, 1979; Posner, 1977). From the perspective of economic welfare in general and contingent valuation in particular, perceived property rights and entitlements may be more important than actual legal ones.

researchers have consistently obtained from their respondents WTA values which are considerably larger than the WTP values for the same amenity.[30] This result, which runs counter to theoretical expectations, has engendered considerable and continuing controversy. The basic issues are the theoretical one of how far apart WTP and WTA Hicksian consumer surplus measures should be, and the methodological one of whether WTP and WTA elicitation questions are equally plausible to respondents. In this section we trace the controversy as it unfolded and develop an explanation for the phenomenon which draws on recent theoretical work by Hanemann.

The first major contingent valuation study to report estimates of both WTP and WTA was Hammack and Brown's (1974) study of waterfowl benefits. They found that their respondents' WTA amounts were a little over four times larger than the same respondents' WTP amounts for the same amenity. In a concurrent development, Willig (1973, 1976) published his seminal theoretical work in welfare economics. Willig resurrected the use of the Marshallian consumer surplus as a welfare measure by showing that the Marshallian consumer surplus lay between WTP and WTA. According to his theoretical analysis, the bound between WTP and WTA was relatively small under most plausible situations.[31] Hammack and Brown's WTP<WTA difference was accordingly dismissed as an artifact of their particular contingent valuation survey instrument.

It was soon noted that Willig's theoretical results were for policies which changed *prices* for consumers with "well-behaved" utility functions. This implied that the Willig bounds might not be appropriate for some commonly estimated but "degenerate" demand functions, for cases where the provision of public goods involves quantity changes, and in particular for policy changes where the quantity of the good provided goes to zero. Randall and Stoll (1980) succeeded in extending Willig-type bounds to general quantity changes.[32] Although Randall and Stoll's results suggested that Willig's basic findings—that WTP and WTA were fairly closely bound if expenditures on the good in question were small relative to total income, and that the Marshallian consumer surplus lay between the two Hicksian measures—were correct, subsequent CV surveys continued to find the same type of large differences between WTP and WTA that Hammack

[30] In this book the term "amenity" is used as a synonym for a change in a public good which is valued in a benefit measurement study.

[31] The Willig bounds suggested that the substitution of the Marshallian consumer surplus for the correct Hicksian WTP or WTA measure would result in an error of 2 percent or less in a known direction.

[32] Willig-type bounds have been clarified or extended in a number of papers, among them Bockstael and McConnell (1980a, 1980b), Hanemann (1980), Takayama (1982), Just, Hueth, and Schmitz (1982), and Stoker (1985).

and Brown had observed. These surveys also revealed another difference in the way their respondents reacted to the WTP and WTA questions: many more respondents gave protest responses or infinite values when asked how much they would accept in return for forgoing the amenity than when asked how much they would pay to receive it. These reactions supported the notion that the large WTA amounts measured in CV surveys were a methodological artifact.

The Randall and Stoll finding that WTA and WTP were closely bound suggested a resolution to the CV researchers' measurement dilemma. On the one hand, questions posed in a WTA format did not seem to elicit meaningful responses from Randall and Stoll's respondents; on the other hand, WTA was thought to be the theoretically correct measure for certain of the changes they wished to value. Why not substitute WTP for WTA in those cases where theory called for WTA question formats (Brookshire, Randall, and Stoll, 1980; Mitchell and Carson, 1981; Desvousges, Smith, and McGivney, 1983; Cummings, Brookshire, and Schulze, 1986), on the grounds that any error induced by this expedient would be minor according to the Randall and Stoll (1980) bounds?

The Randall and Stoll bounds on the quantity changes were a little looser than the Willig bounds on price changes, and involved an additional unknown parameter which they called the price flexibility of income, ξ. Their analysis suggested that measures of WTP and WTA should be within 5 percent of each other under the conditions typical of most contingent valuation studies. Brookshire, Randall, and Stoll (1980) showed how an upper bound on WTA could be calculated from WTP if an estimate of the price flexibility of income were available. They further suggested (1980:484–495) that ξ could be estimated in a manner analogous to an ordinary income elasticity by regressing the log of willingness to pay for the good in question on the logs of the quantity of the good, income, and other possible explanatory variables.

As an illustration of how small the differences among the three consumer surplus measures could be under the Randall and Stoll bounds, consider an example based on the estimates of WTP for national water quality benefits presented in Mitchell and Carson (1981). Mitchell and Carson take a household with an income (Y) of $18,000, a WTP amount of $250 for an increase in the level of water quality from non-boatable to swimmable, and an estimate of the price flexibility of income ξ, for the good of .7 obtained from a regression equation. Using Randall and Stoll's (1980) equations (11) $(\mathrm{M} - \mathrm{WTP})/\mathrm{M} \simeq \xi\mathrm{M}/2Y$ and (13) $\mathrm{WTA} - \mathrm{WTP} \simeq \xi\mathrm{M}^2/Y$, and the information above, we can solve for the approximate values of the two unknown measures, WTA and M, and the Marshallian consumer surplus. The WTP measure is the compensating surplus, WTA is the equivalence surplus, so we have CS ($250) < M ($251.22) < ES

($252.45). Thus the difference between WTP and WTA consumer surplus is approximately 1 percent, or $2.50. Increasing ξ to 1.0 increases this difference to about $3.50. Considering the other possible sources of error in CV studies, differences of this magnitude between the WTP and the WTA formats are small indeed.

Methods of estimating the Hicksian measures exactly, given only an ordinary demand curve, were soon developed for price changes (Hausman, 1981; Vartia, 1983) and for quantity changes (Langkford, 1983; Bergland, 1985). These methods were exact in the sense that, conditional on the ordinary demand curve being known, there was no error in going from that demand curve to estimates of the Hicksian consumer surplus. The papers on exact estimation tended to confirm the validity of the Willig-Randall-Stoll bounds.[33]

As this theoretical work was appearing, a number of contingent valuation studies continued to use both the WTA and WTP formats. Their results showed the same resistance to the WTA formulation and, for those respondents who did not protest or give infinite answers, the same large violations of the implied bounds between WTP and WTA as before.[34] Meanwhile, new evidence began to accumulate which appeared to contradict the conclusion that the WTP < WTA findings were solely a methodological artifact. The evidence came from experiments where researchers used a CV-type format but had their respondents actually buy and sell quasi-private goods using real goods and actual dollars. These experiments included large-scale simulated markets conducted outside the laboratory by Bishop and Heberlein and their colleagues (1979, 1980, 1983, 1984, 1985, 1986) to value commodities highly desired by a select population (special goose- and deer-hunting permits),[35] as well as several innovative laboratory experiments using a variety of different types of goods (Knetsch and Sinden, 1984; Gregory, 1986). Since these studies also found large WTA–WTP differences, it appeared that the WTP < WTA differences found in the CV studies could not be attributed to the hypothetical nature

[33] The "exact" methods have particular advantages when calculating dead weight loss from tax changes, something for which the Willig bounds were not particularly tight. They have the very strong disadvantage (being derived from duality theory [that is, Roy's identity; see Varian, 1984]) that the researcher is required to know (or maintain as an assumption) the functional form of the utility function.

[34] See Cummings, Brookshire, and Schulze (1986), and Fisher, McClelland, and Schulze (1986) for summaries of these studies.

[35] Bishop and Heberlein, in their often-cited 1979 *American Journal of Agricultural Economics* paper, assumed, on the basis of Willig's theory, that the WTA and WTP amounts should be roughly equivalent. This led them to interpret the difference between their WTP amount from a contingent market (in this experiment they were not able to actually sell their goose-hunting permit) and their WTA amount from a simulated market as an indication of the CV method's methodological weakness.

of contingent valuation questions. Attention finally turned from explaining why CV surveys were eliciting incorrect WTA values to why there might be real differences between WTP and WTA, in contradiction to received theory.

A number of hypotheses were put forth to explain the WTP < WTA difference.[36] Some focused on contingent valuation surveys, while others considered a more general class of observed and contingent choice behavior.[37] The proposed explanations may be classified into four main types: (1) rejection of the WTA property right, (2) the cautious consumer hypothesis, (3) prospect theory, and (4) other modifications and reinterpretations of the received economic theory.

According to the first of these explanations, people are motivated to give higher WTA values because they reject the property rights implied by the willingness-to-accept format. As noted, CV studies using WTA questions have consistently received a large number of protest answers, such as "I refuse to sell" or "I want an extremely large or infinite amount of compensation for agreeing to this," and have frequently experienced protest rates of 50 percent or more. It appears that respondents who give protest answers regard the WTA property right as either implausible or illegitimate or both. However, manifest rejection of the WTA format is much less common when actual cash is offered under simulated market conditions (Bishop and Heberlein (1979, 1980), a finding which questions the extrapolation of the protest model to cover those respondents in CV studies who give WTA amounts which are not so high that they have to be regarded as outliers.[38]

The second hypothesis is that consumers are cautious, particularly in contingent valuation surveys. Hoehn and Randall (1983, 1985a, 1987) argue fairly persuasively that consumers who are uncertain, who lack the time to optimize the decision about the good in question, or who are risk-averse will tend to give a lower willingness-to-pay amount and a higher willingness-to-accept amount than they would under conditions of

[36] On this subject, see Hammack and Brown (1974), Gordon and Knetsch (1979), Bishop and Heberlein (1979), Rowe, d'Arge, and Brookshire (1980), Brookshire, Randall, and Stoll (1980), Schulze, d'Arge, and Brookshire (1981), Bishop, Heberlein, and Kealy (1983), Knetsch and Sinden (1984), Brookshire, Coursey, and Radosevich (1986), Fisher, McClelland, and Schulze (1986), Gregory and Bishop (1986), Cummings, Brookshire, and Schulze (1986), Gregory (1986), Coursey, Hovis, and Schulze (1987), Hoehn and Randall (1983, 1985a, 1987).

[37] Some of the early speculation centered on whether income effects could be large enough to generate the WTP–WTA differences in CV surveys; this hypothesis was quickly rejected (Gordon and Knetsch, 1979).

[38] The methodological (as opposed to the theoretical) problems with the WTA format are discussed at more length in chapter 9.

certainty, risk neutrality, and unlimited time to engage in optimal planning.[39] Hoehn and Randall have proposed that these differences will converge if people are given the chance to become familiar with making WTP and WTA value judgments. To test this proposition, Coursey and his colleagues (Brookshire, Coursey, and Schulze, 1986; Brookshire, Coursey, and Radosevich, 1986; Coursey, Hovis, and Schulze, 1987) carried out laboratory experiments using both formats with real goods and money. They interpreted their findings as showing that WTP and WTA amounts do converge as subjects gain experience in repeated trials. However, most of the change comes in the form of decreases in WTA, while the WTP amounts are often quite stable over trials and close to the initial hypothetical WTP responses given by subjects.[40]

The third hypothesis is based on prospect theory (Kahneman and Tversky, 1979),[41] which replaces utility theory's emphasis on final asset positions with a descriptive framework for analyzing preferences based on gains or losses from a neutral reference position. According to prospect theory, the value function is steeper for losses than for gains. Consider an equivalent change in utility under two conditions, *a* and *b*. In *a* the change is a loss, in *b* a gain. In contrast to standard expected utility theory (Schoemaker, 1982), which assumes that people value *a* and *b* the same, prospect theory holds that people's value for *a* will be greater than the value they hold for *b*. Therefore, to the extent that the WTA format implies the relinquishment of a good now held or to which one is entitled (Gregory, 1986), prospect theory predicts a higher amount of compensation will be demanded. The fixed-quantity nature of public goods provision not only tends to accentuate this notion of a difference between a loss and a gain of the same magnitude, but also has the property of frequently implying a one-time discrete choice about whether or not the respondent will ever obtain the good in question.

[39] The behaviorial model proposed by Hoehn and Randall is discussed at more length in chapter 7, where we consider its implications for strategic behavior in CV surveys. We argue there that it is unlikely that strategic behavior significantly influences WTP responses, although the same statement cannot be made about WTA responses.

[40] The Coursey, Hovis, and Schulze (1987) paper concludes that by the final trial, the mean WTP and WTA amounts are not statistically different from each other. Gregory and Furby (1985) reach the opposite conclusion based on a reanalysis of the same data, using somewhat different statistical techniques. The major problem with the Coursey, Hovis, and Schulze experiment is that the N is too small to give enough statistical power to detect any but very large differences between the two measures. We believe that they have convincingly demonstrated a tendency of WTA to converge *toward* willingness to pay, but we would argue that a demonstration of actual convergence between WTP and WTA has not yet been shown to generally occur with experienced subjects.

[41] See also Kahneman and Tversky (1982), Tversky and Kahneman (1982), Machina (1982), Knetsch and Sinden (1984), and Gregory (1986).

The fourth hypothesis to account for the WTP < WTA difference is based on Hanemann's theoretical work (1982, 1983b, 1984b, 1984c), which argues that the one-time discrete choice nature of many contingent valuation surveys has important and unresolved implications for the possible differences between WTP and WTA values. In a few special cases, Hanemann was able to demonstrate that the differences between WTP and WTA could be quite large and yet still be consistent with standard economic theory. Then in 1986 Hanemann put forth theoretical arguments for the general constrained quantity case which showed that the difference between WTP and WTA depended on two unknown parameters, and that in contrast to the Randall-Stoll bounds, quite reasonable values of those parameters could lead to large differences between WTA and WTP values. Hanemann showed that Randall and Stoll's (1980) formulation of the problem was correct, but that the key unknown parameter which they called the price flexibility of income was a very special term which could not be interpreted in a manner comparable to a standard income elasticity. Thus reasonable values for this parameter need not be those in the usual range of income elasticities.[42] Hanemann demonstrated that the price flexibility of income parameter ξ was really the ratio of two other elasticities,

$$\xi = \frac{\eta}{\sigma_0}, \tag{2-7}$$

where η was an income elasticity and σ_0 was an elasticity of substitution between the public good being valued and all other goods in the economic system.[43] If we assume the not-too-unreasonable values of, say, $\eta = 2$, and $\sigma_0 = .1$, ξ now equals 20. Applied to the water quality example discussed earlier, these values result in rather sizable differences among the three measures of CS (\$250) < M (\$300) < ES (\$350). While much larger values of η are somewhat improbable, much smaller values of σ_0 for a number of public goods are quite plausible.

The import of Hanemann's (1986a) findings are that for public goods with many close substitutes, σ_0 will be large, with the result that ξ will be small and WTP and WTA should be close together. For a public good with substitution possibilities such that σ_0 is small relative to η, WTP and

[42] Except for a few very distinct luxury goods, income elasticities are infrequently above 2 and almost never above 5. Houthakker and Taylor (1970) provide income elasticity estimates for a large number of consumer goods.

[43] Technically, Hanemann showed η to be an income elasticity of the direct ordinary demand function for q, while σ_0 is the aggregate Allen-Uzawa elasticity of substitution between q and the Hicksian composite commodity x_0.

WTA can be quite far apart, possibly infinitely so if $\sigma_0 = 0$.[44] Further, the elasticity of WTP with respect to income as typically obtained from a regression estimate of the valuation function (equation 2-5) is not an estimate of ξ, so that it is hard to draw any reliable inference on the magnitude of ξ from the data available in a contingent valuation survey.[45]

It is our view that the WTP < WTA difference may be accounted for by a combination of the factors identified in these hypotheses. Thus, respondents have trouble with the WTA format because they do not find it plausible. At the same time, there are strong theoretical grounds for assuming that WTA cannot be replaced by a WTP measure without biasing the findings. Hanemann has shown that the rationale based on the Randall-Stoll bounds is not valid for public goods with low substitution elasticities (a situation which is probably fairly common); and the prospect or non-expected utility theory is gaining support from within the economics profession. The result is that contingent valuation researchers continue to be faced with a dilemma: asking people to accept payment for a degradation in the quantity or quality of a public good simply does not work in a CV survey under many conditions, yet substituting a WTP format where theory specifies a WTA format may grossly bias the findings. This in turn poses a quandary, since researchers frequently wish to value quantities on both sides of the current level of provision of an amenity, and it is generally agreed that the correct measure for a decrease is the Hicksian compensating surplus WTA measure.

We believe that although it may be possible to design successful WTA contingent valuation questions in some situations, particularly if a referendum-type CV format is used,[46] the task will not be easy. However, there are grounds for believing that a WTP measure is the correct format for valuing decreases in the level of provision of a large class of public

[44] If there are frequent opportunities to obtain the public good in question and if provision of it in one period is a good substitute for it in another period, σ_0 will tend to be large. A large irreversible change in the level of the public good will tend to imply a small σ_0, particularly if the good has few other close substitutes.

[45] The parameter ξ is equal to $\partial \ln \Pi(p, q, Y)/\partial \ln(y)$, which in general will not be equal to $\partial \ln \Pi(p, q, U_i)/\partial \ln(Y)$, the parameter estimated in a valuation function. The relationship between the two parameters is somewhat complex. Deriving ξ from the valuation function coefficient on income requires the assumption that the mathematical form of the utility function be known. This is also the problem with the so-called exact methods (for example, Hausman, 1981). Specification of a particular form for the utility function often implicitly bounds ξ, a priori. Further, the specification of WTP function is subject to all of the problems of specifying ordinary demand functions.

[46] Mitchell and Carson (1986c) advocate the use of a WTA referendum for siting waste disposal facilities, and we have seen such a referendum successfully approved by voters in a Connecticut town (Winerip, 1986).

goods that previously were thought to require a WTA measure—an issue we now turn to.

A New Property Rights Approach

To the WTP–WTA dilemma we propose a partial resolution based on a rethinking of the property right implied in those public goods which require annual payments or their equivalents to maintain a given level of the good.[47] A number of important public goods have this characteristic. Air quality, for example, would rapidly decline if no money were spent by business and government on control measures. For this type of public good, neither ownership nor use per se captures the relevant relationship between the amenity and the consumer.

The implications of this observation are summarized in table 2-2, where column A summarizes the conventional understanding of the correct surplus measures. When private ownership is assumed, as it is in the column A model, the relevant dimensions that define the property right are whether the individual owns the good or not and whether he uses it or not. In this model, CS_{WTA} and CS_{WTP} are indicated for a decrease and an increase in the status quo, respectively.

A different set of dimensions is needed to define the property right to a public good which requires annual payments to maintain a given level. As shown in column B, the first dimension is whether the public good is individually or collectively held. Every member of a relevant collectivity (city, nation, condominium owners' association, and so on) has a common property right to the good, although individual members may be granted differential access (often for a fee, and on an equal basis) by the relevant governing body. For some goods, access may convey exclusive-use rights. Although, from the individual consumer's perspective, it is possible to identify a fairly wide variety of property rights for public goods, our purpose can be served by working with two basic dimensions, collectively held and individually held rights.

Collectively held rights occur where access (or potential access) to the good is available to all members of the collectivity, and individual members cannot sell their access right. Air and water quality are excellent examples of goods to which individual consumers have collective, nontransferable property rights of this kind.[48] If there is a cost to providing

[47] Gordon and Knetsch (1979) appear to have had this type of property right in mind when they said that one of the two questions a benefit-cost analyst could ask was, What is the user willing to pay to continue the present use?

[48] The concept of collective rights must be regarded as an ideal type, however, since even air and water quality are transferable under certain types of administrative arrangements such as those which allow corporations to buy and sell permits to emit specified levels of

Table 2-2. Hicksian Surplus Measures for Private and Public Goods

	A. Private goods		B. Public goods[a]		
	Own	Not own		Individually held	Collectively held
Use	CS_{WTA} (decrease)[b]	ES_{WTP}	Level currently accessible	CS_{WTA}	CS_{WTP} (decrease)[b]
Do Not Use	ES_{WTA}	CS_{WTP} (increase)[b]	Level not currently accessible	CS_{WTP}	CS_{WTP} (increase)[b]

[a] For public goods which require annual payments (or their equivalents) to maintain a given level of the good.

[b] Indicated measure for a decrease (increase) in the amenity from the status quo.

the good at a given quality level, it is normally borne by all consumers through some combination of taxes, higher prices, fees, and the like. If the level of payment is not maintained, the quality of the good will often deteriorate—a condition which holds for many environmental amenities, and which we will assume in the discussion that follows. If a quality increase is desired, higher payments will be needed to cover the cost of providing the new quality level. The less excludable the good, the more likely it will be collectively held, since entrepreneurs cannot efficiently provide it at a profit. The appropriate analogy for this type of public good is not marketed goods, but maintenance fees such as those paid by condominium owners. Purchase of a condominium conveys private property rights to the apartment itself, but condominium owners are also legally obligated to pay fees, whose level is collectively determined, to maintain the property and its grounds. Owners may collectively agree to increase their fees in order to provide a more lavish common amenity. Conversely, nonpayment of fees or a reduction in their level would lower the quality of the common amenity. All owners have equal rights to use these nondivisible collective goods.

Individuals are sometimes granted exclusive rights to use a public good by the collectivity because such a grant is deemed to serve the public interest. Typically, these individually held goods are excludable and subject to congestion. Various allocation rules are used, ranging from auctions to free grants based on competence, first-come-first-served, or other principles. Occasionally, as is the case with mining claims on public land and broadcasting frequencies, the rights are transferable. In these cases the public good has not been fully transformed into a private good because

pollution. The opposition to these arrangements by environmental groups is motivated in part by a belief that rights to these public goods should not be transferable (Kelman, 1981; Bohm and Russell, 1985; Tietenberg, 1985).

the government still maintains an interest in the right and can revoke it. Someone who wishes to purchase a broadcast frequency from an existing license holder, for example, must meet certain governmentally imposed criteria in order to take possession. More commonly, the collectivity grants a nontransferable right. Such rights are granted free to wilderness users through the allocation of wilderness permits, and auctioned or otherwise allocated for fees to those who wish to use public lands to mine coal or drill for oil, to harvest trees, or to graze livestock.

The second dimension for determining the appropriate surplus measure for a public good is whether or not a given level of quality is accessible. We use the concept "accessible" (Bohm, 1977; Gallagher and Smith, 1985; Smith, 1986c) instead of "actual use" because public goods often involve various kinds of existence values in addition to use values (see chapter 3). At the present time water in the nation's lakes, rivers, and streams meets a minimum standard (level) of boatable quality; among the currently unavailable (unmet) minimum standards for freshwater bodies are fishable or swimmable levels of quality. This framework for conceptualizing the property right to public goods has important implications for the choice of the correct Hicksian surplus measure for contingent valuation surveys. Our aim is to measure the benefits of these goods from the consumer's initial level of utility. Where a given quality level of a public good is not currently available, a CS_{WTP} measure is indicated to determine the value of the increased provision, just as it is for determining the consumer surplus for a private good which an individual neither owns nor currently uses. In both cases, the measure is the amount the consumer is willing to pay for the improvement which leaves him just as well off before the change as after.[49]

[49] A conceptual problem with the theoretical underpinnings of our proposed property rights framework bears mention here. As noted earlier, the status quo property right is $U_0 = V(p, q_0, Y_0)$. Another way of expressing U_0 is $V(p, t_0, Y_0|q_0)$, where t_0 is the current tax price of q_0 and the consumer is forced to consume q_0. We would like to be able to measure

$$CS = (Y_0 + t_0) - e(p, q_i, U_0)$$

for an arbitrary q_i. Thus far this is a well-defined measure. The problem comes when we ask: What is the respondents' utility level after taking away q_0 and giving back t_0? If q_0 can still be purchased for t_0, it is clear, using the LeChatelier principle (Samuelson, 1947; Silverburg, 1978), that

$$V(p, t_0, [Y_0 + t_0]) \geq V(p, t_0, [Y_0 + t_0]|q = q_0),$$

because the respondent no longer needs to purchase q_0 if spending t_0 on other goods increases his utility more than the purchase of q_0 does. The function $V(p, Y_0 + t_0)$ is in a sense completely specified without reference to the price (or nonavailability) of q_0, and hence it is unclear what level of utility respondents think they have after t_0 has been given back to the respondent and q_0 has been taken away. We believe that respondents will evaluate $V(p, Y_0 + t_0)$ as approximately equal to U_0, and that the ambiguity in $V(p, Y_0 + t_0)$ will not be troublesome in practice.

In the public good case, however, a CS_{WTP} measure is also indicated for a proposed decrease when a given quality level is currently available. Since the consumer is already paying for the good on a regular basis, the Hicksian compensating surplus for this case is the amount the consumer is willing to pay to forgo the reduction in the quality level of the good and still be as well off as before. This is measured in a CV survey in the following way. The respondent would first be informed that she is already making annual payments in some relevant form—higher prices and taxes, for example—to provide the current quality level of a good such as air visibility.[50] She would then be asked to state the maximum payment (which could be the present payment) that she is willing to make to preserve this quality level before she would prefer a quality reduction. To use a referendum analogy, the consumer is asked to set the highest amount she would be willing to pay annually in taxes for a given program which guarantees to maintain the present level of supply of a good for the next and succeeding fiscal years. It should be noted that the WTA format is clearly inconsistent with the nontransferable character of this property right.

Aggregation Issues

Once a contingent valuation survey has obtained the correct theoretical measure for a sample of individuals, the researcher aggregates these values to obtain the total benefits for the good being valued. This procedure raises two questions: is it possible to add up the WTP amounts obtained for individuals in a CV survey to get an aggregate benefit for the relevant population? and is it possible to aggregate these amounts across subcomponents to obtain the relevant total benefit? Another issue is how to present the disaggregated and distributional information on the public's willingness to pay (or accept), which is often key information for policymakers. This information is also useful in the design of mechanisms for collecting the public's willingness to pay for the good in question. These issues are considered in the following subsections.

Aggregation of Individual Benefits

Samuelson, in his seminal 1954 article on public goods, showed that the correct method of determining the aggregate demand for public goods is

[50] In a CV context, making the respondent aware of her current payments for the good in question and indicating that those payments would no longer be required is potentially troublesome, since a respondent would be reluctant to give a WTP amount for the current level of provision which is larger than the current payment revealed for that level, even if her compensating surplus for the good is much larger. In practice this problem may not be too severe if the respondent is simply told that she would get back whatever it was costing, and both the actual current payment and the perceived current payment were small relative to income.

to vertically sum the individual demand curves, rather than to horizontally sum individual demand curves as one would do in the case of private goods. Bradford (1970) expanded Samuelson's analysis to take into account the fact that public goods are provided in fixed quantity, and proposed the concept of an aggregate bid/valuation curve (sometimes referred to as a total value curve), which he defined as the vertical summation of the individual bid/valuation curves, or the appropriate consumer surplus measure at that level of provision. Bradford was able to demonstrate that over any relevant range, the aggregate bid/valuation curve and its corresponding marginal bid/valuation curve were not necessarily continuous or downward sloping. This is a troublesome finding, since if these conditions do not hold it may not be possible to identify the optimal Pareto level of provision.

In addition to these potential problems with the relevant economic functions, multidimensional scaling or indexing problems exist for many public goods—for environmental quality, in particular.[51] This means that an agreed-upon interval quantity scale or index may not be available for the public good being examined. Beyond that, the costs and benefits of providing a public good are frequently measured on different scales.[52] Getting both the cost and the benefit dollar estimates in terms of the same physical quantity-quality axis is sometimes a very difficult task, if not an impossible one. It is difficult, for example, to translate changes in sulfur emissions into changes in visibility improvement.

If the aggregate cost function is known, the marginal cost curve is derivable from it, and if all of the relevant benefit and cost curves are assumed to be continuous with the appropriate curvature properties,[53] then the Bradford framework resembles the traditional profit maximization framework in which the optimal production of the public good occurs where the marginal aggregate bid/valuation curve and the marginal aggregate cost curve intersect (see figure 2-3). What is being optimized here (assuming a given weighting function) is total social welfare or utility rather than profits. This intersection can be shown to be the point where

[51] These problems are generic to every type of benefit measurement technique, not just CV. For a discussion of the indexing issue with regard to water quality, see appendix E in Desvousges, Smith, and McGivney (1983); for a more general discussion of environmental metrics, see Evans (1984).

[52] For example, the costs of providing clean water are frequently specified in terms of treating a specific level of biological oxygen demand in a water body, while the benefits of clean water are frequently specified in terms of the recreation activity that a water body will support.

[53] These assumptions essentially eliminate all of the potential problems just raised. Starrett (1972) has argued that nonconvexities may be common when externalities exist. Brookshire, Schulze, and Thayer (1985) have found nonconvex marginal valuation functions (increasing WTP for increasing "units" of air visibility) in their CV study of the Grand Canyon.

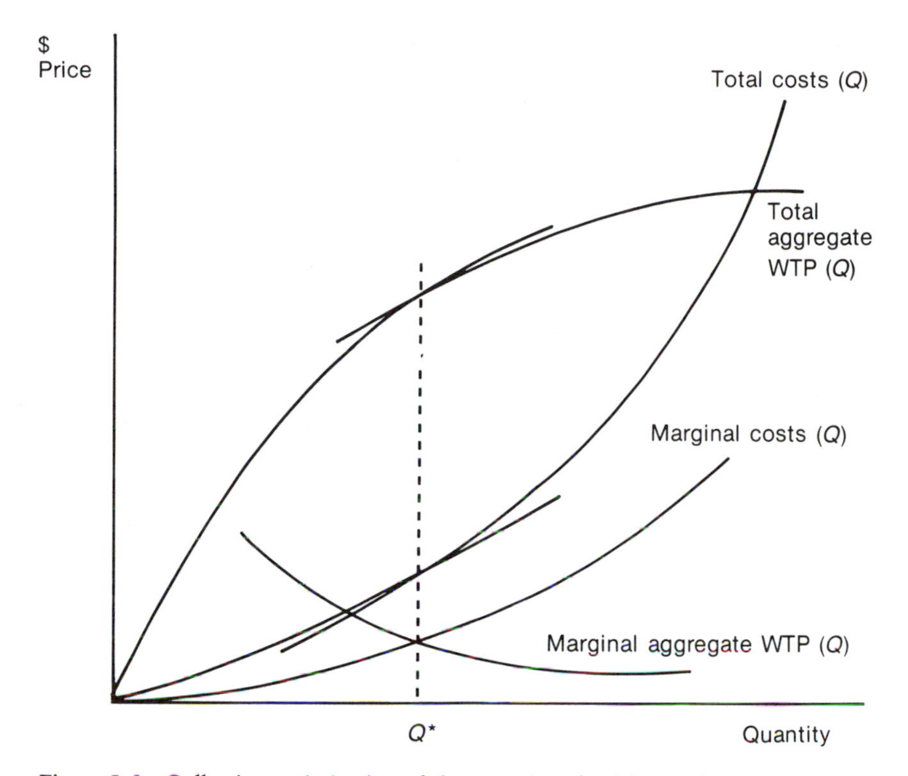

Figure 2-3. Collective optimization of the quantity of public good provided

the rate of commodity substitution equals the rate of technical substitution, which is a necessary condition for Pareto optimality. The equilibrium achieved by providing the public good at this level is known as a Lindahl equilibrium.[54] Randall, Ives, and Eastman (1974) first used and expanded the Bradford framework, which has been part of the theoretical basis for most subsequent contingent valuation studies.

Often what are compared in a benefit-cost analysis are total benefits and total costs rather than the marginal relationships discussed above. This is a result in part of the much greater difficulty, noted earlier, of estimating the valuation (and cost) functions necessary to calculate the

[54] A Lindahl equilibrium has the property, under the assumption of truthful preference revelation, that all individuals in an economy prefer the quantity of the public good provided (given a vector of tax prices, possibly individualized, for each agent in the economy) to any other quantity of the public good and vector of tax prices to which every agent was agreeable. This equilibrium concept is equivalent to a market clearing price and quantity in a market having only private goods. See Johansen (1963), Milleron (1972), or Cornes and Sandler (1986) for detailed discussions.

marginal relationships. The comparison of totals is capable of determining whether a potential Pareto improvement is possible, but it will not tell decision makers whether or not the change represents a move to a Pareto optimal position.

In order to use the findings of a CV study to obtain the correct estimate of either aggregate individual WTP amounts for a specific quantity of the public good or the aggregate valuation function, it is necessary to make several assumptions which are potentially troublesome.[55] First, a weighting scheme for individuals must be chosen. Since WTP (but not WTA) involves an income constraint, the standard weighting scheme is to assume that the current distribution of income is acceptable from a social welfare standpoint.[56] The second assumption typically made is that a suitable payment structure (which may differ from the present payment structure) could be designed to collect all the revenues the respondents in a CV survey indicate they are willing to pay. For example, there may be an upper limit to what respondents would regard as a legitimate charge to gain entrance to a recreational site, a limit that is lower than their true Hicksian compensating surplus for that amenity. A suitable payment structure would make it possible to capture the complete compensating surplus.

Subcomponent Aggregation

Contingent valuation researchers often want to combine separately measured components of a benefit. These components may be different benefits for separate geographical locations such as the Grand Canyon, the Rocky Mountain region, and the eastern United States; or benefits for different parts of a larger program, such as the air and water quality components of the national environmental program. In a series of important papers, Hoehn, Randall, and Tolley (Hoehn and Randall, 1982; Hoehn, 1983; Randall, Hoehn, and Tolley, 1981; Tolley and Randall with coauthors, 1983; Randall and coauthors, 1985) show why, under many conditions, independently measured (by contingent valuation) subcategory benefits cannot be aggregated without overcounting. They also show that when subcategory benefits are measured sequentially in the same study, the order in which the subcategories are presented to respondents influences the values ascribed to each, with the goods valued first receiving higher values than later-mentioned goods, other things being equal. They attribute this behavior to the fact that respondents value each good sequentially

[55] Methodological considerations such as sampling design and execution and response problems, which are relevant to the aggregation of WTP amounts, are discussed in chapter 12.

[56] This assumption is equivalent to saying that each dollar a person is willing to pay has an equal weight in a (equality of dollars) social welfare function. Other weighting schemes are obviously possible.

as if it were a marginal increment to the existing set of environmental amenities they enjoy, instead of valuing each good as an initial possible increment.[57] We present here a heuristic summary of Hoehn and Randall's (1982) theoretical analysis of this problem, and will discuss the practical implications of the problem at greater length in chapter 12.[58]

Take the simplest case of a policy designed to affect three different components (q_1, q_2, q_3) which have an initial level denoted q_1^0, q_2^0, q_3^0) and a goal of q_1^1, q_2^1, q_3^1). Using the expenditure function, WTP for changing q_1^0 to q_1^1, holding the other components at their initial level, is given by $e(q_1^1, q_2^0, q_3^0, p^0, U^0) = Y_1^1$, where similar representations are possible for each of the other two components valued separately. If $e(q_1^0, q_2^0, q_3^0, p^0, U^0) = Y^0$, simultaneous valuation of all three components may be represented as

$$Y^0 - e(q_1^1, q_2^1, q_3^1, p^0, U^0) \tag{2-8}$$

Except in the most unlikely of circumstances, this expression will not be equal to

$$\begin{aligned} 3Y^0 &- e(q_1^1, q_2^0, q_3^0, p^0, U^0) - e(q_1^0, q_2^1, q_3^0, p^0, U^0) \\ &- e(q_1^0, q_2^0, q_3^1, p^0, U^0)^{59} \end{aligned} \tag{2-9}$$

It can, however, be shown that

$$Y^0 - e(q_1^1, q_2^1, q_3^1, p^0, U^0) \tag{2-10}$$

is equal to

$$\begin{aligned} &Y^0 - e(q_1^1, q_2^0, q_3^0, p^0, U^0) \\ &+ e(q_1^1, q_2^0, q_3^0, p^0, U^0) - e(q_1^1, q_2^1, q_3^0, p^0, U^0) \\ &+ e(q_1^1, q_2^1, q_3^0, p^0, U^0) - e(q_1^1, q_2^1, q_3^1, p^0, U^0), \end{aligned} \tag{2-11}$$

where the order of the q_i^1's is arbitrary. These results hold for the aggregation of benefit categories, geographical locations, or policy components.

The basic implication of this result for the CV method is that aggregation of independent estimates of WTP for each q_i will not in general equal

[57] This type of behavior should not be unexpected, since economic theory clearly predicts that one should observe respondents giving decreasing marginal valuations of close substitutes.

[58] Hoehn and Randall's proof relies on a fairly complicated argument using line integrals. The most complete presentation of this proof appears in Hoehn's (1983) dissertation.

[59] Equivalently, $3Y_0 - Y_1^1 - Y_2^1 - Y_3^1 \neq Y^0 - eq_1^1, q_2^1, q_3^1, p_1^0, U^0$ except in special cases.

the correct willingness to pay for all changes simultaneously. Equation (2-11) implies that while the total willingness to pay can be correctly measured by sequential valuation (via the expenditure function), willingness to pay for any particular component or element of the sequence may be different for each sequence except for the two end points. Specifically, these findings suggest that independent estimates of different benefit categories cannot be aggregated together to obtain a combined benefit, that studies done of different locations cannot be aggregated together to obtain estimates of the larger area under consideration, and that the benefits of different policy components cannot be estimated independently. Further, even estimates of individual benefit categories, spatial areas, or policy components made by using the same method and the same sample may be different if they are estimated as parts of different sequences.

Hoehn and Randall's results can be intuitively explained by the notion of complements and substitutes. If the goods are complements, such as mountains and good air visibility, independent estimates of different benefit categories or policy components tend to underestimate the total WTP amount. Hoehn and Randall contend that the general case is more likely to be that where the goods are substitutes—mountain and seaside recreational areas, for example. Here the aggregation of independently derived estimates of different benefit categories or policy components will result in overestimation of the WTP amount. The notion of weak or strong separability is often invoked to justify adding different component benefits together (Freeman, 1979b).[60] Although subcomponent benefits measured by asking respondents to disaggregate their total WTP amounts are somewhat justifiable under this assumption, since double counting is constrained by first obtaining the total WTP amount, the reverse is not true. The basic reason for this seemingly contradictory result is the Hicksian notion (Philips, 1974) that money spent on one good cannot be spent on another, and that all goods are substitutes in this sense.

Tolley and Randall's (1983) study of the value of air visibility in the Grand Canyon empirically demonstrated the existence of the sequence aggregation problem.[61] In 1980, 130 Chicago-area residents were asked how much they were willing to pay in higher monthly utility bills to preserve current air-quality and visibility levels at the Grand Canyon. The visibility levels were portrayed in photographs which were shown to the respondents before they were asked to express their WTP amount. Air quality in the Grand Canyon was the only air quality component valued

[60] Separability implies that the elasticity of substitution between two benefit components is zero.

[61] This was one of several local studies of Grand Canyon air quality undertaken for a larger study directed by Schulze and coauthors (1983). The latters' report contains a copy of the Grand Canyon valuation instrument.

in the study. In 1981, the same researchers replicated the Grand Canyon valuation, but this time respondents did not answer the Grand Canyon question until they had first said how much they would be willing to pay for (1) a reduction in air pollution which would improve from 9 to 18 miles the prevailing air visibility in Chicago, and (2) a 9-to-18-mile visibility improvement in the broader area comprising the United States east of the Mississippi. Photos illustrating the various visibility levels in Chicago were shown to the respondents, in addition to the set of Grand Canyon pictures. The values given to Grand Canyon air quality differed greatly in the two surveys. Valued independently in 1980, the Grand Canyon pollution control program was worth an average of $90 per year to the typical Chicago respondent. Valued by a separate sample of 59 Chicago respondents in 1981, as third in a three-part sequence, the Grand Canyon program was worth only $16 (Tolley and Randall, 1983:93–97).

It is important to note that difficulty in generalizing from parts to the whole and vice versa is by no means a unique property of the contingent valuation method, but is characteristic of the other benefit measurement methods as well.[62] In travel cost studies the spatial aggregation issue is known as the site substitution problem, and in decision theory (Keeney and Raiffa, 1976) it is known as tradeoffs with multiple objectives and sequencing.[63] Hedonic techniques, in order to develop rent gradients for the location under consideration, tend to assume that all substitute sites remain constant. This ubiquitous bias in valuing public goods is perhaps better handled by contingent valuation than by other benefit measurement techniques because of the CV method's flexibility in sequencing policy components (see the discussion of sequence aggregation bias in chapter 12).

Distribution of Individual Willingness to Pay

In deciding whether to undertake a project, a policymaker may be more interested in the distribution of benefits from a policy change than he is in the actual aggregate comparison of cost and benefits. This would be particularly true if it is readily apparent that the change being contemplated would be potentially Pareto-improving and the desired distributional information is directly available from a contingent valuation survey.

[62] The essential problem in all cases is that the substitution elasticities are not known, and are difficult although not impossible to estimate. The aggregation and disaggregation of costs have many of the same problems (Young, 1985).

[63] See Smith and Desvousges (1985) for a discussion and example of how difficult a generalized travel cost model is to estimate and the number of judgments which must be made in the specification stage; see Smith, Desvousges, and Fisher (1986) for a comparison of this model to a CV estimate. The sequencing problem has long been recognized in benefit-cost studies with an engineering orientation; see, for instance, Krutilla (1960).

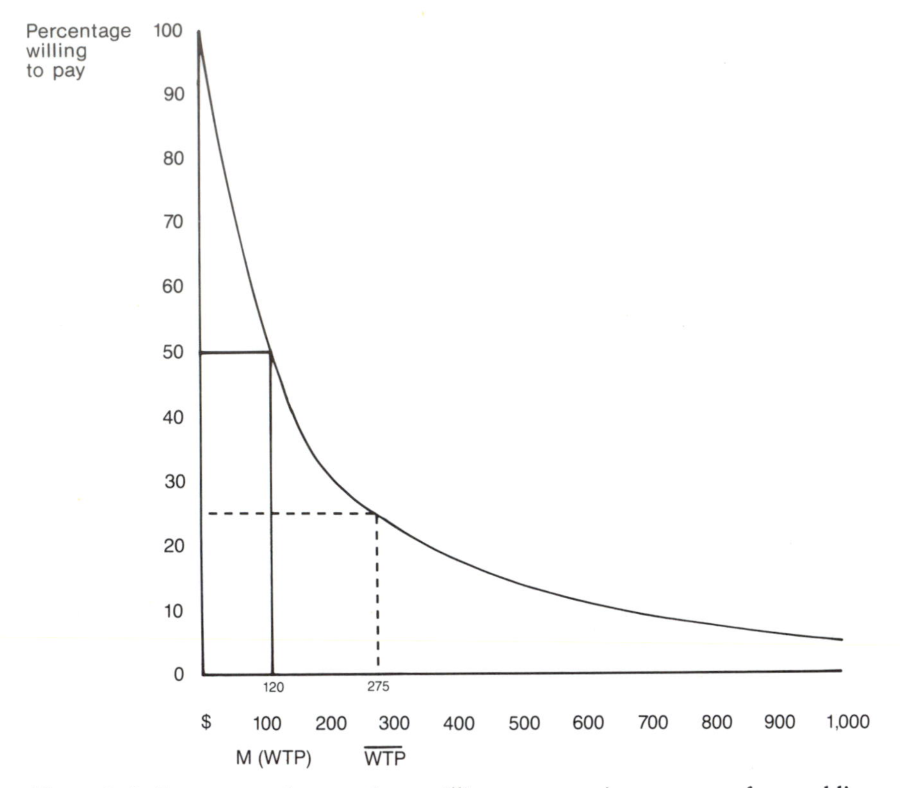

Figure 2-4. Percentage of respondents willing to pay various amounts for a public good

One of the most understandable ways of summarizing this information is to prepare a graph of dollar amounts by the percentage of the public willing to pay each amount or more for the level of the public good in question. This is simply a plot of one minus the cumulative distribution function for the willingness-to-pay responses, as shown in figure 2-4. This graph is based on a smoothed version of one minus the cumulative distribution function for total willingness to pay for swimmable water quality in Carson and Mitchell (1986). Obvious from this graph is the sizable difference between median WTP and mean WTP, a difference which has large implications for the desirability of undertaking the provision of the public good and the method of financing it.[64] On such a graph it is possible to present even more information for the policymaker by plotting separate curves. For example, separate WTP distribution curves could be plotted

[64] For a general discussion of the design of optimal taxation schemes, see Atkinson and Stiglitz (1980), or Mirrlees (1986).

for different income groups, users and nonusers of the good, or different geographical areas.[65]

The Private Goods and Political Market Models

The economic theory presented thus far is for pseudo (public goods) markets which have all of the basic properties of private goods markets except for a slightly different (Lindahl) equilibrium concept based on vertical rather than horizontal aggregation of individual demand curves. For certain types of contingent valuation scenarios, an attractive alternative to the private goods market model is a political market model. Citizens voting in referenda actually make binding decisions about the provision of public goods, which they obligate themselves to pay for through tax mechanisms. The use of such a model in a CV scenario is likely to enhance its plausibility. How might such a referendum model fit into the valuation framework described in this chapter?

Both private goods and political markets exist to coordinate the exchange of goods and services between agents (voters). If one assumes completely selfish behavior,[66] both types of markets are driven by identical individual demand curves for a good and hence should produce identical estimates of individual WTP or WTA amounts for a public good. The two markets imply different approaches to aggregation, equilibrium, and the decision about the quantity of a public good that should be provided.

Outcomes in the private goods market occur at the equilibrium price where demand and supply are equal. Such a competitive equilibrium meets the welfare economic test of Pareto optimality. Aggregation in the private goods market model is over all consumers, so that each has some influence on the market clearing price.[67] The average consumer in a private goods market model is one who purchases the *mean* quantity of the good or who is willing to pay or accept the *mean* amount for a specified quantity of it.

Outcomes in the political market model are determined by a plurality voting rule (usually requiring a simple or two-thirds majority). Large numbers of voters (up to half under a majority approval rule) have no direct influence on the quantity of the public good provided. The average voter is the one who wants the *median* quantity of the public good or who

[65] If the appropriate variables were measured in the contingent valuation study, one could even plot separate curves for different identifiable voter groups such as Democrats and Republicans.

[66] It has long been argued (see, for example, Wilson and Banfield, 1964, 1965) that people may be less self-interested and more public-regarding when they vote in elections than when they participate in the market for private goods. The implications of such behavior are considered in chapter 4.

[67] This influence declines as the number of consumers becomes large.

is willing to pay or accept the *median* amount for a fixed quantity of it.[68] The quantity of the public good that it is desirable for the government to provide is the largest quantity which the appropriate plurality is ready to approve given the proposed cost allocation scheme.

These two ideal-typical market models have tendencies toward almost opposite flaws with respect to the provision of public goods.[69] The potential Pareto criterion (where potential benefits exceed costs) which is coupled with the private goods market model allows a small minority having a high valuation of a public good to force a large majority to pay for it, since that minority need not actually pay for the provision of the public good. In contrast, the political market model, with its plurality voting rules, allows a majority which wants the public good (but does not want to bear all the costs of providing it) to force a minority which does not necessarily want the good to pay for its provision. In many cases more equitable policy outcomes will result if decision makers require that both types of decision criteria be met.[70]

The Implications of Theory for Contingent Valuation Scenario Design

Having reviewed the theoretical basis for the contingent valuation method in microeconomics, it is appropriate to ask, What does theory imply for the design of CV scenarios? In order to obtain data which can be used to estimate equation (2-2) or equation (2-3), CV scenarios must define and communicate to respondents the following:

1. *The reference level of utility.* Usually this is done by defining the level of income available to the respondent and by describing the property rights situation with respect to the particular public good.

   A. *The property right.* In the case of collectively held goods, respondents should understand that they are currently paying for a given level of supply. The scenario should clearly indicate

[68] See Downs (1957), Black (1958), Romer and Rosenthal (1978), Denzau and Parks (1983), and Enelow and Hinich (1984) for discussions of the median voter and other equilibrium concepts in political markets.

[69] The private goods and political market models need not always be in conflict. For example, Bowen (1943) has noted that if preferences were single-peaked and symmetrically distributed and if the voting rule called for a simple majority, provision of the median quantity demanded would be the (potential) Pareto-efficient solution. See Zeckhauser (1973) for an examination of the correspondence between Pareto efficiency and different types of voting systems.

[70] Where equal payments are required from every taxpayer, the decision criterion that the median willingness to pay for a project be greater than its mean cost is sufficient for both types of market models.

whether the levels being valued are improvements over the status quo, or potential declines in the absence of sufficient payments.

B. *Current disposable income.* Respondents should take taxes and fixed long-term obligations into account in giving the WTP amount for the good. If the household is the unit of analysis, the reference income should be the household's rather than the respondent's income. In cases where the respondent is already paying for the provision of the good (in whole or part), the respondent should understand that this payment should be added back to the disposable income.

2. *The nature of the public good.* Unlike ordinary survey questions, which sometimes ask respondents whether they are willing to pay x dollars to improve "air quality," the nature of the good and the changes to be valued must be specified in detail in a CV survey. Although the change is typically described as a specific change in quantity of a public good, it can also consist of a well-defined public policy, along with its intended objectives and probability of success.

   It is also important to make sure that respondents do not inadvertently assume that one or more related improvements are included in the change they are valuing when this is not the case. For example, if people are asked to value only air visibility, it would be important to make sure that they did not include health-related improvements in their WTP amount. Likewise, if they have a tendency to think of environmental improvements in general even when asked about water quality alone, it would be necessary to point out specifically that air quality would remain the same. Conversely, if the intent is to value several public goods, such as the visibility and health improvements caused by air pollution control, this intent should be made clear.

3. *The relevant prices of other goods.* Where the change in the public good will significantly affect the prices of other goods, the impact of this change should be communicated to the respondents. Usually CV studies have assumed these general equilibrium effects to be negligible. However, if private good prices were to be strongly affected by, say, water pollution control policy, and relative price changes were potentially substantial, it would be necessary to make explicit note of this in the scenario. For example, if water pollution control were to increase the price of leather shoes much more than the price of other goods, respondents should be informed of this effect so they could take its impact on their buying patterns into account.

4. *Conditions for provision of the good and payments for it.* When and for how long the good is going to be provided should be clearly

specified if this is not obvious. Respondents should understand the frequency of payments required (monthly, annual, or per trip, for example), and whether or not the payments will be required over a long period of time in order to maintain the quantity or quality change. They should also understand who will have access to the good and who else will pay for it, if it is provided.

5. *The nature of the WTP amount desired.* The scenario should be designed to ensure that the respondents express their consumer's surplus for the good and not some other type of value such as a "fair price."

Summary and Conclusions

Over the more than twenty years since contingent valuation's initial use, the theoretical basis for the CV method has become better understood. In this chapter we have examined the ability of the CV method to measure the correct welfare measures. In principle, the method is fully capable of measuring the appropriate compensating and equivalent measures of welfare changes. However, CV researchers have shown that in practice there is great difficulty in obtaining correct estimates of WTA for decreases in the level of a public good, because respondents tend to reject the implied property right as implausible. We have shown, further, that there is reason to believe that the Willig and the Randall and Stoll bounds—which justified the substitution of the WTP formulation where a WTA formulation was indicated on the grounds that the substitution would not significantly affect the estimate—underestimate the true difference between the measures under conditions which are relevant to many CV surveys. Thus, at least part of the large empirically observed differences between WTP and WTA in contingent valuation surveys may represent real economic behavior rather than being artifacts of the CV method. We have proposed a new interpretation of property rights for public goods which eliminates the need to measure WTA for public goods requiring regular payments to maintain a current quality level.

The conditions for a theoretically valid CV study and for aggregating CV findings have been described and discussed. Aggregation across individuals to estimate the total benefits of specific levels of public goods was seen to be relatively straightforward technically, but involves controversial ethical assumptions. Technically more problematic are the aggregation and disaggregation of subcomponents, such as benefits for separate geographical locations for different types of benefits. Contingent valuation was shown to easily produce the kind of distributional information desired by policymakers. We also compared and contrasted the theoretical properties of the two types of markets that CV scenarios have been used to simulate—private goods and political markets. In conclusion, we have briefly summarized the implications of the theory for the design of contingent valuation scenarios.

In the next chapter we examine the nature of benefits and compare contingent valuation to other methods of measuring those benefits. Overall, we conclude that the CV method stands on firm theoretical ground.

3

Benefits and Their Measurement

Contingent valuation is just one of several methods in current use to measure the benefits of public goods. In this chapter, we review the nature of the benefits which a policymaker might wish to have valued, and compare and contrast the ability of the various existing benefit methods to measure them. We then discuss the properties of the particular contingent valuation formats currently in use.

Property Rights to Quasi-Private and Pure Public Goods

Before looking at their benefits, it is useful to reexamine the public goods concept briefly and to introduce a new class of goods which are intermediate to pure public and private goods. Kopp and Portney (1985) divide the goods entering an individual's utility function into three classes: pure private goods, quasi-private goods, and pure public goods. Pure private goods are bought and sold in organized markets where those participating have an identifiable individual property right to the goods. The process of buying and selling leads consumers to reveal truthfully, at least to some extent, their preferences for these goods if they are to have them. Quasi-private goods, in the Kopp–Portney formulation, are similar to private goods except that they are not freely traded in an organized market.[1] For example, hunting permits issued by states have a purchase price, but that price is not determined solely by the give-and-take of the marketplace. Rather, it is usually arbitrarily determined—often below market price.

[1] When moving from pure public goods to impure public goods, either one or the other of two aspects of that impurity tend to be emphasized: the presence of congestion/rivalry in the use of the good, or the practicality and possible desirability of exclusion from the good. The term "quasi-public" is often used when the emphasis is on the first type of impurity and "club goods" when it is on the second (Cornes and Sandler, 1986) because the latter are often provided by voluntary membership organizations. The goods we designate as quasi-private tend to suffer from both types of impurities.

Thus, while values for these goods are not determined on the basis of competitive prices, it is often possible to observe the quantity of these goods consumed by individuals. Pure public goods, such as air visibility or the national defense program, have no explicitly identifiable individual property rights because consumers cannot be excluded from enjoying them. Since they are not traded directly in any market, we can observe neither a competitive market price for these goods nor the quantity of them desired by consumers. Our version of the Kopp–Portney taxonomy is summarized in table 3-1.

Pure and quasi-private goods are primarily distinguished by the difference between individual and collective property rights. Every member of a relevant collectivity (city, nation, condominium owners' association, and the like) has a common property right to one or more pure public goods. Individual members of the collective may also be granted differential access to other public goods (often for a fee and on an equal basis) by the relevant governing body. Such access may convey exclusive use rights to some goods; we will term those goods involved "quasi-private" goods. Although it is possible to identify a fairly wide variety of property rights for public goods (from the individual consumer's perspective), our purpose can be served by working with the two polar types, collectively held and individually held rights.

Collectively held rights occur where potential access to the good is available to all members of the collectivity and individual members cannot sell their access right.[2] If there is a cost to providing the good at a given quality level, it is normally borne by all consumers through some combination of taxes, higher prices, fees, and the like. As noted in chapter 2, if this level of payment is not maintained, the quality of many public goods may deteriorate. If a quality increase is desired, higher payments will be required to cover the cost of providing the new quality level. The less it is possible to exclude members of the collective from the good, the more likely the property rights will be held entirely by the collective, since the good cannot be efficiently provided at a profit. Air and water quality are examples of goods to which individual consumers have collective, nontransferable property rights of this kind.[3]

[2] For pure public goods, the relevant concept is not whether it is used or not by the individual economic agent, but whether or not a given level of quality is accessible (Bohm, 1977; Gallagher and Smith, 1985; Smith, 1987).

[3] Even with respect to pure public goods such as air and water quality, the concept of collective rights must be regarded as an ideal, since the rights are transferable under certain administrative arrangements such as those that allow corporations to buy and sell permits to emit specified levels of pollution (Bohm and Russell, 1985; Tietenberg, 1985). The opposition to these arrangements by environmental groups is motivated in part by a belief that rights to such public goods should not be transferable (Kelman, 1981).

Table 3.1 Classes and Characteristics of Goods

Class of good	Characteristics	Examples
Pure private	Individual property rights Ability to exclude potential consumers Traded freely in competitive markets	Agricultural products Automobiles Financial services
Quasi-private	Individual property rights Ability to exclude potential consumers Not freely traded in competitive markets	Public libraries Recreation in parks TV frequencies
Pure public	Collective property rights Cannot exclude potential consumers Not traded in any organized market	Air visibility Environmental risks National defense

The appropriate analogy for the price of a pure public good is not that of the price of a privately marketed good, but that of maintenance fees such as those paid by condominium owners. Purchase of a condominium conveys private property rights to the apartment itself. But condominium owners are also legally obligated to pay fees, whose level is collectively determined, to maintain the property and its grounds. Owners may collectively agree to increase their fees in order to provide a more lavish common amenity. Conversely, nonpayment of fees or a reduction in their level would result in a lower-quality common amenity. All owners have equal access rights to "use" these nondivisible collective goods.

In some cases, the collectivity grants individuals exclusive rights to use a public good because the granting of those rights is deemed to serve the public interest. Typically, these individually held goods are excludable and subject to congestion, in the sense that one person's enjoyment of a good may affect someone else's enjoyment of the good. Various allocation rules are used, ranging from auctions to free grants based on principles such as competence and first come, first served. Occasionally, as with mining claims on public land and with broadcasting frequencies, the rights are transferable. In these cases, the public good has not been fully transformed into a private good because the government still maintains an interest in the property right and can revoke it. Someone who wishes to purchase a broadcast frequency from an existing license holder, for example, must meet certain governmentally imposed criteria, or she cannot take possession. The more common case is where the collectivity grants a nontransferable right. Such rights are granted free to wilderness users through the allocation of wilderness permits, and auctioned or otherwise allocated for fees to those who wish to use public lands to mine coal or oil, to harvest trees, or to graze livestock.

One of the advantages of contingent valuation is that the method is particularly suited to measuring the benefits of both quasi-private and pure public goods.[4] Other established methods for measuring benefits, as we will see later, tend to be limited in scope to the class of quasi-private goods. In the next section, we define the concept of benefits and show in more detail how they might accrue, employing a pure public goods example—national freshwater quality—to illustrate the various categories of benefits. The concepts we advance apply to, or are adaptable to, most pure public and quasi-private goods.[5]

The Nature of Benefits

The benefits from the provision of public goods such as pollution control arise from the value individuals assign (Brown, 1984) to improvements in air visibility or water quality and the like. Losses caused by a deficiency of pollution control may be thought of as damages. The distinction between benefits and damages is based upon some reference level of the good, a concept closely related to that of property rights. For ease of exposition, the term "benefits" will be used here to refer to both benefits and the reduction of damages. While the other side of the benefit-cost equation—the "costs"—may also have many meanings, we will try to avoid confusion by defining costs solely in terms of the value of the resources used to produce a public good such as pollution control.[6]

Each type of good has a unique configuration of advantages to confer on an economic agent (individual or household); expressed in another way, there are unique reasons why an agent's level of satisfaction or utility might increase if the good is provided to him. According to the economic doctrine of consumer sovereignty (Penz, 1986), an agent's spending behavior in markets is a sufficient signal of his preferences for various goods, and the reasons why he holds these values are of no economic importance. The question of why someone might value changes in the provision of nonmarketed public goods, however, is of more than passing interest to the researcher who wishes to measure benefits. The researcher needs to identify all the agents who might potentially benefit from a given change in order to have that population represented in the analysis. The identification of agents

[4] Methods similar to contingent valuation are used by marketing researchers to trace out demand curves for private goods—particularly new goods for which market data are unavailable. See, for example, Pessemier (1960) or Jones (1975).

[5] We will often use the term public goods when we do not wish to make a distinction between pure public goods and quasi-private goods.

[6] What we mean by costs is clearly conveyed by the term "control cost function," which is defined by Just, Hueth, and Schmitz (1982:278) as "the cost incurred by a polluter (due to reduced output and other opportunity cost in addition to direct pollution abatement cost) to reduce pollution from the level which would occur in a free (unregulated) market in the absence of control policies." The term hints that the calculation of the cost side of a benefit-cost analysis is not as straightforward as it might first appear.

is fairly straightforward conceptually, but often somewhat more difficult to accomplish in practice.[7] It is also important to identify the full range of possible benefits so that a method (or methods) can then be chosen which offers, all other considerations such as cost being equal, the most promise of measuring those benefits.[8] As we will show later in this chapter, the ability of different techniques to estimate categories of possible benefits varies. Where certain categories of benefits cannot be measured, they should be explicitly identified so policymakers can understand the limitations of the benefit assessment. Finally, an awareness of why and how agents benefit from changes in public goods can be helpful in knowing where to look for traces of the demand in actual private good market transactions, or, in the case of CV studies, in knowing which possible benefit categories to remind people to consider when they determine their willingness to pay for the good.[9]

A Typology of Possible Benefits

A comprehensive assessment of the benefits of a change in the level of a public good should include all of the benefits which will legitimately accrue from a specified change in the provision of a given good. This concept is sometimes known as the "total value" approach (Randall and Stoll, 1982; Boyle and Bishop, 1985). Some kinds of benefits are easier to measure than others, and the failure of economists to measure nonuse environmental amenities has long been criticized by noneconomists. Much of the history of benefits measurement can be written in terms of how researchers have devised ways to measure a larger and larger fraction of the total benefits of providing a public good (Smith, 1986a).[10] The CV method figures importantly in this history because one of its advantages is that it allows the researcher to directly measure the various types of nonuse

[7] For a discussion of problems in defining the appropriate population for a benefits study, see chapter 12.

[8] The need for an exhaustive compendium of benefits is well recognized by market research practitioners who study the potential market for new products (see, for example, Myers and Tauber, 1977).

[9] Agents benefit from the perceived changes in the public good rather than the actual technical changes. Credibility problems occur when actual conditions are more favorable than perceived conditions. The most cost-effective course of action by public officials in such cases would be to raise utility by an educational effort rather than proceeding with additional technical changes (Pauly, Kunreuther, and Vaupal, 1984). See Desvousges, Smith, and McGivney (1983) for a discussion of the chain of effects and responses to water quality regulatory actions.

[10] Unmeasured benefits have often been referred to in the economics literature as "intangibles" (Wyckoff, 1971; Bishop and Cicchetti, 1975; Haveman and Weisbrod, 1975), a term which reflects the propensity of applied economic analysis to define reality in terms of observable market behavior. A more appropriate designation for the unmeasured benefits is Ciriacy-Wantrup's (1952) concept of "extra-market" benefits.

benefits. However, as the benefits measurement frontier moves farther from the traditional economic understanding of consumption, and therefore from the possibility of validation by some form of market behavior, concerns are expressed that inclusion of this broader range of benefits may overstate, perhaps considerably, the value of public goods (Goldsmith and Company, 1986:38).

The typology of freshwater quality benefits in figure 3-1 illustrates how the possible benefits of a change in the level of a public good may be categorized.[11] This typology, from Carson and Mitchell (1986), is intended to be reasonably exhaustive of the possible benefits of freshwater quality improvements. Its use would help to ensure that all the benefits of a given amenity are identified and distinguished from one another so as to avoid double counting. For reasons discussed below and at more length in chapter 12, none of the benefit categories or subcategories should be considered to be additively separable without the imposition of additional and somewhat arbitrary restrictions. This is because unique estimates of the different benefit categories and subcategories generally do not and cannot exist.

In its present form, the typology embodies several major changes from its predecessors (Mitchell and Carson, 1981, 1984; Desvousges, Smith, and McGivney, 1983). Among these are a reorganization of the existence class of benefits, and a deemphasis on the distinction between direct and indirect user benefits. We also drop the use of the term "intrinsic" value (taken to be equal to option value plus existence value) because the common philosophical usage of the word in a related context means that something has value in and of itself (Frankena, 1973; Edwards, 1985; Randall, 1986a), a viewpoint which conflicts with the economic notion that something has value only if an economic agent is willing to give up scarce resources for it. Finally, we will initially rule out uncertainty by assuming, as most contingent valuation studies have, that the respondent buys certain (sure) provision of the amenity. This eliminates the possibility of any kind of option value (Schmalensee, 1972; Bishop, 1982) or quasi-option value (Arrow and Fisher, 1974; Hanemann, forthcoming).[12] Once uncertainty is introduced, there is a switch from defining total benefits in terms of a Hicksian compensating or equivalence surplus to the analogous type of option price. We will follow the line of recent work (Chavas, Bishop, and Segerson, 1986; Plummer and Hartman, 1986; Smith, 1986a, 1987b; Smith and Desvousges, 1986b; Fisher and Hanemann, 1986) to argue that, from the standpoint of contingent valuation, option and quasi-

[11] By "benefit," we refer to the paths through which the changes in the level of satisfaction indicated by an agent's utility function occur.

[12] Hanemann (1984d) shows the relationship between option value and quasi-option value and how option value can arise in a quasi-option framework.

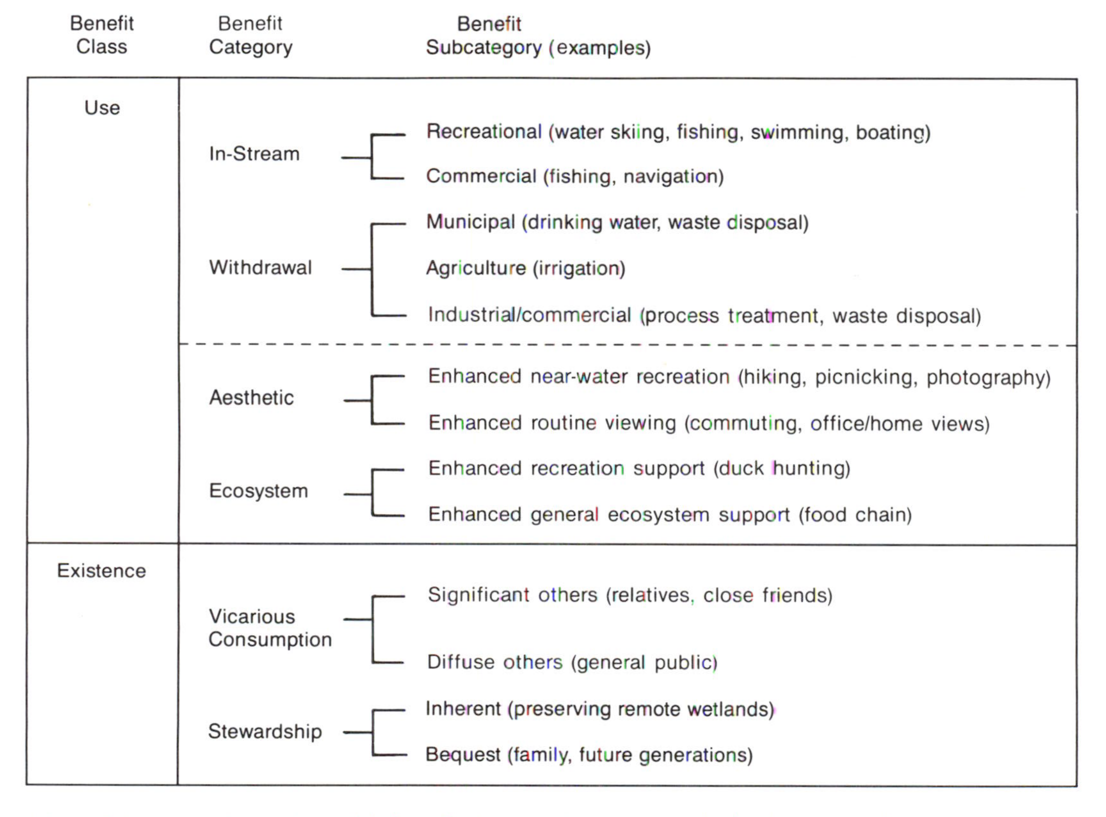

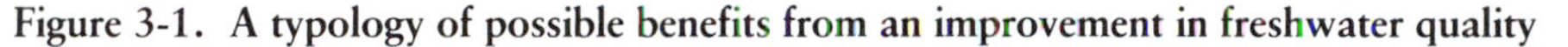

Figure 3-1. A typology of possible benefits from an improvement in freshwater quality

option values are more properly viewed as correction factors to calculations of total benefits from incorrect perspectives than as separate categories of benefits.

The Use Class of Benefits. The use class of benefits consists of all the *current* direct and indirect ways in which an agent expects to make physical use of a public good.[13] The dashed line in figure 3-1 indicates the division between direct use and indirect use categories of benefits. In the case of surface freshwater quality, direct use benefits may arise as a result of recreational or commercial activities that occur on the water, or they may arise from withdrawal activities such as the use of water for agricultural irrigation, for cooling or washing in industrial processes, or (after treatment) for drinking water. This benefit class also has an indirect dimension, created when the water body's characteristics enhance nearby activities. The figure lists two types of indirect use benefits: those which occur because water quality is a vital component of an ecosystem or habitat that supports certain types of recreation—hunting or bird watching, for example; and those which occur because water quality provides an aesthetically pleasing setting for activities like picnicking by a stream or looking out of one's window.

Because they could be measured by observing changes in market prices, commercial activities such as commercial fishing, irrigation, and industrial water treatment were initially of greatest interest to economists. A seminal paper by Krutilla (1967) played an important role in expanding benefits assessment to include the function played by resources (freshwater quality, for instance) in supporting public recreational activities. A new benefit measurement technique known as travel cost analysis (Clawson and Knetsch, 1966) quickly gained favor, as it appeared to be well suited for measuring what had previously been regarded as "intangible" recreation benefits. Freeman's (1979a, 1982) subsequent review of the available studies of water pollution benefits showed that direct recreational benefits were much larger than any of the commercial benefits. More recently, economists have attempted to expand the scope of empirical benefit assessments further to include the indirect use benefit categories. One

[13] For pure public goods, the term "use" is somewhat of a misnomer, as households are really purchasing (and implicitly gaining benefits from) personal access rather than use per se. (The access concept is discussed further later in this chapter.) Note also that CV surveys obtain respondents' willingness to pay for expected use in the future—the measure desired by policymakers—whereas observed behavior measurement techniques (such as travel cost) are based on prior behavior. The CV method rests on the assumption that people will behave in the way they said they would if the hypothetical situation they were placed in were to become real, whereas the assumption underlying the behavior-based measures is that the relationships estimated in the past will continue to hold in the future.

notable effort related to water quality is Hay and McConnell's (1979) travel cost study of nonconsumptive wildlife uses.[14]

The Existence (Nonuse) Class of Benefits. Resource economists often use the term "existence" to refer to nonuse benefits. Here people obtain utility from an amenity for various reasons other than their expected personal use. In contrast to use values, which occur because people are physically affected by an amenity in some way, existence values involve the notion that a person doesn't have to visit a recreational site to gain utility from its maintenance or improvement (Krutilla and Fisher, 1975). Received wisdom in economics would tend to deny the economic meaning of such a concept, holding that if people really did obtain utility from existence values, their behavior would express those values. In fact, people do find ways to express existence values in the private marketplace. A good portion of the millions of dollars in fees and voluntary contributions paid by members of environmental groups (Mitchell, 1979a), and the willingness of environmental activists to volunteer their time to lobby for such legislation as the Alaska Wilderness Bill, can be cited as evidence for the reality of existence values for wilderness amenities. Referenda on environmental programs often receive very strong voter support even among voters whose communities are unaffected by the improvements.

Although behaviors such as these are indicative that existence values are real, complete markets where people can express their existence value with dollars are typically missing. The Sierra Club addresses its direct mail to less than one-tenth of American households, and it cannot guarantee the provision of any amount of a particular environmental amenity to those individuals it does reach. Many types of public goods never reach the ballot. Even pressuring an individual congressman to vote to provide a particular good requires knowledge of the legislative status of individual bills and their content that most concerned citizens lack.

In other fields of economics (labor economics, for example) the term "psychic income/cost" (Maddox, 1960; Thurow, 1978) is often used to describe a concept similar to existence value. It has been demonstrated that some people are willing to accept a lower wage (or other pecuniary benefits) in exchange for the satisfaction of status (job titles) or the satisfaction of doing something worthwhile (Lucas, 1977). It has also long been recognized that there are psychic costs to moving people out of farming (Maddox, 1960), and that many urban workers are willing to

[14] Most empirical efforts to measure indirect use benefits have been directed toward measuring improvements in air visibility, both by the hedonic price method and the contingent valuation method (for example, Randall, Ives, and Eastman, 1974; Tolley and Randall, 1985).

return to a rural setting for wages far lower than a cost-of-living differential would suggest (Deaton, Morgan, and Anschel, 1982). In the cases of both existence value and psychic income, people are influenced in part by preferences for attributes of situations or goods that are secondary to the attribute which is presumed to be the primary source of value, such as the salary paid to a worker or the recreational use of a water body. In neither instance do the benefits result from the process of consumption as it is usually described in economic models, where the commodities are exhausted or used up (Smith 1986a).

Influential early discussions of existence benefits were presented by Krutilla (1967) and Krutilla and Fisher (1975). As the CV method has gained acceptance and benefits estimates have appeared that reflect a significant existence component, a number of economists have sought to clarify the nature of these benefits,[15] and others have attempted to use CV surveys to obtain separate measurements of one or more of the various types of existence benefits (see chapter 12). Despite the vigorous debate and profusion of terminology that characterize this literature, we believe the issues are now sufficiently well understood to permit CV researchers to measure with confidence the total benefits respondents receive for goods which include nonuse components.

What kinds of benefits might people obtain from the provision of a public good, apart from their personal use of the good? In figure 3-1 we identify four types of benefits, which fall under two categories—vicarious consumption and stewardship. In the case of vicarious consumption values, utility is gained from knowing about the consumption of others. These "others" may be generalized, or they may be particular individuals known to the respondent. The motivation behind vicarious consumption values may stem either from a sense of obligation to provide the good, or from a sense of true shared and interdependent utility. In practice, it is often difficult to get respondents to distinguish between these motives,[16] but for the purposes of economic analysis (including the design of a CV survey) it is rarely necessary to do so.

Stewardship values involve a desire to see public resources used in a responsible manner and conserved for future generations (Pigou, 1952; Ciriacy-Wantrup, 1952). We distinguish two types of stewardship values.

[15] For recent discussions, see: Bishop, 1982; McConnell, 1983; Brookshire, Eubanks, and Randall, 1983; Randall and Stoll, 1983; Smith, 1983; Freeman, 1984a; Edwards, 1985; Madariaga and McConnell, 1985; Brookshire, Schulze, and Thayer, 1985; Brookshire, Eubanks, and Sorg, 1986; Smith, 1986a, 1986b; and Hanemann, forthcoming. Driver, Brown, and Burch (1986) discuss the range of motives which may lie behind existence values for wilderness preservation.

[16] Gift-giving, altruism, philanthrophy, and interdependent utility are discussed from various perspectives in Winter (1969), Krebbs (1970), Schall (1972), Becker (1976), Goldman (1978), Mitchell (1979a), Margolis (1982), Sugden (1982, 1986), and Edwards (1985).

Bequest values exist when someone enjoys knowing that the current provision of an amenity will make it available for others—family or future generations—to enjoy in the future. Also part of stewardship are inherent values, which stem from the respondent's satisfaction that an amenity itself—a wilderness area, for example—is preserved regardless of whether it will ever be used by anyone. When people, in the absence of any intention to see harp seals in Canada or any expectation of doing so, contribute money to prevent the harvesting of baby harp seals for their pelts, they are manifesting something close to this type of existence value for that good.[17]

The different types of existence benefits coexist with each other and with use values. For example, while stewardship values do not result from current human use, they may be stimulated by and occur simultaneously with use: someone's stewardship value for wilderness lakes is likely to be enhanced by the experience of fishing in them during wilderness hiking expeditions. Thus, while the several dimensions of existence value are analytically distinguishable, and all enter into a consumer's utility function, they are likely to be very difficult to disentangle and measure separately. The same may be said for certain types of use values, such as the aesthetic (visibility) and health benefits from improved air quality.

While our definition of the existence class of benefits includes inherent values, Brookshire, Eubanks, and Sorg (1986) have recently argued that these values should be excluded from benefit measurements on the grounds that this type of motivation is not consistent with the "efficiency ethic" they believe underlies benefit-cost analysis. They describe the efficiency ethic as a management ethic focused on human welfare. For them, vicarious consumption and bequest values pass the efficiency-relevant test because these values involve efficient use of the resource in the interest of humans, but they do not believe this is true of inherent values. They argue that even though a person may be willing to pay something "simply because he believes we ought to protect wetlands wildlife against human action which would threaten the existence of the wildlife," this is not an acceptable motivation for benefits because it does not contribute to human welfare (Brookshire, Eubanks, and Sorg, 1986:1515). Actions based on ethical considerations, on a desire to do what is right, are "counterpreferences" since they do not increase utility. Brookshire, Eubanks, and Sorg advise researchers to probe individuals' motives underlying statements of willingness to pay in order to avoid confounding efficiency-relevant values and ethical values.

[17] The line between existence values of this kind and use values is blurred when, for example, an interest group attempts to provide a market in which one can purchase photographs and lobbying in behalf of protection of harp seals.

We believe the analysis of Brookshire, Eubanks, and Sorg rests on a number of faulty assumptions. First, it should not be assumed that there is *no* basis in welfare theory for restricting benefits to human use of an amenity.[18] Second, it should not be assumed that ethical motivations are unique to inherent values. A "commitment to do what is right" underlies the vicarious and bequest values which Brookshire, Eubanks, and Sorg accept as legitimate. People gain utility from helping others without expectation of a material reward because they have learned to value this behavior. Third, and most important, it is erroneous to assume that making choices on the basis of ethical beliefs necessarily involves self-sacrifice; in fact, those who make choices of this kind obtain utility from satisfying internalized social norms.[19] For some people such things as preserving wilderness for its own sake or living in a small town among friends and relatives are amenities whose provision genuinely improves their personal welfare as they conceive it. Far from being "counterpreferential," in properly conducted CV studies choices based on these preferences are motivated by self-interested and egoistic considerations.

Some economists accept the economic importance of existence values (including inherent values), but believe that using contingent valuation to measure this category of benefits will inevitably result in flawed benefit estimates. They fear, in particular, that the nonexperiential nature of benefits in this category may lead people to inflate their WTP amounts where the category is invoked. For example, Mendelsohn (1986) has argued against the use of CV to value existence damages in Superfund suits on the grounds that since people have no observable connection to an amenity, "accepting existence value as a potential part of resource valuation consequently opens the pandora's box of potential valuation estimates."

Whether or not the CV method can obtain valid benefit measures of amenities which include an existence component is, of course, a basic concern of this book. We believe CV can do this, provided the potential problems are recognized and overcome. For example, if care is not taken in its wording, a scenario that evokes existence values may stimulate a symbolic response in which the respondent treats the good as representative

[18] Economic theory does reject the notion that "trees have rights." The statement that is consistent with economic theory is that "trees have rights to the extent that humans are willing to pay for those rights."

[19] A social norm is a pattern of expectation shared by members of a social group (Bredemeier and Stephenson, 1962). To the extent that it is internalized, people tend to act in accord with a norm even in the absence of group pressure or the direct threat of coercion. On norms and obligations, see Parsons (1951) and Etzioni (1968) for sociological discussions; Mueller (1986) for a public-choice discussion; and Callicott (1986) for a philosophical discussion.

of a much larger class of goods. Instead of restricting his value to the preservation of the particular park or set of parks described in the scenario, the respondent might include some or all of the value he holds for preserving parks in general. The possibility of symbolic bias, and the fact that stewardship or vicarious consumption values for public goods are often regarded as a "good thing" in our society, thus posing the problem of "compliance bias," suggest that particular caution is needed to avoid inflating existence values by inadvertently inducing one or another of several types of bias (for a detailed discussion of measurement biases, see chapter 11). In the absence of bias, we would expect that the smaller the amenity change, the more local the change, and the less unique the good being valued, the smaller the existence values should be as a percent of total value. On the other hand, programs such as national environmental improvement whose provision enjoys widespread support (Mitchell, 1979b, 1980b; Bloomgarden, 1982), and in which stewardship motives tend to be pronounced, are likely to involve substantial existence benefits.

While we believe CV surveys are capable of measuring benefits that include a nonuse dimension, we are less optimistic about their ability to obtain meaningful estimates of separate component values. Our understanding of respondent behavior in the CV setting is that when people are asked to value an amenity, such as the cleanup of a Superfund hazardous waste site, they make a holistic judgment. Instead of going through a mental process in which they value each of the relevant benefit categories and subcategories and then combine these values in their minds to arrive at a total value, respondents are likely to arrive at a global judgment, based on the conditions described in the scenario, about what the cleanup is worth to them. This judgment will reflect the configuration of benefits they believe will accrue to them from the amenity. It follows from this model of decisionmaking that respondents in the CV context are likely to find it difficult to separately value the component benefits if asked to think about them one by one. (We will return to this issue in chapter 12, where we evaluate the different approaches to estimating existence values.) Fortunately, the total WTP amount is satisfactory for most benefit-cost analyses.

Relationship between the Use and Existence Classes of Benefits

We now turn to the relationship between the two major classes of benefits, use and existence.[20] For ease of exposition we will continue to assume a world in which there is no uncertainty of any kind. We will also assume

[20] This section draws heavily on three recent papers by V. Kerry Smith (1986a, 1987a, 1987b).

that a desirable increase in q from q_0 to q_1 is under consideration, and that a Hicksian compensating surplus measure is requested by the policy-maker. The total value of the change from q_0 to q_1, in such a case, is given by

$$\text{TOTVAL} = e(p_0, q_1, U_0) - e(p_0, q_0, U_0), \quad (3\text{-}1)$$

which is simply equation (2-2) of chapter 2. What are the implications of the relationship

$$\text{TOTVAL} = \text{USEVAL} + \text{EXVAL}, \quad (3\text{-}2)$$

or, more precisely, how can USEVAL and EXVAL be defined so that such a relationship is both true and unique? Two solutions to this question have been proposed. The first is based on the notion of access and the second on the notion of weak complementarity.

The first solution can be seen by asking what happens when the agent for whom we are trying to ascertain maximum WTP is personally denied access to use q_1 if it is supplied. The answer to the question of how much the agent is willing to pay for q_1 under this condition is one possible way to define the agent's existence value for the good, and can be written as

$$\text{EXVAL}^* = e(p_0, q_1, U_0 | A{=}0) - e(p_0, q_0, U_0 | A{=}1), \quad (3\text{-}3)$$

where A equals zero when the agent does not have personal access to the level of q specified in the second expenditure function (but does still have access to the reference level, q_0), and A equals one when the agent does have access to q_1. Use value can now be seen as a residual category defined by

$$\text{USEVAL}^* = \text{TOTVAL} - \text{EXVAL}^*, \quad (3\text{-}4)$$

and could be more properly termed the value of access.[21]

The second solution is based upon the concept of weak complementarity (Maler, 1974, 1986; Freeman, 1979b; Hanemann, 1986c), which implies that the public good in question and some particular private good, X, are consumed together, so that when the private good is not consumed the "use" demand for the public good is zero.[22] The utility of q_1 under this approach can be represented as

$$U(q_1,x) = H[f(q_1,X), g(q_1)], \quad (3\text{-}5)$$

[21] The term "access value" in this general context appears to have originated with Bohm (1977) and Gallagher and Smith (1985).

[22] For papers taking this approach, see Maler (n.d.), Freeman (1981), and McConnell (1983).

where user benefits are those associated with the $f(q_1,X)$ function. The $g(q_1)$ function is said to embody the existence value. TOTVAL is then associated with $U(q)$ and is defined the same as in equation (3-2) above. The benefits associated with use, USEVAL**, are considered measurable by the observed/indirect market-based benefit measurement techniques,[23] such as travel cost, so that EXVAL** can be seen as a residual category defined by

$$\text{EXVAL}^{**} = \text{TOTVAL} - \text{USEVAL}^{**24} \qquad (3\text{-}6)$$

Note that while access is a necessary condition for use, it need not imply use, so that in general USEVAL** ≠ USEVAL*, and EXVAL** ≠ EXVAL*.[25]

The access approach to distinguishing use and existence values can potentially be implemented in a CV survey, but use of the weak complementarity approach to obtain a separate estimate for existence value is cumbersome and methodologically problematic. Implementation of the second approach would require a CV study to obtain an estimate of TOTVAL, and a travel cost analysis to measure USEVAL**. Moreover, it would be very difficult to adequately convey the weak complementarity concept that lies behind USEVAL** to respondents in a CV survey. USEVAL* AND USEVAL** may be close in many instances, and their degree of divergence will depend largely on how the significant others component of vicarious consumption, and the inherent component of stewardship, become manifest under the access and weak complementarity concepts.[26]

Introducing Uncertainty

Thus far we have conveniently assumed that respondents do not harbor uncertainty about whether they will want to have a given amenity provided in the future, or whether a given program will actually be able to supply the desired level of quality. Clearly, uncertainty is an important aspect of

[23] Observed/Indirect benefit measurement techniques are those which use observed purchase of a private market good to infer something about the value of a public good. They are discussed in more detail later in this chapter.

[24] If USEVAL** does not include all use benefits—indirect use benefits are particularly likely to remain uncounted under techniques based on weak complementarity—they will be included in EXVAL**.

[25] The two approaches to defining existence value can be seen to be equivalent when access is defined in terms of the price of X.

[26] Under both restrictions, motives related to vicarious consumption-significant others are likely to be reflected in USEVAL if the benefits are primarily gained by recreating with friends.

many public goods, especially environmental goods. For example, on the demand side, people may be unsure about whether they will ever want to use a currently unused and uncontaminated aquifer. On the supply side, a sewage treatment program may or may not accomplish its goals, depending on whether the estimates of the recovery capacity of a river's biological system are correct or not.

Since Weisbrod's (1964) seminal work on uncertainty, which made specific reference to the possible irreversible consequences of the destruction of a national park, scholars have considered uncertainty in a benefit-cost analysis from two distinct perspectives. The first of these is known as the "timeless" or option value approach (Cicchetti and Freeman, 1971; Krutilla and coauthors, 1972; Schmalensee, 1972; Bohm, 1975; Anderson, 1981; Graham, 1981; Bishop, 1982; V. K. Smith, 1983, 1984, 1987b; Mendelsohn and Strang, 1984; Freeman, 1984b; Plummer and Hartman, 1986). From this perspective, option value is the amount that people will pay for a contract which guarantees them the opportunity to purchase a good for a specified price at a specified point in the future, and may be thought of as a risk premium to compensate for uncertainty about future taste, income, or supply.[27]

The second perspective is known as the "time-sequenced" or quasi-option value approach (Arrow and Fisher, 1974; Henry, 1974; Krutilla and Fisher, 1975; Conrad, 1980; Freeman, 1984a; Miller and Lad, 1984; Hanemann, forthcoming; Fisher and Hanemann, 1985a, 1985b, 1986, 1987; Graham-Tomasi, 1986). In this approach quasi-option value is regarded as the risk premium people will pay to delay an activity which, if undertaken, might foreclose making a better-informed decision at a later time. Quasi-option value assumes that the consumer knows what she wants, that the outcome of a given action is uncertain, and that there is the likelihood that more information will become available at a later time which will permit a better decision. Quasi-option value can also be thought of as the value of information conditional on undertaking a particular action.

Option Value, Option Price, and Expected Consumer Surplus

The nature of option value under the timeless approach continues to be a topic of debate among resource economists. Recent work indicates that it is important to distinguish ex ante and ex post points of view in thinking about the concept. Consider a situation in which the state of the world

[27] Option value has sometimes been confused with the purchase of access rights. Option value represents a relatively small payment now for the right to be able to buy (and pay for) something later, while the purchase of access rights involves full payment now for being allowed to do something later (without necessarily having to do it).

that will exist in time period 2 is unknown in time period 1. Option price (OP) is defined as the ex ante state (period 1) independent willingness to pay for a specified change in the level of the public good in question.[28] Option price is usually considered to be the relevant welfare measure in benefit-cost analysis, because governments must decide what actions to carry out in period 1, and citizens' payments for the change are rarely dependent on what state of the world actually materializes in period 2. These payments can be defined in terms of a planned expenditure function (see chapter 2). In contrast, expected Marshallian consumer surplus (EMCS, the concept that travel cost studies attempt to measure) is an ex post measure.[29] It is found by summing over the probability of each state of the world times the expected consumer surplus in that state of the world as if that state were known to have occurred in period 2.[30]

The relationship between option value, option price, and expected consumer surplus is shown by

$$\mathrm{OV}(f(X)) = \mathrm{OP}(f(X)) - \mathrm{EMCS}(f(X)), \tag{3-7}$$

where $f(X)$ is the probability distribution for different states of the world, X.[31] Option value is a combination of ex post (EMCS) and ex ante (OP)

[28] The use of the term "option price" for this concept is the source of much confusion among those not privy to the intricacies of the option value literature, as the term seems to suggest the price of an option and is used by Wall Street in precisely that way. Ignoring certain subtleties, option price in the present context may be viewed as what a consumer will pay now in order to obtain a good.

[29] We want to clearly distinguish between ex ante and ex post states of the world (Is it going to rain?/It is raining) on the one hand, and ex ante expected utility from undertaking an activity and ex post actual utility gained from undertaking the activity on the other. Only the states-of-the-world use is relevant to our discussion and the option value literature. Difference in utility under the expected and actual utility approaches may be much larger, but are rarely considered in economics because of the impossibility of providing contingent contracts against such differences unless they are a result solely of the realized state of the world.

[30] In practice, expected consumer surplus is measured only in the state of the world that did occur.

[31] For an example that shows the relationships among option value, option price, and expected consumer surplus, consider a one-time production of an outdoor event where tickets are sold only at the gate immediately before the performance and the price is determined by the demand. If it is clear and bright, tickets would sell for $20, and if it is raining tickets would sell for $10. Assuming that there is a 40 percent chance of sunshine and a 60 percent chance of rain and that the demand curves for the tickets in the two states of the world are horizontal, the expected consumer surplus for the gift of a ticket is $14 (.6 * $10 + .4 * $20). Now, allow tickets to be sold the day before the show and assume demand is such that tickets can be sold for $12. This number represents option price. Option value is −$2 ($12 − $14). Another way to represent option value is to ask, How much could a certificate that allowed the bearer to purchase a ticket at the gate for $12 if it was sunny (an option) be sold for on the day before the event? The answer (now from the

welfare concepts (Chavas, Bishop, and Segerson, 1986; Plummer and Hartman, 1986; and Smith, 1987b). Smith further argues that EMCS has no special merit as a reference point for measuring welfare changes, as it is simply what is measured by looking at preferences as reflected in ex post behavior.[32] Equation (3-7) also shows how option value may be viewed as a correction factor which can change an ex post welfare measure into the desired ex ante option price welfare measure.[33] Freeman (1984b) and Smith (1984) argue on theoretical grounds that this correction factor should be small for non-unique resources.[34]

The problem with trying, in a CV survey, to measure option value in the timeless sense is that respondents understandably do not find very plausible the concept of buying an option to be able to buy a specified level of a public good in the future.[35] Nor are contingent contracts meaningful from a policy perspective, as the provider of public goods—government—cannot control people's uncertainty about whether they would use these goods. In contrast, experience shows that what CV surveys almost always offer for sale—certain provision of the amenity for possible use in the future at a specified price—accords quite closely with people's understanding about how public goods are provided. As Smith (1987b) puts it, in a contingent valuation survey one is really selling access to public goods, not use.

Viewed in terms of policy relevance and measureability in a contingent valuation survey, a different situation exists when we turn from the demand side to the supply side. Because government does have some control over how much certainty its policies entail, it may be desirable to measure the benefits of reducing supply uncertainty. Graham-Tomasi and Wen (1987) have proposed a measurement of this kind that is amenable to use in a contingent valuation survey.[36] Their concept, which we will call a

---

seller's perspective) is \$2 (\$14 − \$12). Options have been sold for stocks, agricultural commodities, foreign exchange, and occasionally real estate.

[32] For an extended discussion of this issue with reference to a CV study of hazardous waste risks, see Smith, Desvousges, and Freeman (1985).

[33] As Smith and Desvousges (1986:30) point out, "option value is an index of the importance of the perspective—ex ante or ex post—used to value either changes in the resource, or in the terms of access to it, under conditions of uncertainty."

[34] Chavas, Bishop, and Segerson's (1986) work suggests that an ex ante approach to developing this correction factor may be more fruitful than the option value approach.

[35] We can envision cases involving quasi-private goods where one might pay something now to join a pool of people who had the right to buy a public good in the future at some fixed amount. Brookshire, Eubanks, and Randall (1983) came close to invoking such a situation in their CV study of hunters' willingness to pay for a grizzly bear or bighorn sheep stamp which would entitle them, on payment of a fixed amount, to participate in a drawing for licenses to shoot these endangered species at a given time in the future.

[36] For an example, see Wen, Easter, and Graham-Tomasi (1967).

Hicksian uncertainty bias correction (HUBC) term, is defined by the equation

$$\text{HUBC}\ (f(X)) = \text{OP}(f(X)) - \text{HCS}(\overline{X}), \tag{3-8}$$

where HCS is the relevant Hicksian consumer surplus measure evaluated at the expected value for the $f(X)$, and the latter may be thought of specifically as the distribution of policy outcomes. In contrast to equation (3-7) where the consumer surplus measure used is Marshallian and is evaluated as the average over different states of the world, we now have a Hicksian consumer surplus that is evaluated at a single state of the world.[37] These properties make it possible to measure HCS using contingent valuation. The difference between OP and HCS measured in a CV survey, using either the same respondents or two equivalent subsamples, would provide an estimate of the Hicksian uncertainty bias correction. In principle, option price could be obtained for several different f_i (X), and a curve depicting HUBC could be traced out as a function of the level of uncertainty.

Quasi-Option Value

The essence of quasi-option value lies in the possibility that information can be acquired at a future time which would allow the person valuing the amenity to make a better decision. Quasi-option value is always positive since it is the value of information conditional on undertaking a particular current action. Knowledge can be consciously sought or it can be acquired in a passive manner. Particular policies will have quasi-option value to the extent that they promise to create knowledge (Freeman, 1984a) or make passive learning possible (Arrow and Fisher, 1974), or both. Where this is the case, CV surveys of such policies should inform respondents of these possibilities so that respondents will take them into consideration in determining a WTP amount for the amenity. The value of additional information is likely to be of greatest importance when valuing goods subject to possibly irreversible changes, such as endangered species,[38] aquifers vulnerable to contamination, and the damming of wild rivers.

[37] In the absence of uncertainty, the two are equivalent.

[38] With regard to endangered species (Fisher and Hanemann, 1985b, 1986), we may further distinguish between direct quasi-option value, where use benefits are at risk, and indirect quasi-option value, where the loss of the amenity potentially threatens other valued amenities. Someone may value an endangered species because further knowledge of it may reveal its value for, say, medicine, or because subsequent knowledge may show that the species plays a crucial role in supporting other, valued, aspects of an ecosystem (Norton, 1986).

The complexity of valuing future information can be illustrated by a proposed policy to install scrubbers in power plants or factories to reduce air pollutants believed to cause acid rain. The installation of scrubbers creates quasi-option values to the extent that we learn how they work and that we learn more about the causes and effects of acid rain by delaying the (relatively) irreversible damage to some lakes and trees while knowledge accumulates (Fisher and Hanemann, 1985a). On the other hand, installation of a scrubber involves a large and essentially irreversible investment in its own right (a scrubber has a twenty-year lifetime). Delaying its installation creates quasi-option value to the extent that information is likely to be gained in the near future on whether scrubbers are needed (Graham-Tomasi, 1986). Quasi-option values underscore how important it is for CV surveys to convey to respondents the appropriate future informational stream of the alternative scenarios (status quo versus proposed scenario) that the respondents are asked to value.

Methods of Measuring Benefits

The various methods that can be used to measure the benefits of public goods differ greatly in their data needs and in their assumptions about economic agents and physical environments. They also differ in the types of benefits they are able to measure. Smith and Krutilla (1982) distinguish two major categories of benefit measurement techniques: those based on physical linkages, and those based on behavioral linkages.

Physical Linkage Methods

There is little theoretical basis for the use of physical linkage methods in welfare economics, although these methods are frequently used in practice. They are based on the assumption that some sort of technical relationship (an engineering or a biological relationship, for example) exists between the public good in question and the consumer. Instead of making inferences about the behavioral motivations of trout fishermen or other economic agents, this approach would focus instead on the specific water quality characteristics, such as temperature, which permit trout fishing. Physical linkage methods are frequently referred to as the damage function approach,[39] or, in the case of biological relationships, the dose-response approach. The damage function approach values the estimated effects, using marketplace prices[40]—an attribute which, along with the fact that

[39] For an application of the damage function approach to water quality, see Heintz, Hershaft, and Horak (1976).

[40] For some effect such as human mortality, where market prices do not exist, one of the behavioral linkage benefit measurement techniques must first be used to place a value on human life.

Table 3-2. Behavior-based Methods of Valuing Public Goods

	Direct	Indirect
Observed market behavior	OBSERVED/DIRECT Referenda Simulated markets Parallel private markets	OBSERVED/INDIRECT Household production Hedonic pricing Actions of bureaucrats or politicians
Responses to hypothetical markets	HYPOTHETICAL/DIRECT Contingent valuation Allocation game with tax refund Spend more-same-less survey question	HYPOTHETICAL/INDIRECT Contingent ranking Willingness-to-(behavior) Allocation games Priority evaluation technique Conjoint analysis Indifference curve mapping

the effect is observable (even if only statistically), has lead many economists to favor this approach.

The inadequacies of the damage function approach are widely recognized, however. Its critics (Maler, 1974; Freeman, 1979b) point out that damage functions are not directly related to the consumer's utility function unless averting behavior and price changes are also modeled.[41] Further, damage functions cannot be used to value the existence class of benefits, or even the indirect-use benefits. Freeman and Maler rightly regard use of the damage function as, at best, a first approximation to benefit measurement. If the total benefits of the good being valued are much larger than those comprising the direct-use benefit categories, the damage function approach will result in seriously distorted benefit estimates.

Behavioral Linkage Methods

Contingent valuation and its alternatives are all based on some form of behavioral linkage between a change in an amenity and its effects. Table 3-2 uses two dimensions—how preferences are revealed and type of behavioral linkage—to distinguish four classes of behavior-based benefit estimation methods.

Before assessing each class of methods, it may be helpful to highlight the major issues between the principal competitors for use in making benefit estimates, the Observed/Indirect and Hypothetical/Direct approaches. The Observed/Indirect methods are attractive to economists because they are based, however indirectly, on actual market behavior. Methods in this class are typically two-step procedures. When using the

[41] If the consumer is a firm, the firm's utility function is a profit function which embodies a production function.

household production approach (Bockstael and McConnell, 1983), for example, one estimates a demand curve for purchases of an observed marketed good and then uses a technical or behaviorial relationship between the marketed good and the public good to infer a demand curve for the public good. The Observed/Indirect techniques necessarily rest on unverifiable assumptions when the level of the public good which the researcher wants to value is outside of previously experienced levels. The techniques in this class, because they are based upon observed behavior, measure ex post expected consumer surplus rather than option price, the latter being the correct measure of welfare change in the presence of uncertainty.[42] In contrast, the hypothetical character of the Hypothetical/Direct approach, as exemplified in the CV method, permits the researcher to obtain direct estimates of the correct ex ante measure of welfare changes—option price—and allows great flexibility in the range of benefits and the types of goods it can be used to value. The Hypothetical/Direct methods are based on the assumption that people would behave the same way in a relevant real market, if one existed, as they do in a hypothetical market. All the methods in the two hypothetical classes assume that respondents will give considered answers, that they will not behave strategically, and will not be unduly influenced by extraneous features of the survey instrument or the process of being interviewed.

Observed/Direct Methods. With the Observed/Direct class of methods, preferences are revealed in observed markets and the benefit measures are directly linked with the peoples' preferences. These characteristics represent optimal conditions for valuing goods, and are rarely realized for public or quasi-private goods. For example, in a referendum voters make binding decisions about the provision of public goods by voting on specific propositions. Few of the *public* goods for which benefit information is desired are voted on in such a way that legislators and regulators can use the results of the referenda in a full-fledged benefit-cost analysis. Nevertheless, the methods in this class are useful in defining the characteristics of market models and for validating the values obtained by the methods in the other three classes.

Actual referenda, the first of the Observed/Direct methods, offer an approach to measuring benefits which has some very desirable properties. In a referendum on, say, a water pollution control bond issue, voters indicate whether or not they are willing to support a program to increase the supply of a public good according to a specified payment scheme. Presuming an informed voter, the decision on how to vote is based on the voter's assessment of whether the marginal utility of the program is greater

[42] Ex post in the sense that the state of the world is known. They are ex ante in the sense that the consumer's expected utility may differ from the realized utility of a particular good.

than the marginal utility of the amount she would have to pay. A public goods market is in essence created by the election setting and the wording of the referendum. Unlike the Hypothetical/Direct methods, which rely on hypothetical markets, the amount and method of payment is specified and is politically enforced.

In order to empirically use the results of actual referenda to value a public good, one of three situations must exist:

1. The same electorate must vote in independent referenda for different levels of the good at a fixed tax price or for a fixed level of the good at different tax prices. This might occur in cases where a given referendum fails and its supporters put forth a modified version in a subsequent election.
2. Different jurisdictions must vote on the same level of a good. This occurs when voting data on a single referendum are available for sub-segments of an electorate (precincts, counties, for example).
3. Different jurisdictions must vote on different levels of a good, as when school bond elections are held in different school districts.

The first situation is the most satisfactory for the benefit analyst because it yields a demand curve for the good based on acceptance rates for different levels of provision. Unfortunately, the occurrence of suitable sequences of referenda are extremely rare. The second and third situations require measurable variation in one or more independent variables which are used to estimate a demand curve for the good (Deacon and Shapiro, 1975; Borcherding and Deacon, 1972; Portney, 1975; Freeman, 1979b:102–104). Although referenda of the third type are fairly common, relatively few involve the kinds of goods for which benefit estimates are needed for policy purposes. This is no accident, because the availability of referenda as a decision-making process often precludes the need for other techniques to measure these benefits. There is no need to undertake a CV study of people's willingness to pay for a new school where the citizens can express their binding preferences in a referendum vote. Some referenda of the second type have been held in California during the past decade on such topics as freshwater quality, drinking water quality, land-use issues, nuclear power, parks, prison construction, and support for solar energy (Deacon and Shapiro, 1975; Conway and Carson, 1984).

Referenda are of particular interest to contingent valuation practitioners because they provide an institutional model for asking people to express their preferences for public goods. We will argue in the next chapter that CV scenarios may be usefully conceived of as hypothetical referenda (see also chapter 8). In addition to their role as a model, referenda are a potentially useful behavioral criterion for validating the contingent valuation method. The issue is whether a hypothetical referendum (for example, a CV scenario that uses a referendum format) can predict an actual referendum. We provide an example of the use of actual referenda for this purpose in chapter 9, where we discuss evidence for the validity of the CV method.

Simulated markets, the second of the Observed/Direct methods, are experimental markets set up by researchers in which people actually buy and sell goods under controlled conditions. Bishop and Heberlein (1979, 1986) have constructed several markets of this kind in which they offered hunters the opportunity to buy and sell permits to hunt in state preserves. Ordinarily these permits are only available free of charge to applicants selected in a lottery. By selling the permits to respondents for real money under controlled conditions, the researchers were able to establish a (simulated) market price for the amenity. Simulated markets have somewhat limited applicability as a benefit measurement technique, since they are restricted to quasi-private goods (because of the need for exclusion in order to institute the market) and are difficult and costly to conduct. Where they can be instituted, as in Bishop and Heberlein's work, simulated markets provide a useful criterion for testing the validity of CV values obtained in hypothetical markets for the same good. We review Bishop and Heberlein's use of simulated markets for this purpose in chapter 9.

To the extent that appropriate parallel private goods markets exist for quasi-private goods, such as fishing sites, it may be possible to use these markets to infer the value of the publicly provided commodity. The parallel private market approach has been used to value elementary and secondary education (Sonstelie, 1982) and fishing opportunities (Vaughan, Russell, and Carson, 1981; Vaughan and Russell, 1982).

Observed/Indirect Methods. Observed/Indirect methods rely on data from situations where consumers make actual market choices, as they do in deciding on a trip or buying a house. The value of the nonmarket amenity must be inferred from market data for another good with which it has a known or estimatable linkage. In order to make this inference, the researcher must depend on a large number of assumptions. Freeman (1979b:4) aptly describes the process of determining benefits by methods of this class as "detective work—piecing together the clues that people leave behind as they respond to other economic signals." The methods in this class measure expected consumer surplus rather than option price.

The household production function, advanced by Becker (1965) and Lancaster (1966) and developed to its current state by Hori (1975) and Bockstael and McConnell (1983), is the basis for several of the Observed/Indirect methods. It posits that consumers purchase marketed goods (which have no utility in and of themselves) that are combined with each other, with nonmarketed goods, and with household inputs to produce the goods and services that ultimately generate household utility. A variant of the household production function, the travel cost method (Clawson, 1959; Clawson and Knetsch, 1966), has been used extensively to value site-specific recreation benefits. To estimate the simplest type of travel cost model, one draws concentric circles of different radii around a particular site and calculates the average number of per capita visits to the site by those resident in each distance zone (that is, the area between two circles). The travel information is obtained by survey interviews. These data may be used to trace out a demand curve for the site as a function of distance. If a monetary value is assigned to each mile from the site, a consumer surplus calculation can easily be made for each zone by measuring the area below the demand curve and above the "travel cost" for a resident of that particular zone.[43]

There are several reasons why the travel cost method is problematic. First, travel cost is a highly site-specific technique which generally ignores the possibility that consumers can substitute one site for another. Second, it is not usually possible to explicitly incorporate environmental quality into the travel cost model. People travel to a lake for a variety of reasons, which relate in various ways to the lake's water quality. Some progress has been made in overcoming these two problems in the recent generalized travel cost models which incorporate a first-stage participation estimation with water quality as an argument (Binkley and Hanemann, 1978; Vaughan and Russell, 1982; Desvousges, Smith, and McGivney, 1983; Bockstael, Hanemann, and Strand, 1985; and Caulkins, Bishop, and Bouwes, 1986).[44] The third problem with the travel cost method is the difficulty of knowing how to handle the role of time in a travel cost study. What elements of time are to be interpreted as costs of recreation activity, and what monetary values are to be assigned to these elements (Freeman, 1979b:204)? Although there has been some progress by Wilman (1980) and McConnell and Strand (1981) in addressing this difficult question, it is far from solved.[45] The fourth and fifth problems are even less likely to yield solutions, however. Bockstael and McConnell (1980a, 1980b) have shown that the travel cost method is very sensitive to the functional forms used in estimation, and the work of Vaughan and Russell (1982) confirms

[43] For a basic description of this procedure, see Freeman (1979b). Modern versions of the travel cost method take into account a large number of additional factors; see McConnell (1985).

[44] See also the hedonic travel cost models proposed by Mendelsohn and Brown (1983).

[45] Harris and Meister (1981), Sutherland (1983), and Walsh, Sanders, and Loomis (1985) have used survey approaches to investigate how recreators value their travel time. Interestingly, they find that people would prefer more travel time and would pay for it. Of course, it should be noted that their studies were in highly picturesque places, New Zealand and Colorado. This should stand as a warning, however, to those (Cesario, 1976, for example) who advocate or use one-half or one-third of the individual's wage rate as the measure of the travel time.

this.[46] An even more fundamental problem is the fact that travel cost models can only measure a limited range of benefit categories—the direct recreation benefits and some of the aesthetic and ecosystem-use categories (Hay and McConnell, 1979). The travel cost technique is unlikely to be of much use in valuing commercial direct uses and cannot be used to value existence benefits.

The other major Observed/Indirect method in common use is hedonic pricing (Adelman and Griliches, 1961; Ridker and Hennig, 1967; Griliches, 1971; Rosen, 1974). This method assumes that the price of a marketed good is a function of its different characteristics. Letting X represent a commodity class, and P_i the price of some good i in commodity class X which has characteristics Q_j, we get

$$P_i = P(Q_1 \ldots Q_j \ldots Q_n) \qquad (3\text{-}9)$$

The implicit price of a particular characteristic Q_j is found by differentiating P_i with respect to that Q_j. In a second stage of analysis, this relationship is combined with other restrictions on the demand and supply of the characteristic of interest, and a demand function for that characteristic is estimated. Consumer surplus is then calculated using that demand function. The most common uses of the hedonic pricing technique are in property value studies in which the price of the property is thought to include the value of some environmental service, such as local shorelines or air quality (Brown and Pollakowski, 1977; Harrison and Rubinfeld, 1978; Freeman, 1979b; Brookshire and coauthors, 1982), and in hedonic wage studies (Thaler and Rosen, 1976) where different risk levels are thought to be incorporated into the wages for different jobs.

While the hedonic pricing technique can in principle be used to value all of the use categories, in practice it also suffers from serious problems. First, the data requirements for a valid hedonic pricing study are unusually exacting. It must be possible to control for all relevant characteristics—structural, neighborhood, and environmental; where many or unique resources are already in public hands, this may be impossible. Second, sufficient market data for reliable estimations are difficult to obtain (Just, Hueth, and Schmitz, 1982:294). Housing turnover, for instance, is relatively slow, and it is not easy to locate genuinely comparable houses in relevant neighborhoods. Third, the functional forms of the true underlying hedonic pricing equations are unknown, and the researcher is confronted with a number of competing estimable formulations with quite different

[46] Vaughan and coauthors (1985) also show that the estimation of travel cost models with aggregate-level data rather than individual-level data (as is frequently used) creates additional problems.

implications (Freeman, 1979b; Halvorsen and Pollakowski, 1981). Fourth, people must be aware of the actual physical differences in the levels of the characteristic being valued. To assume such awareness may be unreasonable when dealing with risk levels posed by chemicals or with colorless or odorless air pollutants. Fifth, expectations about changes in the good being valued and other relevant characteristics are generally unobservable, but presumably enter into the determination of prices, especially of property values. For example, people may assume that air quality in a given location is going to improve or decline and incorporate this assumption into their willingness to pay or to accept a given purchase price. Sixth, it is difficult, if not possible, to value simultaneous changes in sites which are substitutes. Finally, as Brown and Rosen (1982) have recently shown, the standard assumptions used to identify systems of hedonic pricing supply and demand equations are tautological and incorrect except in special and unlikely cases.

The last method in the Observed/Indirect class concerns the actions of politicians who make decisions about providing public goods. Some have argued that voting for candidates is a useful source of imputed values for public goods (Barr and Davis, 1966). The assumption here is that political representatives maximize their chances of reelection by identifying and carrying out the preferences of the electorate, in particular the "median voter" (Downs, 1957; Black, 1958; Romer and Rosenthal, 1979). Given a sufficient number of observations on representatives' votes for different program levels, the demand for a specific public good could be derived by making assumptions about the distribution of the electorate's willingness to pay for specific goods and the distribution of expected taxation. Unfortunately, uncertainties about provision and taxation, and the likelihood that most political representatives have multiple objectives when they vote on a given measure, suggest that the tortuous chain of assumptions necessary to obtain the implied value are too fragile to yield valid benefit estimates. For example, studies have shown that political leaders may have inaccurate perceptions of their constituencies' preferences for environmental public goods (Kamieniecki, 1980).[47]

Hypothetical/Indirect Methods. In the case of the Hypothetical/Indirect methods, people are asked to respond to hypothetical markets, but their responses are only indirectly related to valuing the good of interest. Among the methods in this class are contingent ranking (CR), indifference curve mapping, allocation games, the priority valuation technique, and methods

[47] This is a rather consistent finding; see also Mueller (1963), Hedland and Friesema (1972), and Fields and Schuman (1976). However, Lemert (1986) indicates that politicians are consistently more accurate than individual voters at predicting the outcomes of statewide referenda on salient issues.

that ask about a respondent's willingness to engage in a particular form of behavior. The last group of methods can be regarded as hypothetical analogues of the Observed/Indirect behavior methods discussed above. All of the indirect procedures, Observed and Hypothetical, can be viewed as two-step procedures.

Table 3-3 illustrates how two of these techniques—contingent ranking and a hypothetical travel cost method—could be used to value a proposed recreational site. Instead of being asked directly to place a dollar value on the provision of a hypothetical recreational site (A–C), respondents under the hypothetical travel cost method might be asked how far they would drive to use such a site; under the contingent ranking method, they might be asked how they would rank a set of sites having different descriptions. In order to translate these behavioral intentions into an economic value for the site, the researcher would have to posit how they translate into a change in utility that can be expressed in terms of dollars. The justification for undertaking such a two-step valuation process is the belief that respondents are better able to give meaningful answers to related behavior questions than they are to direct valuation questions.

Allocation games and the closely related priority evaluation technique ask respondents to allocate a fixed budget among a specified set of budget categories.[48] They allow little possibility for strategic bias or for part-whole bias (for discussion of these biases, see chapter 11), because the total budget is fixed and respondents are explicitly required to consider the full range of possible allocations. Situations in which an allocation game qualifies as the most useful preference-revelation technique are limited in scope. An example of its usefulness arises when an agency's budget is set and administrators are trying to distribute that budget among several functional areas in an attempt to maximize social welfare.[49]

One methodological problem with the use of allocation games for formal benefit measurement is the difficulty of describing each of a number of budget categories in sufficient detail to allow specific amenity changes to be valued. Whereas a contingent valuation study, for example, would measure the WTP amounts for the provision of specific levels of a public good (such as 30 miles of air visibility in the Southwest versus 60 miles of visibility), an allocation game would have to measure relative WTP amounts for an equally specific set of public good provisions for a diverse set of policy areas (such as education, health, environment, and defense) within a given budget constraint. Another and even more serious problem

[48] For a more extensive discussion of allocation games and the priority evaluation technique, see Sinden and Worrell (1979).

[49] See Hardie and Strand (1979) for an application of this technique to the Maryland State Park Service's capital budget.

Table 3-3. **Examples of Two Hypothetical/Indirect Methods**

Method	A Stimulus →	B Related behavioral intentions →	C Dollar value implied by intention
Hypothetical travel cost method	Building a new recreational site with specified characteristics	How far would you drive to such a site?	Value implied by cost of driving that distance
Contingent ranking method	Description of two or more sites that vary on two (or more) characteristics, one of which is distance from respondent's house	How would you rank-order this set of site descriptions?	Value implied by how distance is traded off for the other characteristic(s)

is that allocation games do not require respondents to make a clear declaration of their willingness to give up a specified amount of money in order to receive the amenity in question.[50] Just because someone allocates a fixed budget among different categories, it does not follow that she is willing to pay the amount allocated to a particular category.[51]

The contingent ranking method (Rae, 1983) requires respondents to rank a set of outcomes consisting of different combinations of goods (and/or their attributes) and associated payment requirements.[52] It has been used to study the potential market for electric cars (Beggs, Cardell, and Hausman, 1981);[53] to value visibility in the Mesa Verde National Park (Rae, 1981a), the Great Smoky Mountains National Park (Rae, 1981b), and in Cincinnati (Rae, 1982); to value reductions in diesel odors from motor vehicles (Lareau and Rae, 1985); and to value recreation and related benefits of water quality improvement in the Monongahela River in Pennsylvania (Desvousges, Smith, and McGivney, 1983).[54] The respondents in the Monongahela study were given four cards, each of which specified one combination of level of water quality and an amount the respondent might be willing to pay for that level of water quality (for example, boatable water for an annual payment of $50, fishable water for $100); respondents were then asked to rank the cards from the one they most preferred to the one they least preferred. The resulting ranked data were used to estimate an indirect utility function for different water-quality changes.

Desvousges, Smith, and McGivney (1983) argue that the potential advantage of contingent ranking over contingent valuation is that CR may obtain more accurate answers because the task of ranking the small set of cards is less demanding of respondents than answering a WTP question for a proposed change in an amenity. This may or may not be true. In the Monongahela study, at least, little discrepancy was found between the

[50] This is also the case with the indifference curve mapping method (Sinden, 1974; Sinden and Wyckoff, 1976; Findlater and Sinden, 1982), where the respondent is asked to trade off one good, such as days spent at park A, for another good, such as days at park B. This method allows the researcher to locate a family of indifference curves, but there is no indication of whether the respondent is interested in recreating in either park.

[51] This is not the case with allocation games that allow a tax refund (see, for example, Strauss and Hughes, 1976), which we regard as belonging to the Hypothetical/Direct class of methods.

[52] A closely related technique used by market researchers is conjoint analysis (Green and Srinivasan, 1978).

[53] For a conjoint analysis of the market for electric cars, see Hargreaves, Claxton, and Siller (1976).

[54] Desvousges, Smith, and McGivney's study was explicitly designed to compare the results obtained by CR, CV, and a travel cost method. The following discussion is indebted to their description and analysis of the CR method.

benefits measured by contingent ranking and those measured by contingent valuation (Desvousges, Smith, and McGivney, 1983:8–20),[55] and there was no evidence that response rates to the CR questions were better than those for CV questions.

The anticipated gain in simplicity of administration is not costless. For the same degree of precision in estimating WTP amounts, CR in general requires many more observations than does a continuous-response CV study.[56] Contingent ranking also requires more sophisticated and less direct statistical techniques to estimate the value of the outcomes than those used for contingent valuation. It can often be difficult to identify the decision rules that respondents have used in completing a contingent ranking exercise. Because the CR format does not permit direct evaluation of the utility change associated with a change in, say, water quality, it is necessary to undertake an indirect evaluation. This is accomplished by interpreting the function as an indirect utility function, taking the total differential of the estimated random utility function with respect to income and water quality and solving for the income change that would be equivalent to any water quality change (Desvousges, Smith, and McGivney, 1983: chap. 6; Hanemann, 1984c). Another drawback to the CR method is that the behavioral model underlying the method and its theoretical properties are at present incompletely understood.[57]

Finally, contingent ranking faces one of the same problems noted above in connection with allocation games: the researcher tends to elicit preferences in the form of attitudes rather than behavioral intentions (see chapter 8). Instead of requiring respondents to declare clearly that they are willing to give up a specified amount of money in order to receive the good in question, the CR technique requires respondents to make rank-order preferences for a set of alternative choices. Desvousges, Smith, and McGivney's (1983) exemplary CR study of Monongahela water quality benefits, with its clear-cut dollar and quality combinations, may serve as a hypothetical example. Some of their respondents were asked to use a set of four cards to rank their preferences for paying \$5 to obtain a river whose water was below boatable quality, paying \$50 for water of boatable quality, \$100 for water of fishable quality, and \$175 for water of swimmable quality. Suppose now that one of these respondents held true maximum WTP amounts for the four levels of water quality of \$1, \$40, \$60,

[55] In the Monongahela comparison, however, the results may have been affected by the fact that both the CV and the CR treatments were administered to the same group of respondents.

[56] This follows from the fact that CV elicits interval/ratio-level WTP responses, while at best CR obtains an ordinal approximation to CV WTP responses, and thus less information.

[57] See Ruud (1986) for a penetrating analysis of this problem with Rae's (1982) study of air visibility benefits.

and $90, respectively. He can rank the cards optimally so as to minimize the disutility of being forced to make such a choice; this is the behavior expected by the researcher.[58] Note, however, that forcing the respondent to make a choice that he would not voluntarily make lowers his initial level of utility, thus violating the assumption underlying Hicksian compensating surplus.[59] Thus the desired unique inverse Hicksian demand function cannot be identified, and at best what can be traced out is a family of indifference curves possessing particular curvature properties.

Hypothetical/Direct Methods. By directly measuring people's valuation of particular hypothetical changes in amenity quality and quantity, Hypothetical/Direct measurement methods shortcut the need to make the large number of assumptions required by the indirect linkage methods. Hypothetical/Direct methods, which include contingent valuation, posit what Smith and Krutilla (1982) term "institutional" links between amenity levels and individual behavior. The institutional assumption is that individual responses to hypothetical markets are completely comparable with individual responses to actual markets, an assumption whose implications we consider at length in the following chapters. Once this premise is granted, however, methods become available which are unique in their simplicity, theoretical justification, and ability to value the entire range of benefit categories.

Since we have already described the contingent valuation method, the discussion of Hypothetical/Direct methods here will be limited to the spend more-same-less type of survey question and to allocation games with the possibility of a tax refund. The spend more-same-less survey approach is based on ordinary survey questions which ask respondents to say whether "we" (the United States) are spending too much, too little, or about the right amount for certain types of government programs (National Opinion Research Center, 1983).[60] As a result of interpreting the trichotomous responses to mean that "spending too much equals a preference for less of the good and lower payments," and so on, researchers have used incompletely developed scenarios as the basis for estimating demand curves for one or more public goods. An early and somewhat simplistic example of the economic analysis of such spend more-same-less questions is Akin, Fields, and Neenan (1973). More recent studies of this

[58] If the respondent regards the offered choices as so far from his preferences that he is unwilling to make the effort to optimally rank the disutility of the various alternatives, his ordering will be meaningless.

[59] Contingent ranking based on observed data (for example, a single choice and a set of possible alternatives) does not suffer from this problem because it is based on voluntary choice.

[60] The spend more-same-less survey question method is sometimes called the "micro-based estimates of demand function" approach.

type have employed increasingly sophisticated discrete-choice statistical techniques (Gibson, 1980; Bergstrom, Rubinfeld, and Shapiro, 1982; Gramlich and Rubinfeld, 1982; Ferris, 1983; and Langkford, 1985).

The obvious risk of the typical spend more-same-less survey question is that the analysis of preferences is based on superficial and uninformed responses. Compared with that in a contingent valuation scenario, the description of the good is sparse. For example, the commonly used question in the National Opinion Research Center's General Social Survey (1983) asks respondents to express their views about "solving the problems of the big cities" and "improving and protecting the environment." There is little effort to create a market or specify a payment obligation. Respondents are rarely given information about what is currently being spent on these programs and how it is being spent. Particularly problematic is the assumption that the respondent knows what his payment obligation would be for each response. Langkford's work (1985) casts doubt on the validity of this assumption.

We include allocation games that offer respondents the possibility of a tax refund among the Hypothetical/Direct methods because this type of allocation game allows the respondent to reject payment in favor of a tax refund for any and all public goods being considered, instead of merely requesting the respondent to allocate a fixed budget among different public goods. An advantage of this type of allocation game is that it forces the respondent to value a large number of goods simultaneously, rather than valuing each in isolation (as is typical of a CV survey). The disadvantages include the superficial description of the different categories of public goods and the fact that if no tax refund is desired, the WTP estimates obtained may not be maximum WTP estimates. The two best known examples of allocation games with tax refund are Strauss and Hughes (1976) and Hockley and Harbour (1983).

The Advantages of the Contingent Valuation Method

The advantages that the hypothetical methods in general, and Hypothetical/Direct methods in particular, offer to the benefits researcher are highlighted by the comparisons shown in table 3-3. Table 3-4 compares the four classes of benefit measurement methods according to five criteria—ability to measure option price, to value goods not previously available, to estimate all existence-class benefits, and to directly estimate the relevant ordinary and Hicksian inverse demand curves.

Of particular importance is the flexibility of the hypothetical methods. As Sen (1977:339–340) has observed, "once we give up the assumption that observing choices is the only source of data on welfare, a whole new world opens up, liberating us from the informational shackles of the

Table 3-4. Key Properties of the Benefit Measurement Methods

Desirable properties	Method			
	Observed/Direct[a]	Observed/Indirect	Hypothetical/Indirect	Hypothetical/Direct
Able to obtain option price estimates in the presence of uncertainty	No	No	Yes	Yes
Able to value goods not previously available	Yes	No	Yes	Yes
Able to estimate all existence class benefits	Yes	No	Yes	Yes
Relevant ordinary (or inverse demand) curve is directly estimable	Yes	No	No	Yes
Relevant Hicksian compensated demand (or inverse demand) curve is directly estimable	No	No	No	Yes

[a] In some cases, only referenda have the desired property.

traditional approach." Within the important constraint that the scenario must have plausibility for the respondent, the CV researcher can easily specify a variety of states of the good to be valued and the conditions of its provision. Moreover, these need not be limited to current institutional arrangements or levels of provision. Brookshire and Crocker, in reference to the inflexibility of methods based on observed behavior, point out that

> the only really sound way of obtaining an estimate of whether the net benefits of a particular property rights restructuring are positive, if one insists upon employing observed everyday behavior, would be to perform the restructuring and observe the results. In some circles, this is simply known as trial and error. Trial and error can be an extremely costly way to perform research because the errors are real rather than hypothetical. (Brookshire and Crocker, 1981:246)

The hypothetical character of contingent valuation, which allows it to obtain ex ante judgments, also permits it (and the other hypothetical-based methods) to obtain WTP amounts that include the existence values, whereas methods that rely on observed behavior can obtain existence values only with great difficulty, if at all. Consider the different values that two people might put on maintaining the current level of air visibility at Mesa Verde National Park. Person A values only that level of visibility which is current during his visit to the park. He is willing to pay some amount to enjoy this attribute during his visit. Person B has no current-use value for the air visibility in the park, but believes that national parks are an important part of America's heritage and represent a public obligation for which he is willing to pay. If a travel cost study of air visibility were made in Mesa Verde National Park, its methodology would assume that travel behavior reveals the respondent's price for the site's services, and that air visibility is jointly supplied along with the site's other attributes. Thus, such a travel cost study might be able to indirectly estimate person A's use value for the park's air visibility, but could not measure the stewardship values held by person B. A contingent valuation study, in contrast, is not hindered by this restriction. The respondents in a CV survey put a price on an amenity after assessing the total change in welfare that it represents to them. If the respondents are reminded of the relevant benefit categories that should be kept in mind,[61] their WTP amounts

[61] This is necessary to ensure uniformity in the respondents' conceptualization of the valuation situation. If they are not so reminded, some respondents may fail to consider nonuse values, either because they did not think of such values in the context, or because they mistakenly assumed that the interviewer wanted them to consider only the use dimension. Desvousges, Smith, and McGivney (1983) were the first to use a "value card" as a visual aid in interviews in order to remind respondents of the potential benefit categories. Under the headings "use," "might use," and "just because it's there," their card listed reasons why the respondent might value clean water in the Monongahela River.

should reflect the full range of their preferences. Thus in a CV study the WTP amounts of persons A and B would include both use and existence benefits (to the extent that they exist).

The Hypothetical/Direct methods (including contingent valuation) also are able to directly measure specific points on an individual's compensated demand curve. This avoids the problem, which plagues the other methods, of potential bias resulting from wrong assumptions about the form of the individual utility functions. For example, instead of imposing separability conditions on individual behavior, based on the researcher's assumptions about consumer tradeoffs, CV studies can allow the respondent to make his or her own tradeoffs in terms of money (Brookshire and Crocker, 1981:246).

Summary and Conclusions

In this chapter we have considered the nature of benefits and the wide variety of methods economists have to measure them. Public goods were divided into two types, quasi-private goods (public goods with individually held property rights) and pure public goods (public goods with collectively held property rights). This distinction has implications for the type of market model appropriate for use in designing a contingent valuation survey. The benefits of both kinds of public goods result from the values individuals place on these goods. These values are subjective and multidimensional. A correct benefits assessment will include all the benefits that legitimately accrue to a given improvement. An inventory of freshwater quality benefits was used to illustrate the principal types of benefits, which include use and existence benefits and their several subdivisions. This inventory did not include option value, on the grounds that option value is not a meaningful benefit category for ex ante welfare measures.

Our discussion of benefit measurement methods focused primarily on the relationships among the various types of behavior-based methods. We classified sixteen different benefit measurement methods into four types according to whether the method relied on preferences revealed in observed or hypothetical markets and on a direct or indirect linkage between the method and the willingness-to-pay value. The Hypothetical/Direct methods, which include contingent valuation, were shown to be the only class of methods simultaneously capable of obtaining option price estimates in the presence of uncertainty, valuing goods not previously available or marketed, estimating all existence class-benefits, and obtaining in a direct fashion the relevant Hicksian demand curves. It was the prospect of realizing these advantages that motivated the CV pioneers to explore the uncongenial (to the well-trained economist) realm of hypothetical markets.

4

Variations in Contingent Valuation Scenario Designs

Having described the relationship between the contingent valuation method and other methods of measuring the benefits of nonmarketed goods, we now turn to three key issues in the design of contingent valuation scenarios that are the subject of current debate among CV researchers. Each involves design choices the researcher must make that may have a large effect on the quality of a CV study's findings and their appropriateness for use by policy makers. The first is whether a CV study should be based on a private goods market or a political market. The second issue concerns which of the several techniques for eliciting the WTP amount from the respondent is most appropriate for the study, and how the chosen technique should be implemented. These two issues are discussed in this chapter. The third set of choices involves how much and what kind of information about the amenity and the hypothetical market the researcher should include in the material presented to the respondent during the course of the interview. These decisions involve tradeoffs on the researcher's part between the need to inform the respondent about relevant features of the hypothetical market and the need to avoid information overload, and between the desire to measure benefits in a manner that offers policymakers the utmost flexibility in using the findings and the difficulty respondents have with scenarios that are too abstract (in the sense that they lack concrete details about the amenity and the conditions under which it would be provided). Discussion of the information issue begins in this chapter and continues throughout the book.

Private Goods Markets and Political Markets

Contingent valuation studies simulate a market for a nonmarketed good. Until recently researchers took it for granted that a private goods market was the appropriate model for CV scenarios. In a well-developed competitive market, where property rights are nonattenuated and the goods and services are in rival supply, unimpeded voluntary exchange moves toward a Pareto-efficient equilibrium. This model embodies the notion of a consumer who has realized tastes and whose purchasing decisions are based on a full understanding of the available alternatives as a result of prolonged experience in the market. Preferences are most credibly expressed through actual behavior in such markets, since the prospect of making a cash payment requires the consumer to take into account the alternative uses to which the money might be put. As Bishop and Heberlein point out, this model of consumer behavior is quite different from how respondents behave in CV surveys, even when the scenario values a quasi-public good such as goose-hunting permits:

> When people buy things in a market, they may go through weeks or months of considering the alternatives. The process will often involve consultations with friends and may also involve professionals such as lawyers or bankers. It may also entail shopping around for the best deal on the product in question. And for the majority of items in the consumer's budget, there is a whole history of past experience in the market to base the decision on. All this is markedly different than spending an hour or two at most with a mail survey or a personal interviewer attempting to discern how one might behave in a market for a commodity for which one has never actually paid more than a nominal fee. (Bishop and Heberlein, 1979:927)

One attempt to address this disjunction was the proposal made by the authors of the 1986 state-of-the-art report (Cummings, Brookshire, and Schulze, 1986) to limit the use of the CV method to those situations which best emulate consumer markets. Two of their original "reference operating conditions" for a valid CV scenario are explicitly based on the consumer goods market model:[1]

1. Subjects must understand, be familiar with, the commodity to be valued.
2. Subjects must have had (or be allowed to obtain) prior valuation and choice experience with respect to consumption levels of the commodity. (Cummings, Brookshire, and Schulze, 1986:104)

Taken literally, these conditions would limit the CV method to goods that are currently marketed in some fashion, such as hunting permits.

Despite its strong theoretical appeal to economists, there are persuasive grounds for questioning the appropriateness of this type of consumer

[1] The book edited by Cummings, Brookshire, and Schulze (1986) consists of a report, which was formally reviewed by a number of scholars, and a reply in which the authors revised or qualified a number of their original conclusions. The reference operating conditions referred to here were qualified in the editors' conclusion.

goods market model for CV studies. In the first place, Bishop and Heberlein's private goods market is an idealized market which is not always attained in the real world of consumer behavior, even when expensive purchases are involved. Market researchers have long recognized that many purchases are infrequently made and that the information people gather before making purchasing decisions differs greatly, depending on the purchase situation, the type of good, and the consumer's past experience (Bettman, 1979).

Second, market behavior is not always a superior indicator of people's preferences. Many people who smoke or engage in compulsive behavior such as gambling do not regard their behavior as expressing their real desires (Rhoades, 1985:166). When it comes to public programs, private goods market behavior is even less relevant. It has long been argued that people may be less self-interested and more public-spirited when they vote in elections than when they participate in the market for private goods (Buchanan, 1954; Wilson and Banfield, 1964, 1965).

The strict application of a private goods market model ignores any but self-interested consumption behavior and therefore downplays the "public-regardingness" behind existence values.[2] These values can affect political behavior.[3] Crenson (1971), for example, compared community political support for air pollution programs in the 1960s and found that political parties in communities characterized by a "private-regarding ethos" tended to ignore pollution issues, whereas in communities where a "public-regarding ethos" prevailed such issues were likely to reach the agenda. Mitchell and Carson (1986b) have observed similar behavior at the individual level in a study conducted to measure the benefits of reducing drinking water risks. In a pretest, the same respondents gave lower amounts when asked how much they would pay to have a pollution control device installed and maintained in their homes by the city water company than when asked how much they would pay in higher water bills to have the town's water plant install and maintain new equipment that would achieve the same risk reduction, but for everyone in the community. These respondents valued the latter program more because they perceived that it protected others besides themselves. There was no indication that they subordinated their private desires when they took the broader public interest into account (see Wildavsky, 1964).

These considerations suggest that political markets are a more appropriate analogue for CV surveys that value public goods than are private

[2] For tests of the "public-regardingness" hypothesis, see Attiyeh and Engle (1979) and Martinez-Vazquez (1981).

[3] Lane (1986) discusses the public's perceptions of fairness in political and private goods markets. See also Kahneman, Knetsch, and Thaler (1986).

markets.[4] A considerable body of theory on political markets has been developed by economists and political scientists (see, for example, Deacon and Shapiro, 1975; Bergstrom, Rubinfeld, and Shapiro, 1982; Enelow and Hinich, 1984; Langkford, 1985). As we have seen, the form most relevant to the contingent valuation method is the referendum, where the voter is faced with a one-time (or at best very infrequent) choice of voting yes or no to a predetermined policy package.[5] Here the behavior to be predicted by a CV study is how informed voters would actually vote if a proposition to provide an amenity was in fact on the ballot. The voting decision suggests a more complex, and some would say a more realistic (Morgan, 1978, for example), model of decisionmaking than the one implied by the idealized private goods market model. Instead of assuming that people express preexisting, well-realized preferences, the referendum model assumes that people make choices which are influenced by multiple motives, by contextual factors, and by less than perfect information.

Advantages of the Referendum Model for CV Surveys

Several aspects of referenda recommend them as an appropriate model for contingent valuation surveys. One is that public goods are paid for collectively, once the decision is made to supply them. This is why past CV studies have tended to use payment vehicles that imply coercion, such as the imposition of higher prices or taxes if a proposed regulatory program were to be implemented. Posing the elicitation question in the form of a referendum clearly implies such a payment scheme. Another is that, as noted earlier, referenda are actually used as a mechanism to enable citizens to make binding decisions about the provision of public goods, such as a new school building or a water pollution control program to be financed by a bond issue. Respondents are therefore likely to be familiar with their method of operation and their use in the political system. In addition, the voter's decision in a referendum has clear economic implications for his household, which will have to bear its share of any cost implied by the proposal if it passes. We saw in the preceding chapter that this feature makes it possible to use actual referenda behavior to estimate demand curves for the good offered in the referendum. Finally, the referenda model lends itself well to the survey setting. A ballot, after all, is similar to a multiple-choice questionnaire, and political polling to predict election outcomes is a well-recognized feature of public life.

[4] This alternative model has received support from Ridker (1967), Cummings, Cox, and Freeman (1984), Lareau and Ray (1985), Carson, Hanemann, and Mitchell (1986), Mitchell and Carson (1986c), Randall (1986b), and Cummings, Brookshire, and Schultze (1986).

[5] In almost all states voters may vote on binding propositions placed on the ballot by state legislatures, and some states provide for advisory referenda as well (Magleby, 1984:1).

This raises the issue of whether the information that can be provided to the respondent in a CV survey is adequate to simulate a referendum. The ideal political market assumes perfect information, an assumption which Hayes (1981:157), with reference to political markets in general, has termed "the most important prerequisite . . . and almost certainly the least realistic." Magleby's (1984) recent study of direct legislation documents how actual referenda, as informed decision-making mechanisms, suffer from numerous weaknesses. The agenda of issues offered to voters often does not represent the issue concerns of the public, and the issues are often presented in a way that may seem rigid and inflexible. The information pamphlets sent by states to voters in advance of referenda may be difficult for even high school graduates to understand, and the wording on the ballot typically is even more complex. Because of low voter turnouts and, among those who vote, lower participation rates on ballot propositions than on candidates, decisions made by referenda disproportionately represent the views of better-educated, better-off, white citizens. Beyond that, instead of being the considered judgment envisioned by the advocates of direct democracy, the vote on individual propositions is "typically the result of snap judgments based upon superficial emotional appeals broadcast on television" (Magleby, 1984:188).[6] These considerations led Magleby to suggest that "for many voters, direct legislation [that is, referenda] can be a most inaccurate barometer of their opinions" (1984:144).

There is no question that referenda are clumsy and imperfect devices for eliciting public preferences in the real world. However, compared to consumer markets where contingent valuation surveys fall short of the real-world counterpart, CV surveys have several features that permit them to overcome these problems. One advantage is that a CV survey can be more representative than its real-world referendum counterpart. The poor, for example, are less likely to engage in various forms of political participation, including voting, than those who have more income (Beeghley, 1986); CV interviewers, instead of relying on voters to register and come to the polls, go to the respondents' homes and work hard to cajole people's participation. Random sampling procedures and repeated visits to not-at-home respondents can ensure that the respondents are truly representative of the relevant population. To the extent that people with lower education or incomes are underrepresented among respondents because they are harder to find or because they are unable to give meaningful WTP amounts, statistical techniques can be used to weight the data to compensate at least in part for this shortfall.

[6] See Oppenheimer (1985) for a discussion of why it is often "rational" for voters to be poorly informed when making election decisions.

A major factor in limiting voters' knowledge about the candidates and issues on which they vote is the cost of acquiring information (Hayes, 1981). Contingent valuation surveys lower the cost of acquiring information in several ways. The choices offered to CV respondents are framed much more clearly with regard to the nature of the good being valued and the conditions under which it would be provided than is usually the case with referenda as they appear on the ballot. Also, CV surveys, particularly those conducted by an interviewer, are likely to be more successful in communicating information about the amenity than the communication procedures used in actual referenda. In CV surveys the information is usually presented verbally, thus encouraging respondent interest and participation. The survey designer generally puts extensive effort into the pretesting phase of the project to make the instrument clear and to provide the information people want and need to make their decisions. CV instruments can employ visual aids when necessary, and can provide background material and comparisons relevant to the choice situation—an advantage that voters rarely have made available to them, even in states that distribute voter pamphlets. Finally, CV interviewers can be trained to provide special clarifying responses to frequently asked questions.[7]

These features of CV referenda help to create the appropriate conditions for people to express their "volitions"—the essence of political choice, according to Lindblom (1977), who describes voting as a process in which the voter arrives at a state of mind and will that did not before exist (1977:136). A contingent valuation scenario offers respondents the opportunity to make such a judgment under relatively optimal conditions. To the extent that this occurs, a CV survey is a form of "super-referendum" in which individuals give meaningful values for nonpriced goods.

What criteria for a CV survey are implied by the referendum model? In order for a contingent valuation study to adequately simulate a referendum, respondents should understand the amenity and how it will be provided and paid for. Since participation in referenda is voluntary, CV studies should encourage, but not "require," answers to the WTP questions from respondents who would prefer not to answer. The referendum model also implies a voting criterion for making decisions. Since referenda involve a majority or two-thirds rule for deciding whether or not a public commodity is to be supplied, a few extreme (outlier) responses should not be permitted to unduly influence the aggregate value.[8]

For contingent valuation surveys, we do not advocate exclusive use of the referendum model. Both the private goods and the referendum models

[7] All information conveyed in this manner must be restricted to predetermined statements in order to maintain comparability across interviews.

[8] Extreme outlier responses are impossible if the survey asks for a single discrete yes/no response to fixed quantity/tax price combinations.

have their merits. The choice of which to use depends upon the nature of the good and the type of benefits that the study intends to measure. The private goods market model is appropriate for certain types of quasi-private goods which can potentially be provided to consumers for a price, such as access to parklands and game species for recreational purposes. These are situations where user values predominate and where the imposition of a permit or access fee is plausible. In the case of public goods, however, the referendum model may be preferred, as it invokes the correct payment context and the full range of values appropriate to public goods.

Elicitation Methods

The CV researcher's objective is to obtain the respondents' consumer surplus for the amenity—the maximum amount the good is worth to the respondent before he would prefer to go without it. It might be thought that the best way to do this would be to ask the respondent what maximum price he is willing to pay for the described good, and record the answer. Unfortunately respondents often find it difficult to pick a value out of the air, as it were, without some form of assistance,[9] just as they tend to be hard pressed to determine the highest price they are willing to pay for items at a garage sale in the absence of price tags. As a consequence, the open-ended format tends to produce an unacceptably large number of nonresponses or protest zero responses to the WTP questions (Desvousges, Smith, and McGivney, 1983).[10] This problem has led researchers to experiment with elicitation techniques which attempt to facilitate the respondents' valuation process by simplifying the choice process or by offering a context in which to value the good. These techniques help reduce the number of nonresponses and, according to interviewers, make it easier for respondents to successfully complete the valuation process.

A look at the manner in which these elicitation methods vary in their familiarity to respondents, in their potential for biasing WTP amounts, and in their ability to obtain the optimal amount of information provides an excellent introduction to the difficult tradeoffs that CV researchers face in constructing hypothetical markets that are meaningful to the respondents, yet free of bias. Understanding of these tradeoffs has been enhanced by a number of studies which have tested the properties of one or more

[9] In some cases the open-ended question works smoothly, as in a study (Mitchell and Carson, 1986c) of drinking water risks, where the respondents (residents of a small town in southern Illinois) were familiar with the concept of paying for drinking water quality through their water bills.

[10] In protest zero responses, respondents give a $0 WTP amount even though the good does have some value for them.

Table 4-1. A Typology of CV Elicitation Methods

	Actual WTP obtained	Discrete indicator of WTP obtained
Single question	Open-ended/Direct question Payment card Sealed bid auction	Take-it-or-leave-it offer Spending question offer Interval checklist
Iterated series of questions	Bidding game Oral auction	Take-it-or-leave-it offer (with follow-up)

elicitation methods,[11] or have systematically compared the methods with each other.[12]

Table 4-1 lists nine elicitation methods, which are categorized along two dimensions: (1) whether the actual maximum WTP for the good in question is obtained, and (2) whether a single WTP question or an iterated series of questions is asked (for a single level of the public good being valued). The properties of the different elicitation methods follow directly from the two dimensions of our typology.[13]

The first dimension concerns the amount of information collected from the respondent about her preferences. Does it obtain the respondents' actual WTP amount or a discrete indicator of WTP, such as whether the respondent is willing or not to pay a single amount proposed by the researcher? Some researchers advocate the use of the discrete-choice format because they believe this technique makes it easier for respondents to give a meaningful value for the good. If it were not for this methodological

[11] See, for example, Thayer (1981); Mitchell and Carson (1981); Roberts, Thompson, and Pawlyk (1985); and Boyle, Bishop, and Welsh (1985).

[12] See, for example, Randall, Hoehn, and Tolley (1981); Desvousges, Smith, and McGivney (1983); Sellar, Stoll, and Chavas (1985); Bishop and Heberlein (1986); and Johnson, Shelby, and Bregenzer (1986).

[13] Examples of the different methods are found in these sources: direct question, Desvousges, Smith, and McGivney (1983); sealed bid auction, Bishop and Heberlein (1986); payment card, Mitchell and Carson (1984); bidding game, Randall, Ives, and Eastman (1974); oral auction, Bohm (1972); checklist, Loehman and De (1982); take-it-or-leave-it offer, Sellar, Stoll, and Chavas (1985); take-it-or-leave-it offer with follow-up, Carson, Hanemann, and Mitchell (1986); spending question offer (more-same-less), Bergstrom, Rubinfeld, and Shapiro (1982).

consideration, CV researchers would always prefer to obtain the respondent's maximum WTP amount because it provides much more information about the respondent's value for the amenity, and therefore allows the use of relatively straightforward statistical techniques. The logit or probit techniques used to estimate WTP from discrete-choice data require the researcher to make strong assumptions, which are difficult to verify, about the mathematical form of the valuation function.

Regarding the second dimension, CV researchers disagree about whether a single question or an iterated series of questions (such as the bidding game, where an initial amount is proposed and bid up or down, depending on the respondent's acceptance or rejection of the starting amount) is most appropriate for a contingent valuation scenario. Some CV researchers prefer the use of a series of elicitation questions on the grounds that this procedure is needed in order to get respondents to search their preferences with the necessary thoroughness (see, for example, Hoehn and Randall, 1983). Other researchers, including ourselves (Mitchell and Carson, 1981), have argued against iterated questions on the grounds that this type of probing tends to induce various forms of compliance bias: the respondent gives higher values for an amenity, not because these values represent his true WTP amount, but because he feels pressured by the follow-up questions to give more than he really is willing to pay.

The issues involved in the use of particular elicitation methods may be seen by examining four of the principal methods used by CV researchers in the field. These are the bidding game, the payment card, the take-it-or-leave-it offer, and the take-it-or-leave-it offer with follow-up approaches. In addition, problems entailed in choosing an elicitation format for a contingent valuation survey serve to introduce the complex interplay between method and theory and to highlight the partial nature of our knowledge, at this point in time, about how respondents make and express their preferences in the CV setting.

The Bidding Game

The oldest and the most widely used elicitation method in contingent valuation surveys until relatively recently has been the bidding game (Davis, 1964). The bidding game is modeled on the real-life situation in which individuals are asked to state a price—the auction (see the description of its application to the Brookshire, Ives, and Schulze [1976] CV survey in chapter 1). Although the bidding game imitates an auction, and therefore is likely to be familiar to respondents, the characteristic that Davis thought most important was the simple nature of the choice it required the respondent to make. Is the respondent willing to pay a particular price for the good, yes or no? Additional virtues, according to others, are the likelihood that the bidding process will capture the highest price consumers are willing to pay, thereby measuring the full consumer surplus (Cummings, Brookshire, and Schulze, 1986), and the likelihood that the process of iteration will enable the respondent to more fully consider the value of the amenity (Hoehn and Randall, 1983). These desirable features are achieved at what is now regarded as an unacceptable cost by many CV researchers (Cummings, Brookshire, and Schulze, 1986). The starting bid tends to imply a value for the good. Studies have shown that even if a respondent rejects the initial bid, starting points well above the respondent's true WTP amount will tend to increase the revealed WTP amount, while starting points well below it will tend to decrease it (Roberts, Thompson, and Pawlyk, 1985). (For further discussion of starting point bias, see chapter 11.)

The Payment Card

The payment card method was developed by the present authors (Mitchell and Carson, 1981, 1984) as an alternative to the bidding game. We sought to maintain the properties of the direct question approach while increasing the response rates for the WTP questions by providing respondents with a visual aid which contains a large array of potential WTP amounts, ranging from $0 to some large amount.[14] This procedure circumvents the need to provide a single starting point, yet offers the respondent more of a context for her bid than the direct question method provides. In the benchmark version of the payment card (Mitchell and Carson, 1981, 1984), the context is enhanced by identifying some of the dollar amounts on the card as the average amount households of the respondent's income category are currently paying for other public goods.[15] The question posed to the respondent is: "What amount on this card or any amount in between is the most that you would be willing to pay" for the level of good being proposed? Randall, Hoehn, and Tolley's (1981) experimental comparison of several elicitation techniques suggests that the use of nonrelated public good benchmarks does facilitate a respondent's valuation process.[16] Al-

[14] Similar to the payment card is the checklist method, where respondents indicate which of a list of payment *ranges* includes their WTP amount. Hammack and Brown (1974) used this technique in a mail survey, and Binkley and Hanemann (1978) used it in a personal interview survey. See Loehman and De (1982) for a treatment of data from a checklist as discrete choices in a logistic regression framework, and Cameron and Huppert (1987) for a treatment of payment card data as a grouped interval response.

[15] Respondents were grouped into five income categories; the payment cards used in this study are reproduced in appendix A (for a full description of the payment card method, see Mitchell and Carson [1984] and Desvousges, Smith, and McGivney [1983]). Note that in order to avoid bias, the other goods should not be directly related to the good a study wishes to value, for otherwise respondents may base their values on the related goods without giving sufficient thought to the value held for the good in question.

[16] This does not appear to be the case, however, with "trivial" private goods such as toothpaste, which are likely to inflate willingness to pay by imposing implicit moral pressure on respondents.

though payment cards appear to pose less of an "anchoring" problem than the bidding game, the payment card method is potentially vulnerable to biases associated with the ranges used on the cards and the location of the benchmarks. We will return to this issue in chapter 11, where we discuss bias induced by implied value cues and how it may be minimized.

The Take-It-or-Leave-It Approach

Another important CV elicitation method was developed by Bishop and Heberlein (1979, 1980): the "take-it-or-leave-it" approach, as they call it, uses a large number of predetermined prices, t_j, chosen to bracket expected maximum WTP amounts of most respondents for the amenity. Each respondent is asked if she is willing to pay a single one of these prices for the amenity on an all-or-nothing basis, with no further iteration. The prices are randomly assigned to respondents so that each price is administered to an equivalent subsample.

The take-it-or-leave-it method has several strengths. It simplifies the respondent's task in a fashion similar to the bidding game, without having the iterative properties of that method. The respondent has only to make a judgment about a given price, a type of judgment performed frequently by consumers and by those who vote in referenda. In this respect, the take-it-or-leave-it method may be especially suitable for mail interviews or, presuming the scenario does not require the use of visual aids, telephone interviews. This approach is also incentive-compatible in that it is in the respondent's strategic interest to say yes if her WTP is greater than or equal to the price asked, and to say no otherwise (Zeckhauser, 1973; Hoehn and Randall, 1987).

Other elicitation methods are more vulnerable to strategic behavior in those situations—fortunately not found in many CV surveys (see chapters 6 and 7)—where the scenario offers the respondent a clear opportunity to promote a desired policy outcome by exaggerating her WTP amount. The incentive-compatible property inherent in a single binary choice under a plurality voting rule is reduced if the respondent is given a choice among three or more alternatives, or a choice that is conditional upon an earlier choice. Open-ended elicitation methods that ask the respondent to state her maximum WTP amount have no incentive-compatible property.

The take-it-or-leave-it approach, however, suffers from several drawbacks. Relative to other elicitation methods, take-it-or-leave-it is inefficient in that many more observations are needed for the same level of statistical precision in sample WTP estimates, because only a discrete indicator of maximum WTP is obtained instead of the actual maximum WTP amount. Also, take-it-or-leave-it may be subject to a nonzero background level of yea-saying. This problem, which is discussed in chapter 11, is the discrete choice analogue of starting point bias and is somewhat harder to detect.

Perhaps the most serious drawback of the take-it-or-leave-it elicitation method, however, is that one must make assumptions about how to parametrically specify either the valuation function or the indirect utility function to obtain mean WTP. Because of the growing popularity of the take-it-or-leave-it approach, recent work on this important problem deserves further comment.

Bishop and Heberlein (1979, 1980) made the original estimates of mean WTP (and WTA) from take-it-or-leave-it responses. They noted that a logistic or probit regression curve could be fitted to the percentage of respondents willing to pay (or accept) each of the randomly assigned prices. The area below (above) this curve is equal to mean WTP (WTA). This can be a somewhat messy procedure to implement and raises questions about behavior outside the range of the t_j used in the survey.

Cameron and James (1987) have recently shown how mean WTP (or WTA) can be obtained directly from the parameters of a probit equation.[17] This is possible because of the special structure of the take-it-or-leave-it situation. The stimulus variable t_j is measured in the same unit—dollars—as is the underlying latent variable WTP. To see this more clearly, assume that the valuation function takes the form

$$\mathrm{WTP} = X\beta + u, \tag{4-1}$$

where WTP is an $n \times 1$ vector, X is an $n \times k$ matrix containing a constant and possibly other variables such as income and taste attributes, β is a $k \times 1$ vector of unknown parameters, and m is an $n \times 1$ vector of random terms distributed $N(0, \sigma^2 I)$. Let I be an $n \times 1$ indicator vector whose ith element equals one if WTP is greater than or equal to t_j, and zero if WTP is less than t_j. If we estimate the probit regression,

$$\mathrm{Prob}\ (I = 1) = 1 - \phi(-[t\ \mathrm{X}]\ [\alpha\ \gamma]), \tag{4-2}$$

where t is an $n \times 1$ vector whose elements consist of the t_j assigned to each of the i respondents, and β and γ are parameters to be estimated. Cameron and James (1987) show that $\alpha = -1/\sigma$ and $\gamma = \beta/\sigma$, so that $\beta = -\gamma/\alpha$. If X is simply a vector of ones, then β equals median WTP and by the symmetry property of the normal distribution β also equals mean WTP.[18] If X consists of a constant plus other variables, then WTP_i is estimated by $X_i\hat{\beta}$. Cameron and James (1987) derive expressions for obtaining the correct standard error for β.

[17] For a discussion of the logistic regression case, see Cameron (forthcoming).

[18] It is straightforward to show that β is the value of t_j which solves Prob $(I_i | t_j) = .5$.

While this approach is intuitive and easy to implement, it tends to obscure the importance of the normality assumption on u (and hence WTP). This assumption is much more important than is typically the case with a probit regression, particularly if X consists of only a constant term. If, for instance, WTP is distributed log-normally, as often appears to be the case, then $\log(t_j)$, rather than t_j, should be used as a regressor. When $\log(t_j)$ is used, mean and median WTP may diverge quite dramatically as σ (of which mean WTP is a function) becomes large.

Hanemann (1984c) has examined the take-it-or-leave-it elicitation question from the perspective of the indirect utility function rather than the valuation function. Here it is clear that assumptions about the random component drive the estimate of mean WTP. A few respondents who are willing to pay the largest proposed t_j may almost completely determine mean WTP.[19]

Median WTP is much less sensitive to distributional assumptions. With a large sample size and the right choice of the t_j, median WTP can be calculated accurately using a response surface approach, thus avoiding, for all practical purposes, the need to make assumptions about the nature of the valuation or indirect utility function.

Take-It-or-Leave-It with Follow-up

The fourth elicitation method under consideration is the take-it-or-leave-it offer with a follow-up approach, recently proposed by Carson, Hanemann, and Mitchell (1986) as a possible way to overcome the inefficient nature of the standard take-it-or-leave-it offer. In this fourth approach, the respondent is asked a question requiring a yes or no answer about whether he would pay a specified price. If the respondent says yes, another WTP question is asked using a higher price randomly chosen from a prespecified list. If the answer is no, the follow-up question proposes a randomly chosen lower price. Although this method offers the potential for considerable gains in efficiency, all of the other problems with the take-it-or-leave-it offer still hold. The efficiency is gained by exploiting the revealed preference postulate in combination with Neyman double sampling. Essentially, the implicit percentage of respondents approving at each of the t_j design points is calculated and then used in a logistic regression. Carson, Hanemann, and Mitchell (1986) recommend that only one, or at most two, follow-ups be asked, that the follow-up price should represent a large, discrete jump

[19] This problem tends to be even more pronounced when dealing with willingness-to-accept questions, where mean WTA is often infinite under particular specifications. Bishop and Heberlein (1979, 1980) originally dealt with this problem by setting the upper limit of integration for the area above their logistic regression curve at the highest t_j offered. See Sellar, Chavas, and Stoll (1986) for an extension of the original Bishop and Heberlein approach.

from the original price, and that it should be phrased in a "what if" fashion so as to maintain the incentive-compatibility property, which is lost if the tax price and quantity of the public good offered are not viewed by the respondent as exogenous to the choice situation. Carson and Mitchell (1987) show that survival analysis statistical techniques developed for irregular interval inspection time failure data can be used to handle more complicated assignment schemes for the t_j's which exploit revealed preference relationships between different levels of the good being valued. Under certain conditions, especially when a sizable number of true zero responses are expected, it may be useful to employ a continuous rather than a discrete follow-up question. This approach was used to value reduced risk from drinking water contamination (Mitchell and Carson, 1986c), where the respondents were first asked whether or not they were willing to vote in favor of paying at least a small amount of money to receive a specified reduction in risk. Those who said yes were then asked to give the highest amount the program could cost them before they would vote no.

Summary and Conclusions

Those who are critical of the ability of surveys to provide meaningful valuation information frequently compare the hypothetical market in contingent valuation surveys unfavorably to a private goods market model where consumers make informed purchases of familiar goods. In this chapter we have called attention to market research findings which suggest that this idealized model is not an accurate description of actual consumer behavior in most situations. We have argued further that a particular type of political market model, the referendum, is more appropriate for use in valuing pure public goods than is a private goods market model. CV surveys using the referenda approach and an open-ended elicitation method would ask respondents for the highest amount a specified amenity could cost them before they would vote no in a binding referendum on whether or not that good should be provided by the government. The correspondence between CV scenarios and referenda are close, and CV surveys are able to obtain expressions of preference under more representative and informed conditions than is usually the case with actual referenda.

Contingent valuation researchers have used a variety of techniques to elicit the WTP amount from respondents. We have examined briefly the open-ended question, the bidding game, the payment card, and two take-it-or-leave-it elicitation methods. For most purposes the bidding game technique is not recommended because it is prone to starting point bias. Each of the other techniques requires the researcher to be sensitive to their potential drawbacks. The take-it-or-leave-it methods have gained in favor in recent years because they simplify the respondents' valuation choice and lend themselves to use in mail and telephone surveys. Although they are congruent with the referendum model—actual referenda imply a take-it-or-leave-it cost to the voter's household—take-it-or-leave-it methods are independent of the referendum model. Other methods, including the open-ended technique, have been used to elicit WTP amounts in contingent valuation referenda, and studies based on the private goods model have employed the take-it-or-leave-it approach.

5

The Methodological Challenge

In the earlier chapters we placed the contingent valuation method in the framework of welfare economics and benefit estimation, and concluded that there are strong theoretical grounds for using surveys to measure benefits if truthful revelation of preferences can be obtained. We now turn to the methodological challenge of using sample surveys to accomplish this goal. Can surveys obtain reliable and valid willingness-to-pay amounts from random samples of people? In this chapter we briefly review the nature of surveys and raise the methodological issues that are addressed in detail in the subsequent chapters. Because many readers may not be trained in survey research techniques, we will not assume a familiarity with this methodology in the following discussion.

Survey Research

Rossi, Wright, and Anderson (1983) define sample surveys as "relatively systematic, standardized approaches to the collection of information . . . through the questioning of systematically identified samples of individuals." Thirty years ago Rensis Likert (1951) predicted that survey methodology, then a relatively new research technique, would have an increasingly wide application in all the social sciences. Proof of his prescience is as close as the nearest major journal in social psychology, political science, sociology, or economics, where, depending on the discipline, currently 20 to 56 percent of the articles use survey data (Presser, 1984).

The modern survey is the result of two key methodological developments. The first is probability sampling, which enabled survey findings to be accurately projected to larger populations. Sampling itself is centuries old, but sampling based on strict probability principles has a much shorter history. In the United States it was first applied to survey research in the 1930s and came of age after the 1948 presidential election, which most of the major polls had confidently predicted would go to Dewey rather than

Truman, the winner. Although the pollsters' use of the inappropriate quota sampling method was only one of the factors behind it, that debacle provoked national survey organizations to adopt the area probability methods first developed by the census bureau (Rossi, Wright, and Anderson, 1983). These techniques give each element in the population a known non-zero probability of being selected into the sample, thus making possible the use of statistical inference to project the results to the relevant population.[1] If rigorously implemented, findings based on sample sizes as small as 600 to 1,500 people can be representative of the entire United States population (or any other population) with a high degree of confidence.[2]

The second development on which the modern survey rests is "the art of asking questions" (Payne, 1951). This phraseology nicely captures the essentially qualitative nature of the achievement, which is based on a great deal of experience and a relatively small number of controlled experiments, most of them of recent vintage (Schuman and Presser, 1981, for example), to test the effects of different ways to word questions. The result is that there is no single "correct" version of a questionnaire.

Such diversity should not be taken to imply that anything goes. Writing survey questions appears deceptively easy to many people. In reality, it is difficult to convey even relatively simple ideas so they are uniformly and correctly understood by respondents who come from diverse backgrounds, have differing levels of education, and may or may not feel comfortable in the interview situation. Words that seem clear to the researcher may be ambiguous to some or many respondents. Seemingly slight changes in word order may convey unexpected meanings.[3] Fee (1979) found that his respondents held at least nine distinct understandings of the term "energy crisis." Sudman and Bradburn tell the delightful story of two priests who discuss whether it is a sin to smoke and pray at the same time. Each consults his superior on the matter and when they next meet each relates the outcome:

> The Dominican says, "Well, what did your superior say?" The Jesuit responds, "He said it was all right." "That's funny," the Dominican replies, "my superior said it was a sin." Jesuit: "What did you ask him?" Reply: "I asked him if it was all right to smoke while praying." "Oh," says the Jesuit, "I asked my superior if it was all right to pray while smoking." (1982:1)

Another aspect of the art of asking questions is determining the best sequence in which to ask the questions. Experience has shown, for exam-

[1] Kish (1965), Sudman (1976), Cochran (1977), Yates (1980), and Frankel (1983) provide useful overviews of sampling theory and practice.

[2] The same sample size will be equally effective for large or small populations.

[3] Many striking examples are given in Payne's (1951) classic book on question wording, *The Art of Asking Questions*.

ple, that questions about the respondent's personal characteristics—the background questions—are best left to the end of the questionnaire, when the respondent is more relaxed about being interviewed and less likely to take offense at having the interviewer probe into his private life. A sequence of questions in a questionnaire that "works" will flow from one topic and question to another so that the transitions are smooth, the interrelationship between the parts is perceived by the respondent to be logical, and the pace is varied enough to retain the respondent's attention.

In-Person, Telephone, and Mail Surveys

Survey instruments may be read to the respondent in person or over the telephone, or they may be sent in the mail with a request to fill out and return the questionnaire. Bradburn (1983:294) states the current consensus in the survey research community when he writes: "Contrary to the common belief favoring face-to-face interviews, there is no clearly superior method that yields better results for all types of questions." In recent years both the high cost of in-person surveys and methodological developments in telephone survey technology have led the major academic survey research centers to experiment successfully with telephone interviews, a methodology which commercial polling houses had used for many years (Groves and Kahn, 1979). The sampling problems presented by unlisted telephone numbers have been overcome by the use of computer-based random digit dialing techniques.[4] An even less expensive survey method is the mail survey, which, unlike telephone interviews, permits the use of visual aids. Here, too, methodological advances have improved the technique. It used to be thought that low response rates were inevitable in mail surveys, but techniques are now available that result in respectable (70 percent or larger) response rates for mail surveys under some circumstances.[5]

What characteristics of contingent valuation questions should influence the choice of method? At least three come to mind. First, CV questions often involve complex scenarios that require careful explanation, and which benefit from the use of visual aids and from close control over the pace and sequence of the interview. Second, the need to obtain dollar values requires a method which can motivate respondents to make a greater-than-usual effort. Third, the need to extrapolate from the sample to make benefit estimates for populations necessitates the use of survey methods which support techniques to compensate for missing data.[6]

[4] See Frey (1983) for a discussion of random digit dialing and other aspects of telephone survey methodology.

[5] See Dillman (1978, 1983) and Tull and Hawkins (1984) for useful and somewhat different perspectives on mail and telephone survey practices.

[6] See chapter 12 for a discussion of the relationship between sampling problems and survey methods.

On the basis of these criteria, the in-person survey, where the interviewer conducts the interview in the respondent's dwelling place, is the method of choice for most CV studies. The physical presence of the interviewer offers the greatest opportunity to motivate the respondent to cooperate fully with a complex or extended interview and allows the interviewer to probe unclear responses and to provide observational data (Schuman and Kalton, 1985). In-person interviews also lend themselves to the use of visual aids, or display cards, to help convey complex ideas or bodies of information, and they support missing data techniques.

The large potential cost savings in using telephone and mail surveys has not gone unnoticed by contingent valuation researchers, however. Several have used mail surveys (Bishop and Heberlein, 1979; Schulze and coauthors, 1983; Walsh, Loomis, and Gillman, 1984; Bishop, Heberlein, Welsh, and Baumgartner, 1984; Bishop and Boyle, 1985), and others have conducted CV surveys by telephone (Oster, 1977; Roberts, Thompson, and Pawlyk, 1985; Carson, Hanemann, and Mitchell, 1986; Sorg and coauthors, 1985; Mitchell and Carson, 1986b; Sorg and Nelson, 1986). But costs apart, what are the tradeoffs between the mail and telephone methods and the more expensive in-person technique?

The greater impersonality of telephone surveys compared with in-person interviews reduces the ability of the interviewer to motivate the respondent. The absence of visual cues during the telephone survey makes it harder for the interviewer to adjust the interview to the respondent's circumstances, and the telephone interview precludes the use of visual aids to help communicate the scenario.[7] The result is that respondent attention spans for descriptive material are much lower in telephone surveys than in surveys where the interviewer is present and the description can be reinforced by visual aids. In telephone surveys it is thus difficult, if not impossible, to maintain respondent interest and attention while communicating even moderately lengthy CV scenarios.

Although mail surveys have an advantage over telephone interviews in being able to use visual aids, and an advantage over both in-person and telephone interviews in avoiding the possibility of interviewer bias, they suffer from several important shortcomings from the CV point of view. In the first place, they require the respondent to read and understand the description given in the scenario, and unfortunately the reading level of a surprising number of Americans is quite low. According to the National Assessment of Educational Progress, which conducted a study of literacy among a national sample of 3,600 young adults between the ages of 21 and 25, 6 percent were unable to read a short sports story in a newspaper,

[7] It may sometimes be possible to mail materials to households before conducting the telephone interviews; see Sorg and coauthors (1985) for an example.

20 percent could not read as well as the average eighth grader, 37 percent could not summarize the main argument in a newspaper column, and only 43 percent could use a street map (Kirsch and Jungeblut, 1986). These data understate reading comprehension problems, since this sample had a higher level of education than comparable cohorts of older people. It would appear that unless the scenario in a mail questionnaire is very short and simple, or the respondent is reasonably well educated and also highly motivated, the respondent may miss important details or misinterpret one or more aspects of the scenario.

A second set of problems results from the self-administered character of mail surveys, which makes it difficult to use skip patterns (where the choice of follow-up questions depends on the respondent's answer to the previous question) or to tailor the interview to the individual respondent's needs. A well-trained interviewer can pace the interview according to the circumstances, can repeat questions when the respondent indicates puzzlement or uncertainty, and can (within the limits imposed by the interview protocol) answer the respondent's questions.[8] The self-administered character of mail surveys also means that there is no way to keep respondents from browsing through the questionnaire before they start to fill it out. Thus precluded is the use of multiple scenarios, where it is desired to have the respondents answer each one in a fixed sequence and without knowledge of the following scenarios.

Finally, from the sampling point of view mail surveys are vulnerable to serious sample nonresponse bias problems which cannot easily be corrected by available weighting techniques. These problems occur because those in the sample who fail to fill out and return the questionnaire are typically those who have the least interest in the amenity being valued. (See the discussion of sampling and nonresponse bias problems in chapter 12.)

While in-person interviews are clearly the technique of choice for contingent valuation surveys,[9] experience with telephone and mail CV surveys suggests that, with the exception of the sample nonresponse bias problem

[8] Accepted survey practice forbids interviewers to provide ad hoc explanations or answers to respondent questions. Interviewers are instructed to read to respondents only the material provided, which may, however, include answers prepared by the researcher to questions which pretesting has shown to be of concern to some respondents. This additional material is used only if the respondent specifically raises the issue.

[9] On the basis of their study of the national aggregate benefits of air and water pollution control, which obtained relatively similar findings for mail and in-person interviews, Randall and coauthors (1985) conclude that the in-person interviews were not superior to their mail questionnaires. Unfortunately, the response rates they achieved for each methodology were too low (44 percent for in-person interviews, and 36 percent for mail questionnaires) to make a definitive judgment on this issue, and they did not consider the serious sample nonresponse problems mentioned above and discussed in chapter 12.

in mail surveys, their shortcomings may be overcome if the respondents are familiar with the amenity,[10] or if the scenario is relatively simple. For example, the offshore recreational divers interviewed by Roberts, Thompson, and Pawlyk (1985) over the telephone were also familiar with the type of diving amenity they were valuing, and the goose hunters who received the Bishop and Heberlein (1979) survey in the mail were well acquainted with the hunting opportunity they were asked to value. Bishop and Heberlein also had an unusually low nonresponse rate (only 6 percent did not return the questionnaire), thanks to a concerted follow-up effort and the use of a $5 incentive.

As the survey material becomes more complex and less familiar to the respondents, however, the results are less satisfactory. When we used a relatively simple referendum format in a telephone survey of people's values for reduced risks of contracting giardiasis from San Francisco's water supply (Mitchell and Carson, 1986b), use of the telephone method involved a very clear tradeoff between cost and precision. For despite considerable instrument development by an academic survey research organization experienced in conducting difficult telephone interviews, we found it necessary to omit a number of important aspects of the hypothetical situation from the scenario presentation which could have been easily incorporated into a personal interview.

Data Comparability

Irrespective of how it is administered, a major requirement of a survey is to ensure that the data it obtains are comparable—that is, that the information is gathered in a standardized fashion so one person's answers can be compared with those given by another. To this end, survey organizations devote considerable care and resources to pretesting questionnaires and training interviewers. Pretesting is the survey equivalent of the test flight. Just as no plane manufacturer would go into production without rigorously testing its latest design, so no survey writer would assume that a questionnaire on a new topic—especially if the questionnaire is complex—could be sent directly to the field without careful tryouts under fieldlike conditions. Even experienced survey practitioners are often surprised when certain questions work better than they had anticipated, and others, which they thought were winners, turn out to be fatally ambiguous. Pretests normally consist of an extended period of trial-and-error experimentation with draft versions of the questionnaire to see which wordings and orderings of the questions work best. If the topic is novel, the pretest process may include preliminary in-depth research, perhaps

[10] This is why mail and telephone interview techniques are likely to work best with recreational users.

using focus groups, to learn how people conceptualize and talk about the topic (see Desvousges, Smith, Brown, and Pate, 1984; Randall and coauthors, 1985; Mitchell and Carson, 1986b).

Comparability also imposes demands on how interviewers conduct themselves in surveys. As David Riesman (1958) once observed, the basic task of the interviewer is to "adapt the standardized questionnaire to the unstandardized respondents." Except for mail surveys, questioning is a social process. Each interaction between an interviewer and a respondent is unique owing to the particular circumstances in which the interview occurs and the personal characteristics of the two participants. In order to adapt the questionnaire without distorting or changing it, the interviewer has to motivate the respondent to enter into a special kind of relationship. Sudman and Bradburn describe how interviews differ from ordinary conversations:

> The survey interview . . . is a transaction between two people who are bound by special norms; the interviewer offers no judgment of the respondents' replies and must keep them in strict confidence; respondents have an equivalent obligation to answer each question truthfully and thoughtfully. In ordinary conversation we can ignore inconvenient questions, or give noncommittal or irrelevant answers, or respond by asking our own question. In the survey interview, however, such evasions are more difficult. The well-trained interviewer will repeat the question or probe the ambiguous or irrelevant response to obtain a proper answer to the question as worded. (1982:5)

It is precisely at the point of probing and handling respondent queries that comparability can be lost unless the interviewer rigorously follows instructions not to offer any information or explanations other than those described in the handbook for the study.[11]

The Problematic Nature of Surveys

Despite fifty years of experience with surveys, survey methodologists agree that our knowledge about how respondents answer survey questions is still relatively primitive (Bishop, 1981; Schuman and Presser, 1981; Dijkstra and van der Zouwen, 1982; Bradburn, 1983; Jabine and coauthors, 1984; Turner and Martin, 1984). We know a great deal about how surveys can go wrong and much less about how to keep them from going wrong. Fortunately, there is currently considerable interest in the latter topic,[12]

[11] The Research Triangle Institute's (1979) *Field Interviewer's General Manual* offers an informative overview of the interviewer's role and training.

[12] See, for example, the reports, prepared under auspices of the National Research Council, of the Panel on Survey Measurement of Subjective Phenomena (Turner and Martin, 1984) and of the Advanced Research Seminar on Cognitive Aspects of Survey Methodology (Jabine and coauthors, 1984), as well as the research currently being undertaken by Roger Tourangeau and his colleagues (Tourangeau and coauthors, 1985).

with the result that our understanding of response effects is growing more rapidly than ever before. Nevertheless, the state of the art is sufficiently problematic that Kalton and Schuman (1982), two of the leading academic survey methodologists, advise against taking the marginal distributions of answers to nonfactual questions "too seriously," advocating instead that researchers concentrate their attention on some form of correlational analysis. Since the distribution of WTP amounts is the primary focus of attention in CV surveys, Kalton and Schuman's observation deserves notice as a warning against methodological hubris.

Response effects are a type of nonsampling error which can distort a survey's results. They occur when one or more characteristics of the question, the respondent, the interviewer, and the context in which the interview takes place unduly affect response behavior. The many dimensions involved make the opportunities for such effects legion. For example, when asked a question, respondents may misunderstand it, or their answer may be influenced by the order in which it appears or by the interviewer's manner. One team of hardy researchers (van der Zouwen and Dijkstra, 1982) arrayed causal factors by various types of survey outcomes and determined that an adequate nonsampling errors model must take into account at least 300 potentially important interaction effects. The state of response-effect theory is suggested by the fact that "theory" does not appear in the index of their book, *Response Behaviour in the Survey-Interview* (1982), which is devoted to examining the 104 bivariate propositions they developed about respondent behavior.

Why is the survey interview vulnerable to response effects? The answer lies in the social nature, motivational mysteries, and cognitive capacity of human beings. In-person and telephone interviews are social situations and humans respond to these situations in complex ways—in part idiosyncratic, in part programmed by learned rules of behavior or norms, and in part in reaction to the prevailing norms. The ideal respondent would be motivated to devote as much time and energy to the interview as required and to answer every question truthfully. The actual respondent's motivation is likely to depart from this ideal in ways that fluctuate according to his current obligations, his sense of what the interviewer wants to hear, his current degree of self-confidence, his self-image, and his reaction to the interviewer.

As for cognitive capacity, survey researchers have long recognized that surveys must adjust to human frailties. These frailties include the difficulty some people have in understanding seemingly simple questions and instructions, and the difficulty many respondents have in accurately recalling even recent and concrete events (such as a visit to the doctor a week earlier). Survey researchers have recently become interested in the findings of cognitive psychologists (Kahneman, Slovic, and Tversky, 1982; Nisbett and Ross, 1980, for example) who have suggested, on the basis of their laboratory experiments, that humans tend to use certain rules of thumb (heuristics) in ways that can affect their responses to survey questions. For example, the "availability" heuristic (Tversky and Kahneman, 1974) holds that people tend to give the answer that is most immediately available to them in memory when asked to give a verbal report of any kind. Applied to surveys, this response implies that "the behavior itself of answering questions may act as a 'primer' which makes some cognitions more *accessible* or salient than others" (Bishop, 1981). In the case of the "anchoring" heuristic, another tendency identified by cognitive psychologists, people make estimates by starting from an initial value, which may be suggested by the formulation of the problem, and then adjust that value to yield the final answer (Tversky and Kahneman, 1974). For example, when subjects were asked to estimate the frequency of death from each of 40 different causes, their estimates tended to be higher or lower depending on whether they were first told that about 50,000 people die annually in motor vehicle accidents or that about 1,000 people a year die from electrocution (Slovic, Fischhoff, and Lichtenstein, 1982).

Some people look to these cognitive research findings for a possible basis for a theory of response effects (Bishop, 1981; Jabine and coauthors, 1984). Others, however, are skeptical about the payoff from this source. The survey methodologists Schuman and Presser (1981) found that cognitive research findings offered surprisingly little theoretical guidance for the solution of the question-wording and context-effect problems they studied in their split-sample survey experiments. They attribute the lack of relevance to the great difference between controlled laboratory experimentation and "the encounters with ordinary people that characterize surveys" (1981:313). This is a perceptive comment, which may be augmented by the observation that the cognitive research enterprise is devoted to finding factors that create differences in respondent outcomes, whereas survey methodologists endeavor to identify ways to avoid or to minimize differences caused by such factors.

The difference in research agendas has thus far made the findings of cognitive psychologists less useful to CV researchers than they would otherwise be. For example, if the difference between the intervals used in two sets of payment cards is made large enough, an experiment comparing the two payment card formats will surely identify a response effect. Except for demonstrating that badly designed payment card formats are vulnerable to "range bias" (see chapter 11), this finding is not particularly important for the design of CV studies if the possibility of such bias is well recognized. From a practitioner's point of view, the most useful experiments are those that attempt to determine the conditions under which such differences do and do not occur. For our part, we have found the heuristics described above to be a useful source of hypotheses about potential bias problem areas, and we draw on them in our review of biases to which CV surveys may be prone, in chapter 11.

Contingent Valuation and Conventional Surveys Compared

Surveys are diverse in what they try to measure and the uses to which they are put. In comparison with other uses of surveys, the measurement requirements and the aims of contingent valuation surveys are particularly demanding. This is not because the kinds of information sought by CV surveys differ from those sought by other surveys, as table 5-1 demonstrates. Even the raison d'être of the CV survey—generation of willingness-to-pay data—has its counterpart in the attempts of political pollsters to predict elections and the efforts of market researchers to predict consumer behavior. Nor are contingent valuation surveys unique among surveys in posing hypothetical situations to their respondents; similar "what-if" questions are used in other settings to divine the prospects of potential candidates or to discover which characteristics of prospective products will most endear them to consumers: "If Ted Kennedy was the Democratic candidate in the 1988 presidential election and George Bush was the Republican candidate, which one would you vote for?" "If the new toothpaste I described to you was green in color would this make you more likely to try it, less likely, or wouldn't it affect your decision one way or another?"

The differences between contingent valuation studies and the other attempts to measure future behavior in surveys lie in (1) the novelty of the situation most CV surveys pose to the respondent, (2) the need in CV surveys to construct a market in which the good can be bought, and (3) the effort required for a respondent in CV surveys to arrive at a meaningful answer to many WTP questions.

Contingent valuation surveys are novel. While respondents are generally familiar with having to make judgments about referenda propositions or expressing opinions about whether or not they will buy a car this year, the request to set a dollar value on what they would be willing to pay for a particular hypothetical change in a public or quasi-private good is unfamiliar to most people. The novelty is enhanced when respondents lack direct experience with the good being valued, such as a park they have never visited. In our experience, professional survey researchers who encounter a CV instrument for the first time tend to assume that these types of judgment are beyond the capacity of many respondents in a survey setting because of the novelty, and are surprised at the quality of the valuation data which can be obtained from a well-designed instrument.

Respondents in CV surveys value the good in a specific setting. Survey researchers have long recognized that respondents' answers to general

Table 5-1. Comparative Examples of Information Obtained by Surveys

Type of information	General surveys	CV surveys
Self-reported behavior	Current employment status; voting behavior	Recreational participation; consumption of bottled water
Demographic information	Sex, education, income	Sex, education, income
Personal knowledge	Names of respondent's senators; awareness of the accident at Three Mile Island	Awareness of risks to drinking water quality from PCBs
Opinion attitudes	Satisfaction with job; concern about crime; "Are we spending too much, too little, or the right amount for education in this country?"	Concern about air pollution; confidence in government
Expected future behavior	Voting intentions; expected future purchases of consumer goods	Willingness to pay for specified amenities; voting intentions in a hypothetical referendum

questions are often poor predictors of how they would react to specific proposals. In a survey taken shortly after World War II (Cottrell and Eberhart, 1948:8), large majorities supported the idea that the U.S. government should do everything it could to stimulate world trade, yet opposed a plan to extend large credits to Great Britain for this purpose. Because they create a hypothetical market in the good, CV surveys are required to be even more specific than this. They must, in the words of Randall, Hoehn, and Brookshire (1983:637), confront "the respondent with a well-defined situation and elicit a circumstantial choice contingent upon the occurrence of the posited situation." The "posited situation" typically includes such factors as the current level of the provision of the amenity, the amount of increase or decrease in provision the respondent is to value, how this will be provided, how the respondent will pay for it, and who else will pay for it. Considerable description is sometimes required to construct a plausible market and depict the good in adequate detail.

In sharp contrast, conventional surveys offer comparatively few details about situations the respondent is asked to value. Consider the following public opinion question:

> Now, I'd like to find out how worried or concerned you are about a number of problems I'm going to mention: a great deal, a fair amount, not very much, or not at all. . . .
>
> . . . How worried or concerned are you about the presence of toxic chemicals such as pesticides or PCBs in the environment?
>
> Cleaning up our waterways and reducing water pollution?
>
> The disposal of industrial chemical wastes that are hazardous? (Mitchell, 1980b)

The question covers three environmental amenities. The description of each is very brief, there is no attempt to describe different levels of provision, and the only response requested of the respondent is a simple choice between four levels of concern on an ordinal response scale. In comparison, a contingent valuation scenario for just one of these goods—reduced risk from toxic chemicals—would have to identify the chemicals, their uses, consequences of their use, and current level in the environment. It would also describe several levels of chemical contamination in enough detail so the respondents could understand the changes they are asked to value in dollars, and would provide information about how these changes would occur and how the respondent would pay for them. Compared with the minute or two it would take to ask the conventional survey question on this topic, a contingent valuation version would require anywhere from ten to thirty minutes.

Contingent valuation surveys tend to require a greater effort from the respondent than most conventional surveys. Few respondents bring well-realized values for unfamiliar amenities to the CV interview. Yet they are asked to pay attention to the sometimes lengthy description of the market, search their preferences, take their income constraint into account, and determine a dollar amount which represents the most they would pay for each level of the good the survey attempts to value. As Fischhoff, Slovic, and Lichtenstein (1980) point out with reference to such extreme cases as CO_2-induced climatic change and genetic engineering, people's opinions about such topics are likely to be "labile" or subject to change because they may not have thought through the implications of their views, they may bring contradictory values to bear on the situation, or they may vacillate between incompatible, but strongly held, positions. In situations like these, respondents will be tempted to minimize effort by resorting to strategies that ease the burden of decision, such as giving an off-the-cuff answer or an answer suggested by an aspect of the scenario that is not intended to convey value.

The measurement problems faced by CV surveys become especially important when their intended use is taken into consideration. With the exception of market surveys and political polling for candidates, most opinion surveys are not directly used in making decisions; if they are so used, a high level of precision is not assumed.[13] The avowed purpose of contingent valuation surveys, on the other hand, is to obtain benefit estimates for benefit-cost decisions where an efficiency criterion is employed to allocate resources to their highest-valued uses. In order to fulfill this mission, the WTP amounts obtained in a particular study are typically aggregated across the relevant population and presented as an authoritative representation of the good's benefits. In such an enterprise, questions about the quality of the data, their validity, and the precision of the estimates will be raised by policymakers, and as has already happened in several instances, by lawyers and judges.[14] This places a special burden on CV studies to use the best available methodology.

[13] Exceptions to this generalization exist (such as the Census Bureau's employment and cost-of-living surveys, the results of which may trigger important decisions), but these are large, routine, data-gathering efforts which measure relatively well-defined, self-reported behaviors. Smith (1986a) notes, however, that even these questions require a fair amount of judgment on the respondents' part.

[14] In 1986 benefit estimates based on CV surveys figured in an administrative hearing between the Northern States Power Co. and the Minnesota Pollution Control Agency (Welle, 1985; Carson, Graham-Tomasi, Rund, and Wade, 1986), and in several court cases brought against corporations by the Colorado Department of Law under the Comprehensive Environmental Response, Compensation and Liability Act of 1980 (known as the CERCLA or Superfund legislation) (Department of the Interior, 1986).

The Challenge to the CV Researcher

Each method of valuing the benefits of quasi-private and public goods has particular methodological challenges which its users must address. It is worth repeating that *the principal challenge facing the designer of a CV study is to make the scenario sufficiently understandable, plausible, and meaningful to respondents so that they can and will give valid and reliable values despite their lack of experience with one or more of the scenario's dimensions.* The difficulty of writing CV scenarios which accurately communicate the intended meaning to respondents who have varying levels of education, life experience, and interest in the topic is often underestimated by researchers who have had little experience in survey research. Unless respondents understand all the components of a scenario in the way that the researcher intends them to be understood, there is no assurance that those surveyed will properly value the good. And even if an instrument is understandable, the market it portrays must also be plausible. Respondents are unlikely to take seriously relationships or situations that do not seem credible, such as the use of an electric utility bill payment vehicle for a study of risk reductions from toxic waste dump sites. Finally, unless respondents are able to relate the scenario to their personal knowledge and experience in such a way that the market is genuinely meaningful to them, they will not be motivated to expend the effort necessary to determine their personal value for the good. Part of the challenge in conducting a contingent valuation study is to get the respondents to accept a tradeoff between money and the public good which maintains the same level of utility. Economists usually assume that asking people to make this tradeoff is not problematic, whereas research conducted by social psychologists suggests that people have difficulty with tradeoffs (Abelson and Levi, 1985:287).

Sources of Error in Contingent Valuation Studies: An Overview

It is not possible to evaluate the contingent valuation method without considering the reliability and validity of its observed responses—the WTP amounts. Our framework for assessing reliability and validity is based on measurement theory.[15] The basic concern addressed by measurement theory is the relationship between the observed response and the underlying unobserved "true" value that generated it. The difference between the two is defined as "measurement error." Our understanding of this relationship in the contingent valuation case is complicated by the fact that as there is

[15] This discussion is based on Bohrnstedt (1983).

usually no external criterion against which we can compare a WTP amount,[16] "truth" may be inherently unobservable. The results of a market survey of consumer intentions to buy Toyotas may be matched against subsequent Toyota sales figures, whereas surveys of people's willingness to pay for SO_2 emission reductions to reduce acid rain generally have no market outcome by which to judge their accuracy.

Although the concepts of validity and reliability are commonly referred to in the social sciences, they are applied to nonpsychological data with surprising infrequency. Indeed, as Bohrnstedt (1983:69) notes, "historically survey researchers have paid scant attention to examining the reliability and validity of their measurements." Perhaps this is why these concepts have rarely been applied to CV studies in a way that is consistent with measurement theory. The sources of potential error in contingent valuation surveys, from the measurement theory perspective, can be briefly outlined.

We begin by positing that the *j*th individual's true willingness to pay for a specified level of a particular public good (TWTP) can be described by

$$\mathrm{TWTP}_j = f(X,\alpha), \tag{5-1}$$

where X is a matrix of the *j*th individual's attributes such as income and environmental attitudes, and α is a vector of unknown parameters.[17] It is impossible, however, to observe TWTP_j. Potentially available to the researcher instead is the individual's revealed willingness to pay (RWTP_j). RWTP_j can be represented as

$$\mathrm{RWTP}_j = h[f(X,\alpha), g_1(W,\beta), g_2(R,\psi), g_3(Z,\delta)], \tag{5-2}$$

where $f(X,\alpha)$ is simply TWTP_j; $g_1(W,\beta)$ describes the random error process as a function of the matrix of variables W (possibly a sub or super set of X, as are Y and Z) and a vector of unobserved parameters, β; $g_2(R,\psi)$ describes the systematic error process as a function of a matrix of variables R and vector of unknown parameters ψ; $g_3(Z,\delta)$ is a function which describes how likely RWTP_j is to be actually observed; and $h[\cdot]$ is an aggregator function with unknown properties. The $g_1(\cdot)$ function is closely related to reliability issues discussed below and in chapter 10; the $g_2(\cdot)$ function is closely related to validity issues and is discussed below and in

[16] See, however, chapter 9 for a review of the handful of studies that compare contingent valuation WTP amounts with the external criterion of payments in parallel markets.

[17] "True value" is a useful abstraction that is conventionally employed in discussions of survey measurement error. For a discussion of the true value idea in this context, see Turner and Martin (1984:104).

chapters 9 and 11; and the $g_3(\cdot)$ function is closely related to the nonresponse and sample selection issues discussed in chapter 12, which affect aggregate WTP across respondents. We briefly discuss each of these potential sources of error in turn below.

Reliability

Reliability usually refers to the extent to which the variance of a response or estimate is a result of random sources or "noise." Measurement reliability can be cast either in terms of classical test theory (that is, a test-retest situation) or in terms of sampling theory. Both perspectives are relevant to contingent valuation studies. The quantity that one typically estimates from a contingent valuation survey is $\overline{\text{RWTP}}$. The usual measure of the reliability of $\overline{\text{RWTP}}$ is the standard error of the mean which is given by

$$\hat{\sigma}_{\overline{\text{RWTP}}} = \hat{\sigma} / \sqrt{n} \qquad (5\text{-}3)$$

$$\hat{\sigma}^2 = \frac{1}{n-1} \sum_{j=1}^{n} (\text{RWTP}_j - \overline{\text{RWTP}})^2 \qquad (5\text{-}4)$$

If $E(\overline{\text{RWTP}}) = \overline{\text{TWTP}}$, then the standard error of the mean contains no bias component, and in conjunction with n can be used to describe a confidence interval for the difference between the $\overline{\text{RWTP}}$ from any particular sample and $\overline{\text{TWTP}}$.[18] It follows directly from equation (5-3) that increasing the size of the sample will increase the reliability of the WTP estimate, other things being constant. No matter what sample size is used, the estimate of $\overline{\text{TWTP}}$ can be expressed in terms of a confidence interval, which is preferable to the use of the standard point estimate alone, since the confidence interval also conveys some sense of the measure's reliability.

The variance of the RWTP_j is a result of two factors: a deterministic component, hopefully reflecting only the underlying (and sometimes quite large) variance of the TWTP_j, and a random error component. This random error component is a function of sampling variance, which is controlled by the researcher through the choice of the sample size and design, and of instrument and interviewer effects, which can occur at any point in the process. Since the procedures to control sampling variance in surveys are well understood (although not always practiced), in later chapters we will focus on the control of random error induced in the process of obtaining the respondents' answers to the WTP questions. The less

[18] One can say (exactly, if *normality* is assumed, and asymptotically for other distributions) that $\overline{\text{TWTP}}$ falls in the interval $\overline{\text{RWTP}} \pm t^{*}\sigma/\sqrt{n}$, with the probability represented by the choice of t, the Student's t variate used.

well-formed the respondent's value for a good, and the less understandable and meaningful the WTP questions are to the respondent, the greater the chance that his answers will represent guesses rather than considered responses. Ambiguity in question wording, and scenarios too unrealistic to be meaningful to respondents, are therefore important contributors to the random error in contingent valuation studies.

Validity

To the extent that an instrument measures the concept under investigation (is unbiased, in the terminology of statisticians and economists), it is valid (Bohrnstedt 1983). The absence of systematic error is implied by

$$E(RWTP_j - TWTP_j) = 0, \forall_j \qquad (5\text{-}5)$$

Systematic errors are perhaps the more serious threat to the accurate measurement of respondents' WTP amounts, and are more difficult to assess and adjust for, than are many types of random error.[19] Unlike random error, which is amenable to assessment by sampling and test theory, there is no applicable body of theory by which validity can be assessed (Carmines and Zeller, 1979; Bradburn, 1982) because we lack an explanatory model of the cognitive processes which underlie respondents' verbal self-reports (Bishop, 1981:591). In these circumstances the prevention of systematic error necessarily has an ad hoc character, although survey researchers have developed rules of thumb, based on experience and a growing body of survey experiments, that serve to minimize bias.

The question of bias is complicated in contingent valuation surveys by the general absence of a measurable true WTP value for public goods which can be used to assess the validity of a given study. Bias must therefore be inferred from our partial understanding of respondent behavior—if you ask the question this way, people will be likely to distort their answers—or from evidence in the survey which shows that changing the wording of the scenario in ways that are not expected to affect the WTP amounts in fact does so. "Not expected" is a key phrase here, because such differences may be legitimate contingent effects.

This observation requires some explanation, for until recently there was some confusion in the CV literature on this point. It was earlier assumed that only the nature and amount of the amenity being valued should influence the WTP amounts; all other scenario components, such as the

[19] In a recent review of the general survey research methodology literature, George Bishop concluded that the magnitude of systematic error "greatly exceeds that stemming from random sampling or non-sampling error" (1981:591).

payment vehicle and method of provision, should be neutral in effect (Rowe, d'Arge, and Brookshire, 1980:6). Therefore, according to this view, an experimental finding that the WTP amounts for a given study differ according to whether a utility bill or a sales tax payment vehicle is used was evidence of "information bias." More recently, Arrow (1986), Kahneman (1986), and Randall (1986b) have argued against that view, holding that important conditions of a scenario, such as the payment vehicle, *should* be expected to affect the WTP amounts. In their view, which we accept, respondents in a CV study are not valuing levels of provision of an amenity in the abstract; they are valuing a *policy* which includes the conditions under which the amenity will be provided, and the way the public is likely to be asked to pay for it. The notion that a public good does not have a value independent of its method of financing goes back at least to Wicksell, and is fully consistent with economic theory.[20]

The Relationship Between Bias and Variance

Since both bias and variance refer to how far, on average, an observed value will be from the true value, the researcher obviously wants to minimize both bias and variance. There are many design features of a contingent valuation survey which can reduce both. At some point, however, the researcher is faced with design choices that may reduce either bias, but only at the risk of increasing variance, or vice versa. Toro-Vizcarrondo and Wallace (1968) have shown in a statistical context that bias and variance can be traded off against each other, although to do this some decision criteria or loss function is needed to describe how the researcher is prepared to make the tradeoff. The mean square error (MSE) criteria, or quadratic loss function, is one such decision criterion. The MSE criteria minimizes

$$\frac{1}{n}\sum_{j=1}^{n} (\mathrm{RWTP}_j - \mathrm{TWTP}_j)^2, \qquad (5\text{-}6)$$

which is equivalent to minimizing the sum of the variance plus the square of the bias. This is the most commonly used decision criterion in the econometrics literature, and is often appropriate for contingent valuation studies.[21] In designing a CV instrument it is sometimes desirable to accent

[20] One important implication of this position is that without further evidence it cannot be assumed that the results of a given CV study can be generalized much beyond the conditions specified in its scenario.

[21] MSE is a symmetric loss function which penalizes over- or under-estimation equally. If this symmetry condition is not reasonable, another loss function should be used. It may also be desirable to use a symmetric loss function that puts less weight on outliers, such as minimizing the sum of the absolute deviations. The median is generally a more robust estimator of location than the mean. The family of trimmed mean estimators, which includes both the mean and median as special cases, is discussed in chapter 10.

a feature that increases bias somewhat (particularly if the direction of the bias is known) if it makes the CV scenario much more realistic, thus reducing the random error (that is, variance) in the responses.

Summary and Conclusions

The sample survey as a research technique has a problematic nature, which we discuss here because social scientists unfamiliar with the method are sometimes deceived by its apparent simplicity. Yet survey researchers, out of their wealth of experience, have learned the art of asking questions and have revealed not only many subtle problems, but also sensible solutions. Contingent valuation surveys are less-travelled terrain, with fewer solutions precisely when they are most needed because of the nature of the subject matter and the uses to which CV studies are put.

All benefit estimation methods are vulnerable to error, including those based on actual market prices. Our intention in pointing out the difficulties involved in using survey research to measure benefits is not to counsel despair, but to encourage a mature skepticism and promote methodological sophistication. Once the potential problems are identified, it is generally possible to overcome them, or at least to minimize their effects. If problems persist, knowing what they are will permit the researcher to qualify the findings appropriately.

Although discussion of measurement issues is commonplace in the contingent valuation literature and CV practitioners have conducted a number of interesting methodological experiments, there has been relatively little effort to systematically relate discussions of reliability and validity to measurement and sampling theory. In order to avoid the resulting conceptual confusion and to set the stage for an analysis of the reliability and validity of CV measures, the last part of this chapter has been devoted to a brief overview of measurement theory.

Applied to a specific amenity, measurement theory posits an underlying, unobserved (true) willingness-to-pay amount which is imperfectly captured by the observed representation measured by the researcher. Some of the causes for the discrepancy between the observable and unobservable variables are random, while others are systematic and bias the measure in one direction or the other. The former affect the reliability of the estimate, and the latter its validity or the extent to which an instrument measures the concept it is intended to measure. Assessing the validity of WTP estimates is inherently difficult because most CV studies measure benefits of amenities for which no market measure is available.

One source of error that many economists have believed would invalidate the willingness-to-pay amounts given by consumers in CV surveys is strategic behavior, where respondents deliberately attempt to influence the provision of the amenity by intentionally distorting the amount they say they are willing to pay. In the next two chapters we examine the empirical evidence for strategic behavior, some of which comes from studies that use experimental designs and simulated markets to assess the validity of hypothetical willingness-to-pay measures.

6

Will Respondents Answer Honestly?

As a method of measuring the benefits of goods for which there are no markets, asking people directly to value public goods has many desirable properties. Nevertheless, the beliefs about human behavior embraced by economists make them deeply suspicious of data gathered by the contingent valuation techniques. This attitude contrasts sharply with their respect for studies that use indirect valuation methods, despite the long chain of assumptions made between the data gathered by these methods and the benefits they seek to measure (Mendelsohn and Brown, 1983; Smith, 1986b). As noted, the crucial difference between the indirect and the direct methods is that indirect methods are based on actual behavior, whereas direct questioning necessarily relies on the respondents' willingness to reveal their true preferences.

The most influential statement about why direct questioning is suspect is that given by Paul Samuelson in his 1954 paper (and its reiterations in 1955, 1958, and 1969), where he states: "It is in the selfish interest of each person to give false signals, to pretend to have less interest in a given collective activity than he really has."[1] In the penultimate paragraph, Samuelson pursues this point by making specific reference to questionnaires:[2]

One could imagine every person in the community being indoctrinated to behave like a "parametric decentralized bureaucrat" who reveals his preferences by signal-

[1] The assumption that people will lie if it is in their narrowly defined selfish interests appears to have originated early in the development of economic theory with the work of David Hume and Knut Wicksell. For a useful survey of the assumption in contemporary economic thought, see McMillan (1979a).

[2] It is interesting, historically, that Ciriacy-Wantrup (1947,1952), in work foreshadowing much of Samuelson's 1954 paper, strongly advocated the "direct interrogation" of the relevant public to determine willingness to pay for various public goods.

> ling in response to price parameters or Lagrangean multipliers, to questionnaires, or to other devices. Alas, by departing from his indoctrinated rules, any one person can hope to snatch some selfish benefit in a way not possible under the self-policing competitive pricing of private goods. (1954:389)

The assumption that people will engage in strategic behavior by intentionally misrepresenting their preferences for public goods appears in virtually every public finance text, beginning with the first edition of Musgrave (1959), and is integral to almost every discussion of the public goods problem in textbooks on microeconomics.[3] It is not surprising, therefore, that respondents' strategic behavior is invoked by economists as a reason for rejecting the contingent valuation method (for example, Seneca and Taussig, 1974:95; Tietenberg, 1984:73), and that its spectre lies behind the wariness with which other resource economists accept the method (such as Freeman, 1979b; Fisher, 1981; and Just, Hueth, and Schmitz, 1982).

The form of strategic behavior which has received the greatest attention in the literature is "free riding," where someone pays less than a public good is worth to him in the expectation that others will pay enough to provide it nevertheless.[4] If applied to CV surveys, it follows that free-riding respondents would underbid if they believe they will actually have to pay the amount they reveal, and believe also that there is a good chance the good will be provided even if they understate their true WTP amount. Reversing these beliefs would motivate a less-discussed form of strategic behavior that we will call "overpledging." Respondents would be expected to overbid if they believe they will not actually have to pay the amount they state, yet believe also that the stated amount can influence provision of the amenity.

In exploring the question of whether respondents will answer honestly, we examine the non-contingent valuation literature (theoretical and empirical) bearing on the assumption that strategic behavior will hinder the ability of surveys to measure people's preferences for public goods. We will see that there are important factors which militate against strategic behavior, such as information costs and adherence to social norms of altruism and truthfulness, and that these play an important role in the preference revelation process. We will also see that in many respects the properties of contingent valuation studies themselves militate against strategic behavior. On the basis of these literatures, and evidence from CV studies themselves (considered in chapter 7), we conclude that Samuelson's pessimism is not warranted. Instead of representing an insoluble problem, the possibility of

[3] Interest in the conditions, motives, and implications of lying extends to other disciplines, including political science (Hardin, 1982) and philosophy (Bok, 1978).

[4] The term "free riding" is often used in ambigious ways (Cornes and Sandler, 1986), and sometimes is used to refer to any type of strategic behavior.

strategic behavior is just one of the biases that CV studies must take into account.

Theoretical Developments

Samuelson, in his formulation of the preference revelation problem, did not foresee circumstances where people would *not* have incentives to lie about their demand for public goods. Nor did he envisage any counterbalancing incentives for people to voluntarily cooperate and tell the truth.[5] In its free-riding form, Samuelson's formulation of the strategic behavior problem is essentially the same as the dominant strategy in a simple static prisoner's dilemma.[6] The most fundamental difficulty with this formulation is that, as long as one person can make his own contribution to the public good contingent on everyone else also contributing, the outcome of Samuelson's prisoner's dilemma game will not produce free riding as a dominant strategy, much less the weaker stable Nash equilibrium.[7]

Incentive-Compatible Demand Revelation Theory

The large theoretical literature on incentive-compatible demand revelation was inspired in part by attempts to design preference revelation mechanisms for public goods which would avoid free riding and result in the optimal decentralized provision of public goods. The incentive-compatible demand revelation devices (ICDRDs) proposed by theorists create situations in which it is in the person's selfish interest to choose to reveal his or her true preferences for a good. These instruments include surveys, voting schemes, auctions,[8] and real markets that incorporate penalties for false revelation and rewards for honest revelation. A decade's theoretical work challenges Samuelson's assumption that free riding is *always* the optimal strategy under any questioning format.

The first breakthrough in incentive compatibility theory came in papers by Malinvaud (1971) and Dreze and Vallee Poussin (1971), which put forth the concept of a benevolent planner who, in a continuous incremental dynamic Walrasian tatonnement process, used responses on marginal WTP from consumers to obtain a Lindahl-Pareto efficient equilibrium. They demonstrated that even if consumers did not initially give truthful

[5] See Coase (1974), Stigler (1974), Bolnick (1976), and Sudgen (1982,1986) for observations on the voluntary provision of public goods in the absence of government intervention.

[6] The standard example is a 2 × 2 game matrix where consumers are assumed to be of two different types. More complex representations maintain most of the same properties.

[7] A Nash equilibrium occurs when each actor takes the action of the other actors as given. A stable Nash equilibrium occurs if, upon the last agent making his or her decision, no other agent has any motivation to change earlier strategy.

[8] Of particular note was the demonstration that the particular form of the second price sealed bid auction proposed by Vickrey (1961) has many properties of an ICDRD.

responses, this process converged to the optimal provision of public goods as long as the consumers followed a strategy of maximizing the minimum gain—a strategy shown to lead eventually to truthful responses. One class of ICDRDs, first suggested by Clarke (1971), Groves (1973), Groves and Loeb (1975), and Groves and Ledyard (1977), has the property that under the conditions of the device it is in the selfish interest of the consumer to state his or her true WTP. This class of devices utilizes taxes or subsidies (or side payments) to change the level of provision of the public good.[9]

At about the same time, Hurwicz (1972) proved that true preference revelation was not the dominant strategy in private goods markets. This result was extended in two separate proofs (Gibbard, 1973; Satterwaite, 1975) to show that it was impossible to design Pareto-optimal, nondictatorial social decision rules for both public and private goods if no restrictions were placed on individuals' announced preferences. Thus, a counterpart to Arrow's impossibility theorem was formed. And as Roberts (1976) showed, it follows from the Gibbard-Satterwaite theorem that, as the number of consumers increases, free riding becomes less effective in the case of private goods, while in that of public goods incentives for free riding tend to increase as people realize their contributions will have only a minuscule effect on the supply of the good. Roberts' results, though still suggesting that free riding is a problem in the large-number situation, represent a significant weakening of Samuelson's free-rider postulate and do much to explain the ability of small organizations, such as trade associations and charities, to produce public goods.

Because the various ICDRDs proposed by theorists are generally too complex to be implemented outside of the laboratory, much of the ICDRD literature is not directly relevant to contingent valuation surveys. But the properties of demand-revealing situations that this literature identifies as militating against strategic behavior are relevant, because they provide insight for CV practitioners who wish to avoid strategic bias. These properties are:[10]

1. small numbers or sample size (ideally, few enough to be noticed but not so few that an individual is able to influence provision of or payment for the good for the group as a whole);[11]

[9] Because of the nonbalancing taxes or subsidies that must be collected, none of the ICDRDs yield Pareto optimums, although under fairly general conditions they will converge arbitrarily close to Pareto optimality.

[10] There is not complete agreement on these points. Much of the work on incentive-compatible demand revelation is contained or summarized in Green and Laffont (1979), Laffont (1979), a symposium in the *Review of Economic Studies* (April 1979), and a review essay by Groves and Ledyard (1986). A less technical treatment may be found in Clarke (1980) or Cornes and Sandler (1986).

[11] The researcher can tell individuals they are part of a random sample, the results of which will be used to determine the outcome for the entire population. Green and Laffont

2. reaction functions which do not take the other economic agents as fixed (so that a stable Nash equilibrium exists when no one wants change, because everyone realizes that others would also withdraw their contributions if the agent in question did not pay a fair share);
3. restriction of the domain of acceptable responses (thus limiting the potential gains of strategic behavior);
4. a dynamic context for the supply of and payment for the public good;
5. observability of the actions of the other economic agents either in sum or, ideally, individually;
6. a situation where there are small expected gains from free riding and/or large expected gains from cooperative (non-free riding) behavior by all economic agents.

Other Theories

Although ICDRD theorists disagree with Samuelson about the inevitability of free riding, they share his assumptions that respondents will behave noncooperatively and will tend to give dishonest responses in preference revelation situations if there is any gain to be had by doing so. This is why they believe that the only way to overcome strategic behavior is to use an elaborate incentive-compatible device incorporating the properties listed above as well as side payments. Other theorists, however, challenge Samuelson's basic assumptions about noncooperation and honesty. McMillan (1979b) and Evans and Harris (1982) investigate the noncooperative behavior assumption in a dynamic context.[12] Building on the concept of a meta- or supergame advanced by Luce and Raiffa (1957), they show that even in the absence of an ICDRD, noncooperative behavior, and hence free riding, is at best a suboptimal strategy.

In examining the role of feedback mechanisms in inducing cooperative behavior for supplying public goods, McMillan finds that the greater the gains from dynamic cooperative playing, relative to the static gains from free riding, the greater the incentives for cooperative behavior. The cooperative solution emerges as a stable Nash equilibrium as long as agents do not discount future utility too heavily. In contrast, the noncooperative solution emerges as a stable Nash equilibrium only if agents completely discount future utility. While McMillan shows that incentives for cooperative behavior are present when behavior is replicated over time, he did

(1978) suggested the practical implementation of the ICDRD approach through such a sampling scheme.

[12] There is growing consensus among economic theorists (for example, Kreps and coauthors, 1982) that the naive, static, one-period expected utility models are inadequate for developing predictions about any forms of behavior, especially those with dynamic elements.

not take up the question of how one might pass from noncooperative to cooperative behavior.

This issue is addressed by Evans and Harris (1982), who indicate how the cooperative solution arises from noncooperative behavior through a Bayesian learning process. They argue that to induce cooperative behavior it is not necessary for the sequence of games to actually be played, but only that the play be anticipated by the consumer of a public good. This behavior is consistent with the well-known game theory result (Shubik, 1982) in which the optimal strategy in a prisoner's dilemma game with a stochastic number of plays is a cooperative one.[13] Thus, dogmatic free-rider behavior is not rational, even in a prisoner's dilemma setting. Evans and Harris find a learning lag before the supply of public goods falls to the point where participants feel deprived, and they propose a number of interesting hypotheses about the factors determining its length. Where large numbers are involved, they see longer learning lags and free riding by a significant but not large number of agents. Their analysis is essentially a disequilibrium argument, and they envision a kind of continuous process with some small percentage of agents unlearning (free riding) before relearning as the supply of the public good drops. Cooperative behavior becomes the norm after the initial learning process.

Another line of attack on the honesty-cooperative assumption is provided by the recent work of Akerlof (Akerlof, 1983; Akerlof and Dickens, 1982), which incorporates cognitive dissonance and other psychological concepts to show why shifts in economic behavior are not costless.[14] Building a model of honesty and cooperative behavior, Ackerlof demonstrates that honest behavior results in positive long-run *economic* gains which are jeopardized if any dishonest or noncooperative actions are exhibited.[15] This result contradicts an implicitly maintained assumption

[13] As Evans and Harris do not cite McMillan, these papers appear to be independent of each other. Evans and Harris, however, represent a definite extension of McMillan's results in the Bayesian direction. The origins of Evans and Harris's approach lie in Taylor's (1976) attack on the free-rider hypothesis from the political science perspective, and Rapoport's (1971) and Shubik's (1970) attacks on the superficial nature of the static prisoner's dilemma concept. Hardin (1982) and Axelrod (1984) provide good reviews of much of the political science work in this area. Axlerod, in particular, has become well known for his demonstrations that in the long run a tit-for-tat strategy appears to win out over any type of "strategic" behavior.

[14] Cognitive dissonance has long been recognized as a powerful force acting on consumer behavior in the marketing literature (Holloway, 1967). There is an extensive psychological literature on the development of morals and their effects on later life; see Carroll and Nest (1982) for a discussion.

[15] A number of alternative theoretical approaches, yielding similar results, are not discussed here. Brief outlines of four such approaches—rational altruism, Bayesian decision theory, utilitarian ethics, and natural selection—may be found in articles by Kurz, Harsanyi, and Schelling in the session on "Economics and Ethics: Altruism, Trust and Power" in the

of most ICDRD theorists that dishonest behavior is costless outside the immediate choice situation, and will be engaged in any time net positive direct incentives for false revelation exist. It also suggests that people weigh the *cost* (outside of the choice situation) of false revelation against the gains directly related to the choice situation.

In this brief review of post-Samuelson theorizing about strategic behavior we began with the assumption that free riding is a rational dominant strategy, but then arrived at the position that free riding is an irrational suboptimal form of behavior.[16] The motivation to tell the truth when revealing preferences for public goods and the social norms of fairness and honesty may now be seen as mutually reinforcing. Important empirical support for these ideas is found in the work of experimental economists to which we now turn.

Experimental Results

The experimental economist's laboratory provides a useful place to test the extent of cooperative behavior and the honesty of responses in preference revelation situations. In the laboratory the researcher can give and require payments, repeat trials, introduce complex scenarios, and vary a wide number of stimuli under different conditions—none of which can be done easily, if done at all, in contingent valuation surveys or in real life. Thus far, most of the experimental work of relevance to the contingent valuation method has been restricted to the free-riding form of strategic behavior. Nevertheless, the findings illuminate strategic behavior in general, and many of the experiments have included treatments comparable in certain ways with contingent valuation scenarios.

ICDRD Experiments

The first set of experiments of interest are those that tested the ideas of the ICDRD theorists about mechanisms to reduce free riding. An unexpected and notable finding was that the ICDRD mechanisms worked no better than simply asking people voluntarily to say how much they were willing to pay.

1978 issue of the *American Economic Review* containing the *Proceedings of the American Economics Association.* Harsanyi, commenting on lying, notes that people have inhibitions against killing their own species, that "most people have a natural tendency to tell the truth and [that], even though they are usually perfectly able to lie, they tend to be rather inefficient liars and often give themselves away by inconsistencies or by involuntary behavioral signs of embarrassment, etc. . . . It is easy to see why such natural inhibitions are in general socially useful; and they may also be the results of natural selection."

[16] Recently a number of economic theorists—Scitovsky (1976), Becker (1976, 1981), Sen (1977), Simon (1982), and Margolis (1982)—have gone beyond the free-rider postulate to challenge the fundamental assumption of "selfish" utility maximization.

The initial experimental test of an ICDRD mechanism to value public goods was reported by Scherr and Babb (1975), who used three treatments—the Clarke tax (Clarke, 1980), another ICDRD, and voluntary demand revelation—to obtain individuals' demand for a concert series and for additional books for the college library. The voluntary demand condition was included as a control against which to measure the expected improvements in revealing true WTP made by each of the ICDRD mechanisms. Subjects were given a payment of $10 for participating in the experiment; they could keep it or donate some or all of it to the concert and book funds. Participants were obligated to pay the amount they offered for each good, while the goods purchased through the group's contributions were to be shared equally among all participants, whatever their payments. Contrary to expectations, the researchers found that the voluntary demand revelation treatment elicited larger real payments for the public good than did either of the ICDRD mechanisms. According to Scherr and Babb,

> the proposed pricing systems (ICDRD) may not have inhibited free-rider behavior because there was not a great deal of such behavior to inhibit. The debriefing suggested that few subjects attempted to totally free-ride. Although 75 percent of the subjects wanted to keep over four of the ten dollar allotment for themselves, they indicated the following reasons for not keeping the rest: (1) not wanting to feel cheap, (2) the funds were worthwhile, and (3) altruistic reasons. (1975:42)

Theorists working on ICDRDs have tended to ignore the Scherr and Babb experiment or to discount its results by attributing them to the experiment's novel conditions, to induced altruism, to insufficient incentives, or to the respondents' lack of experience in using the rules of ICDRD.[17] These reactions have prevailed in spite of Johansen's (1977) admonitions, in the *Journal of Public Economics,* that in the literature there was little or no empirical evidence of free riding and that it was difficult to construct a consistent logical argument for free-riding behavior. He warned the economics profession that it had put "misplaced emphasis on the free rider problem."

During the 1970s references were made in the ICDRD literature to a series of experiments being conducted by Vernon Smith that were more realistic and would rectify the learning problems associated with the unfamiliar nature of the ICDRD. When the results of Smith's experiments were published, however, they revealed that the early experimental findings were neither flukes nor the products of unique conditions. In a 1977

[17] Clarke (1975:52), in his comments on Scherr and Babb's experiment, stated, "I would maintain that the most important explanation [for the lack of free riding] lies in their [subjects'] lack of understanding of incentives to strategic behavior."

paper, Smith offered these comments, strengthening Scherr and Babb's original observations:

> But why do real people making real decisions for real money, in the experiments reported here and in the cited experimental public good literature, behave predominantly in accordance with the competitive postulate? Why do they not exhibit the more 'sophisticated,' 'strategic' behavior postulated by Hurwicz and Ledyard-Roberts? I think it is because there are significant direct (and indirect opportunity) costs of thinking, calculating, and signaling which make strategizing uneconomical. In the Auction Election the opportunity cost of failing to reach agreement is the loss of the more valuable best proposition. Strategizing not only consumes costly time and thought, it increases the risk of group disagreement. Disagreement means exclusion from better wealth states on the sobering principle that nobody gets more if there is not more available to get.
>
> Can all the hard scientific evidence from experiments to date be dismissed as irrelevant, too simplistic, or based on particular experimental parameters? I think not, for the reason that such results are consistent with a great deal of field evidence. Thousands of buildings have been purchased by religious organizations, clubs, art associations, and private societies through member voluntary contributions. (1977:1136)

Vernon Smith, Grether, Plott, and their colleagues subsequently demonstrated cooperative behavior many times under a wide variety of conditions.[18] The general conclusion of their work is that cooperative behavior is almost always reached with both voluntary revelation and ICDRD treatments, and usually is reached as quickly (efficiently), if not more quickly, through voluntary revelation. *In other words, in reaching the Pareto-optimal state the performance of voluntary demand revelation is usually as good as and is often better than that of the ICDRDs.* It appears that the following assumptions of the ICDRD theorists do not mirror reality well: (1) that there are zero costs involved in the process of determining optimal strategies; (2) that zero costs result from the act of giving dishonest responses; and (3) that the individual perceives zero risk that her actions might keep the economy from reaching the optimal competitive or cooperative solution.

It is useful at this point to distinguish between the strong and weak versions of the free-rider hypothesis. The strong version, usually associated with Samuelson (1954) and Olson (1965), states that people will not voluntarily contribute any money to the public good without other incentives. In the absence of selective incentives, a contribution rate of less than 25 percent of the optimum is generally considered support for the strong

[18] In particular see Vernon Smith (1977, 1979, 1980), Grether and Plott (1979), and Vernon Smith and coauthors (1982). A good description of the methodology of experimental economics may be found in Plott (1982).

free-rider hypothesis. The weak version states that voluntary contributions may be significant but suboptimal.[19] Over the course of the past decade experimentalists have consistently disproved the strong version, while their findings in regard to the weak version indicate a crucial difference between divisible public goods and those that are indivisible (for example, those where the whole group is excluded from enjoying the good if the group's total stated WTP is not equal to or greater than some specified amount). Under the indivisible condition—the one implicitly used in most CV scenarios—even the weak free-rider hypothesis receives little support.

Bohm's Experiments

In the early 1970s Peter Bohm, a Swedish economist, conducted a path-breaking set of experiments on people's willingness to pay to see a preview of a Swedish television show (Bohm, 1972). Not only was his study one of the first tests of the free-riding hypothesis and the only experimental test to date of the overpledging hypothesis, it also is one of the handful that compare willingness-to-pay behavior in simulated markets with WTP behavior in hypothetical markets (on this aspect of Bohm's study, see chapters 7 and 9). Bohm's results were sufficiently persuasive to lead Atkinson and Stiglitz (1980), in their recent advanced text on public economics, to cast doubt on the empirical significance of free riding.

Bohm's subjects, a random sample of Stockholm adults, were invited to the Swedish Broadcasting Company to answer questions about television programs. Once assembled, they were given a fee of 50 kronor and, on the condition that either the group or the individual meet a payment criterion, were offered the opportunity to bid money to see a closed-circuit preview of a half-hour program by two famous Swedish television personalities. They were led to believe that they were part of a larger body of people simultaneously being offered the same opportunity. Bohm allocated his subjects to six experiment treatments. Five treatments excluded the entire group from seeing the show if the individual bids did not reach a specified total; these treatments differed only in the method of payment they required of participants to see the show once the subjects had revealed their WTP amounts. The provision conditions included actual payment of the bid offered by each individual; indeterminacy of whether payment would be required; payment of a fixed amount regardless of the revealed WTP amount; and no payment requirement. In table 6-1, which summarizes Bohm's treatments, the strategy column gives Bohm's predictions for each treatment, based on his a priori utility maximization assumptions.

Bohm designed his experiments so that treatments I and II provided incentives to free ride, while treatments IV and V motivated subjects to

[19] Cornes and Sandler (1984) refer to weak free riding as "easy riding," and examine in detail why it might occur.

Table 6-1. Bohm's Tests of Strategic Behavior

Treatment if 500 kr. total bid is reached	Provision contingent?[a]	Payment condition			Bohm's predicted outcome[b]	N	Mean bid (kronor)
		Actual	Uncertain	Fixed			
I Pay amount bid	No	X			Free ride	23	7.6
II Pay amount proportional to bid	No	X[c]			Free ride	29	8.8
III Lottery chooses payment method	Yes		X		None	29	7.3
IV Fixed payment of 5 kr.	Yes			X	Overbid	37	7.7
V No payment required	Yes			X	Overbid	39	8.8
Treatment without group exclusion							
VIa Hypothetical WTP amount requested	Uncertain		X[d]		Slightly overbid	59	10.2[e]
VIb Highest 10 bids of 100 accepted	Yes	X			Slightly underbid	59	10.3

[a] "No" means that respondents had reason to believe that understating their willingness to pay would not affect the provision of the good; "Yes" means it was likely that participants believed the provision of the good was dependent on their revealed values.

[b] According to the economic incentives embodied in the scenario.

[c] When giving their responses subjects knew they would have to pay something, but did not know the exact amount of their payment obligation.

[d] Although the scenario for this treatment did not imply payment of any kind, respondents could have perceived the experiment as a market test of a product for the government television network for which they, as taxpayers, would eventually pay.

[e] This amount became 9.45 kr. if one outlier was dropped.

overpledge. A participant in treatment I, for example, could reasonably expect that bids offered by the rest of the "large number" of participants would reach the 500 kronor payment goal, making it possible for her to underbid and still see the show. Conversely, since individual participants in treatment V were told that they would not actually have to pay anything, no matter what they said they were willing to pay, a strategy of overbidding to reach the goal was presumably encouraged. Bohm designed treatment III to minimize strategic incentives. Respondents were told that any one of four payment methods might be required, depending on the results of a random draw. To make sure that the subjects were aware of the incentives, Bohm told the subjects in each treatment what their optimal strategy would be and presented them with counterarguments: possible loss of the good in free-rider cases, and a morality favoring true preference revelation in overpledge cases.

Treatment VI differed from the others in that the good was treated as a private, not a public, good (there was no group exclusion), and its participants were subjected to two forms of preference elicitation. In the first—treatment VIa—they were asked to state "the highest admission fee you would be willing to pay, *if* you had been asked to pay an admission fee for watching the program here" (Bohm, 1972:129; emphasis in the original). Although at this stage it was not explicitly said that the subjects would have to pay, or whether they would be able to see the show, it seems reasonable that some or all of the group would have some positive expectation, but not certainty, about these possibilities. After responses to the hypothetical situation were collected, members of this group were asked to take part in a real auction to see the show. In the second form of preference elicitation—treatment VIb—each person bid to see the program, with the understanding that the 10 highest bidders out of a presumed 100 participants would win the viewing.

Bohm originally thought treatments VIa and VIb would, in combination, reveal the true aggregate demand, with responses to the hypothetical version slightly overestimating it (because some subjects, trying to be "nice" to Swedish Broadcasting, might say the show was worth more than they thought it was), and responses to the auction slightly underestimating it (because some would bid amounts calculated to be just sufficient to be in the top 10 percent). He changed his mind, however, once he saw that the mean bids for treatments I–V were all significantly below those for VIa and VIb. This convergence, his belief that treatment III contained no incentives for strategic behavior,[20] and the possible distortions introduced

[20] We will argue that if a risk-neutral subject did a simple expected-payment calculation where he would pay the stated amount with some positive probability and pay nothing with some positive probability, then the expected payment would be less than the stated payment, and there would be an incentive to overpledge. A sufficiently risk-adverse subject might have an incentive to free ride in such a situation.

by treating the good as a private good led him to conclude that group III's values best represented the true value. According to Bohm, the distortions were twofold. First, treatment VIb was suspect because it was contaminated by treatment VIa. That only 18 of 54 people changed their statements (half increased and half decreased their original bid) between the two treatments suggested to Bohm that people may have felt reluctant to admit they had "lied" in the first round. Second, because the good in treatment VIb was a private good, it introduced elements of competition and "auction fever" into the valuation situation, which rendered the treatment inappropriate as an indicator of true WTP for a public good.

While we acknowledge Bohm's misgivings about treatments VIa and VIb, we think that of all his treatments VIb probably provides the closest measure to the true WTP, for three reasons. (1) The form of treatment VIb was the best of the seven for determining maximum WTP; it was similar to auction techniques successfully used in other experiments for this purpose (described below). (2) Treatment VIb was worded in such a way that respondents should have felt free to change their earlier bids. (3) It is plausible that treatments I through V underestimated the true WTP amount because participants may have been influenced by the then-current price of attending a movie in Stockholm, or may have based their WTP amounts on a presumed group size that they divided into the 500 kronor goal.

We are now in a position to interpret the findings of these important experiments. First, they show that participants were not sensitive to the strategic incentives in Bohm's treatments. The amounts revealed in response to treatments I to V are not statistically different at any accepted significant level even though economic theory would predict some to be higher and some to be lower than the true WTP amount. Second, when treatment VIb is used as a criterion, the amounts revealed by I to V capture 71 to 85 percent of the true value. It is notable that overpledging does not occur here, and the strong free-riding hypothesis does not receive support. Third, the hypothetical treatment (VIa) is closest to the true WTP. These findings bode well for the validity of the contingent valuation method, and will be considered at more length in chapter 9.

Free-Riding Experiments

We now turn to a series of experiments which use nontrivial rewards to test the prevalence of the free-riding form of strategic behavior under realistic fieldlike conditions designed to produce free riding, according to standard assumptions. Their results tend to confirm our interpretation of Bohm's findings. They demonstrate that strategic behavior occurs much less often than standard utility maximization assumptions would predict, except where the person is assured that he will get the good no matter what he says he will pay. Even under this condition, free riding occurs far less than most economists would predict.[21]

Marwell and Ames (1981) conducted a series of experiments in which participants were given tokens, exchangeable for money, that they could allocate to either an individual exchange or a group exchange. The individual exchange returned money directly to the respondent. Tokens given to the group exchange paid out money at a higher rate, but the group sum was divided among all its contributors, irrespective of their donation to the fund. In all of these experiments except one, the payoff from the group exchange was money to each individual, a divisible good. With the exception of an experiment whose subjects were economics graduate students,[22] none of these divisible-good experiments confirmed the strong free-rider hypothesis. The quantity of tokens allocated to the group exchange was in the 30 to 60 percent range. These findings were invariable, even in experiments involving nontrivial amounts of money. They confirm the weak free-rider hypothesis in the case where the payoff is divisible among the group members. However, in the one experiment that used an indivisible payoff, even the weak free-rider hypothesis was not supported. Here the average quantity of tokens allocated to the group exchange rose from 43 percent in an experiment with divisibility to 84 percent in an experiment that was identical except that the money could only be used to buy a group good. This result contradicted Marwell and Ames's expectations, and foreshadowed the results of three other experiments, which we now examine.

Experiments by Schneider and Pommerehne (1981), Brubaker (1982), and Christainsen (1982) used almost identical treatment designs to test the

[21] For related experiments whose conditions are less similar to contingent valuation, see Harstad and Marrese (1982), Kim and Walker (1984), Tideman (1983), Issac, Walker, and Thomas (1984), Issac, McCue, and Plott (1985), and Orbell, Schwartz-Shea, and Simmons (1984). The experiments by Issac and his colleagues represent a counterattack against the experiments discussed in this section. They tried deliberately to find, and were somewhat successful in finding, conditions that motivated people to behave strategically. We believe not only that such conditions exist, but that they are largely avoidable.

[22] Before conducting these experiments, Marwell and Ames gathered a priori beliefs from a small panel of economists who, after examining the experimental design, stated that theory would predict zero public contributions to the group exchange. Asked to give their views about the actual extent of public contributions, the consensus was that between 0 and 20 percent would be contributed, the 20-percent figure being ascribed to risk-taking, altruism, and irrationality. It is interesting that the 20-percent figure was exactly the level of contributions made to the group exchange by the economics graduate students. Marwell and Ames found that attitudes toward fairness were good predictors of the percentage of a person's resources invested in the group exchange. In contrast to the other participants, economics graduate students had great difficulty in answering an open-ended question about what contribution to the group exchange could be regarded as fair. Many of those students who did answer said little or no contribution was fair, whereas three-fourths of the other respondents replied that 50 percent or more of their initial allocation was fair.

free-rider hypothesis. Each experiment involved up to three treatments, which we will label E_1, E_2, and E_3:[23]

- The E_1 treatment solicited WTP offers with a guarantee that the *n* highest bidders would receive the good. E_1 was presumed to elicit the true willingness to pay for the good, since any underrevelation placed participants at strong risk of being excluded from it.
- E_2 solicited WTP offers that obliged respondents to pay their stated amounts and assured that all members of the group (whatever the individual contribution) would receive the good if the group total was equal to or greater than a specified amount. Where total contributions exceeded that amount, each individual's contribution would be proportionally reduced. This treatment represented group exclusion, a concept implicit in many contingent valuation studies of environmental goods. E_2 elicited the true willingness to pay minus the incentives for free riding on others' contributions. The incentives not to free ride in this treatment were the possibility that the contribution of the others would not be sufficient to obtain the good, and internalized norms of fairness and honesty (each individual's contributions were not made public to the other subjects).
- E_3 solicited WTP offers having the condition that payment was required, but all respondents giving any positive amount were guaranteed to receive the good. In treatment E_3 the incentives to free ride were very strong, inhibited only by societal norms.

Schneider and Pommerehne conducted their experiment in a real-life context in which economics graduate students used their own money to bid for a prepublication copy of a new economics text that would be uniquely useful for an upcoming preliminary examination. The students did not know that the market was created for research purposes. Brubaker's subjects, a sample of townspeople in Madison, Wisconsin, bid on certificates good for a $50 shopping spree at any of several local grocery stores. Christainsen's respondents, also Madison townspeople, bid on vouchers good for $200. For obvious reasons, Christainsen did not include an E_3 treatment in his study.

As expected, $E_1 > E_2 > E_3$. In Schneider and Pommerehne's experiment $(E_1 = \$27.62) > (E_2 = \$26.57) > (E_3 = \$16.83)$, with no significant difference between the results of E_1 and E_2. For Brubaker, $(E_1 = \$33.99) > (E_2 = \$27.07) > (E_3 = \$23.96)$. Here the difference between E_1 and E_2 is just significant at the 95 percent confidence level. Christainsen's experiment yielded $(E_1 = \$194) > (E_2 = \$153.05)$, which

[23] For consistency with Schneider and Pommerehne's and Christainsen's ordering, we have applied the labels E_3 to Brubaker's group II and E_2 to his group III.

were significantly different. In the three experiments E_2 as a percentage of E_1 varied from 96 to 79 percent, a relatively minor difference between true WTP and revealed WTP when the usual vagaries of cost-benefit analysis are considered. It is remarkable how consistent these results are with the theoretical argument advanced by McMillan (1979b) and Evans and Harris (1982), discussed earlier.

The E_3 treatments shed light on the degree to which people reveal preferences in the absence of any direct economic incentives to do so. They provide evidence against the strong free-rider hypothesis, because the respondents offered much higher amounts than the lowest positive price for which they were guaranteed the good (that is, \$1.00). There are two interesting interpretations of the E_3 results, neither of which is consistent with the standard free-riding postulate. The first is that

$$NG = MG - WTP_{AL}, \tag{6-1}$$

where NG is the net gain from lying or giving distorted demand revelation, MG is the maximum gain available from giving the minimum WTP necessary to obtain the good in question, and WTP_{AL} is willingness to pay to avoid lying to get MG.[24] This explanation is consistent with the possibility that large enough incentives (gains from dishonest actions) could induce large distortions between true and revealed WTP. For Schneider and Pommerehne WTP_{AL} is \$15.83, and for Brubaker WTP_{AL} is \$22.96, both of which are approximately 60 percent of the optimal free-riding strategy.[25] In this interpretation, lying involves costs and there is a substantial WTP_{AL} to completely avoid lying, at least under the conditions presented in these experiments.[26] The ICDRD framework implicitly excludes the costs of lying. It can easily be shown that introducing such costs in the ICDRD framework would give radically different results.

Brubaker offers an alternative interpretation of the E_3 results—one we find fairly persuasive. He makes the strong assumption that people will always tell the truth in preference revelation situations such as these, but that the "truth" consists of a range within which people's valuations of the good fall. He further assumes that the effort required to narrow that range to the true expected value is so great that people will not be motivated to undertake it without some threat of exclusion from the provision of the

[24] Operationally, MG equals E_1 minus the minimum necessary payment, and WTP_{AL} equals E_3 minus the minimum necessary payment.

[25] A larger range of values would be necessary to make an argument as to whether subjects were giving a constant proportional amount.

[26] Unless Brubaker's hypothesis below, which implies that people tell the truth all the time, is true, we can think of no other reason why someone would pay more than \$20 for a good they could have had for \$1.

good. In the absence of a threat, he believes people will express a WTP at the lower bound of their WTP range (rather than at the mean) in order to minimize the risk of overvaluing the good. To the extent that Brubaker's hypothesis is correct, revealed WTP will be a low estimate, unless the consumer can be encouraged to give the valuation exercise sufficient attention. This is precisely the result Bohm obtained in his I–V treatments.[27]

Types of Strategic Behavior

Before summarizing the experimental findings just considered and relating them to contingent valuation surveys, it will be helpful to identify in more detail the types of strategic behavior that economic thought would predict in a CV setting. We earlier stated that according to standard economic assumptions (Schoemaker 1982; Varian, 1984, for example), strategic behavior in CV studies will be a function of (A) the respondent's perceived payment obligation and (B) the respondent's expectations about the provision of the good. Table 6-2 cross-classifies those factors *within* (A) and (B) which we regard as most relevant to contingent valuation surveys. The key distinction within (A) is whether or not the respondent believes provision of the good is contingent upon the WTP amount he reveals to the interviewer. Regarding (B), we posit three possible expectations about the payment obligation: that the respondent will have to pay the actual amount he offers, that the amount he will have to pay is uncertain (it may be more or less than his revealed WTP amount), or that he will have to pay a fixed amount, which may be nominal or even $0. The joint effect of these factors yields six motivational states. These are also identified in table 6-2, as are the presumed direction of the revealed WTP amount relative to the respondent's true value for the good, and the presumed strength of the motivation.

According to this framework, true preference revelation (TP) is motivated when the respondent believes provision is contingent on the revealed preference, and that he will have to pay the full amount he offers. Under these conditions it is in the person's interest to state his true WTP amount, and strategic behavior corresponds with true preference revelation. A very different strategic motivation, minimize effort (ME), occurs when the respondent believes the provision is inevitable and that there will be no relationship between what is offered and what will have to be paid. Here there is no motivation other than to avoid lying to reveal or not to reveal the person's true preferences, and the rational respondent will make as little effort as possible to search her preferences. It is quite possible for CV scenarios to unwittingly create this situation.

[27] Hoehn and Randall (1983, 1985a, 1987) develop this concept in a CV context; see the discussion of their work in the next chapter.

Table 6-2. A Priori Expectations of Strategic Behavior in CV Settings

	Obligation to pay perceived as:		
	Amount offered	Uncertain amount	Fixed amount
	Provision of good perceived as contingent on revealed preference		
Motivation	True preference revelation (TP)	Variable (SB_1)	Overpledge (SB_2)
Direction	True value	Uncertain	Overbid
Strength	Strong	Weak to moderate	Strong
	Provision of good perceived as likely, regardless of revealed preference		
Motivation	Free Ride (SB_3)	Free Ride (SB_4)	Nonstrategic Minimize effort (ME)
Direction	Underbid	Underbid	Random
Strength	Strong	Weak to moderate	Moderate

The four remaining types of strategic behavior (SB) are all thought to occur because the conditions imposed by the CV scenario are sufficient to motivate respondents to under- or overstate their WTP amounts for the amenity. It follows from the standard assumptions that the more confident a respondent is that a good will be provided regardless of the amount he offers, and the more a respondent believes that he will actually have to pay the amount he offers, the greater is the tendency to underbid or free ride in a CV survey. The behavior that is expected to result from this strong motivational state is free riding (SB_3), where the respondent under-represents his true WTP in an effort to reduce his financial obligation for the provision of the public good.

Conversely, the more a respondent believes that the amount she reveals will influence the provision of the good, and the less she believes her payment obligation will determine the amounts she will actually have to pay, the greater her tendency to overbid. This situation results in strategic behavior$_2$ (SB_2), in which the WTP amount is unconstrained by the responsibility to pay, and the optimal strategy for an individual who values the good is to pledge to pay the highest credible amount in the hope that policymakers will be encouraged to provide more of the good.

Strategic behavior$_4$ (SB_4) results in a weak form of free riding. Although the respondent believes provision of the good is likely, no matter what, the incentives to free ride are dulled by uncertainty about whether or not he will have to pay what he offers.

Strategic behavior$_1$ (SB_1) is of particular importance because the assumptions behind it—that the amount of payment is uncertain and that provision of the good is contingent on the respondent's revealed preference—correspond most closely to the incentives created by most CV scenarios. Contingent valuation researchers avoid (or should avoid) any hint that the good they are trying to value is certain to be provided. To respondents, this position should be credible because the provision of public goods in American society is uncertain. With regard to the payment obligation, however, it would be difficult if not impossible for a CV survey to convince respondents that they would have to pay exactly what they offer in the hypothetical market. However, the idea that their payments might be in some way related to what they offer—the uncertainty condition—can be conveyed by CV surveys.

Of the six types of strategic behavior, the SB_1 type is unique because here the direction strategic behavior would take depends on where the respondent perceives the payment uncertainty to lie. If the respondent's uncertainty is over whether or not she will have to pay the whole amount offered, her expected value (probability times amount) of the amenity will be less than the offered amount and the incentive will be in the direction of overpledging. If, for example, the respondent values the amenity at $30 and figures the odds are around 3:1 that she will either have to pay the whole amount or nothing (a 75 percent chance), then the expected payment would be about $23 if true preference reevaluation occurred. According to logic of the kind used by Samuelson, by offering an amount somewhat higher than her true WTP amount the respondent can exert more influence on the provision of the good without being exposed to an expected value payment greater than her true WTP amount. On the other hand, if the respondent's uncertainty is over the percentage of the offered amount that she would eventually have to pay, a different motivational situation ensues. Here, depending on such factors as the person's adverseness to risk, it is possible to imagine situations in which motivation to free ride or to overpledge could be induced.

In using this typology of strategic behavior, it is important to bear in mind the following three points. First, this model is a summary of the a priori theoretical expectations about strategic behavior in contingent valuation settings, and thus does not take into account all the factors that militate against strategic behavior. Second, the payment and provision *perceptions* in the typology are just that: they involve the respondent's subjective understanding of the scenario proposed in the interview. What ultimately matters, to mention a theme we will emphasize later, is not whether the researcher's scenario is formally correct, but whether the scenario as understood by the respondent is correct. Finally, the predicted direction of the strategic behavior in the four SB cells is based on the assumption that the respondent has a positive value for the good—an assumption that holds in regard to many public goods. Different motivational patterns for SB_1 and SB_2 would be expected from respondents who were indifferent to or opposed to providing the good being valued.

Interpretation of Experimental Findings

We are now in a position to evaluate the degree to which the experiments considered in the first part of this chapter support the assumptions about strategic behavior that are embodied in standard economic theory. The first set of experiments we examined tested the ability of incentive-compatible demand revelation devices to restructure incentives so that the subjects' strategic response would be true preference revelation (TP) instead of free riding (SB_3). They were conducted under controlled laboratory conditions where counterstrategic incentives, such as altruistic or honesty norms, were deliberately minimized. The finding of interest is that where no special incentives were used to induce people to reveal their true preferences, voluntary revelation was as successful as the various ICDRDs. The explanations offered for this phenomenon range from the cost imposed by strategizing to the presence of strong counterstrategic behavior motives.

The findings of those experiments on public good decisions for which we can estimate the percentage of true WTP measured are summarized in table 6-3. Most of the experiments induced SB_3, or the strong free-riding condition. Bearing in mind that these SB_3 experiments were deliberately designed to motivate strategic behavior, the failure of a single experiment to confirm the *strong* free-riding hypothesis is striking. Although in none of the experiments did respondents reveal 100 percent of the true amount they were willing to pay, none revealed less than 61 percent of that amount. Particularly noteworthy are the findings of the two treatments (Schneider and Pommerehne, 1981; Brubaker, 1982) which explicitly guaranteed individual participants that they would receive the good no matter what they bid. The participants in these experiments seemingly faced no information cost, and the effort required to devise a strategy was surely minimal, yet they offered between 61 and 71 percent of the true value for the good.

These data also show that the introduction of a group exclusion condition clearly raises the percentage of the true optimum revealed. Being tied

Table 6-3. Percentage of True WTP Measured in Experimental Studies

Type of behavior	Behavioral motivation	Experiment or treatment		Percentage of true WTP[a]
Strategic $behavior_1$	Variable	Bohm III[b]		71
Strategic $behavior_2$	Overpledge	Bohm IV		74
		Bohm V		85
Strategic $behavior_3$	Strong free ride	Bohm I		74
		Schneider and Pommerehne	c	96
		Brubaker	c	80
		Christainsen	c	79
		Marwell and Ames	c	84
		Schneider and Pommerehne	d	61
		Brubaker	d	71
Strategic $behavior_4$	Weak free ride	Bohm II		85

[a] In every case but one, the true willingness-to-pay criterion was measured in an auction treatment where the winning bidder(s) received the good.

[b] The uncertainty Bohm induced was a random chance that the person would either pay nothing or the full amount. As noted in the text, the motivation for this type of uncertainty is overpledging.

[c] In these experiments provision of the good was dependent on the group meeting the goal. In the Marwell and Ames study, the criterion for no free riding was 100 percent contribution to the group good.

[d] In these experiments provision of the good was guaranteed.

Sources: Bohm (1972), Schneider and Pommerehne (1981), Brubaker (1982), Christainsen (1982), Marwell and Ames (1981).

to the fortunes of a group in the maximum free-riding condition increases the amount revealed by an average of 15 percent.

Bohm is the only researcher to have tested for the presence of strategic behavior under any of the other conditions believed to motivate free riding or overpledging. It should be noted that while the payment and provision conditions of his treatments I through V provided incentives for various types of strategic behavior, Bohm did his best to induce the respondents to disregard the strategic incentives through the use of moral exhortations and by making public the amount that people bid to see the TV program. This effort was successful, as the amounts revealed by his respondents did not vary significantly in the direction predicted by expected utility assumptions. Particularly suggestive of people's relative insensitivity to strategic behavior motivations is Bohm's finding for treatment V: that overpledging did not occur when people were explicitly told (1) they would not have to pay anything, but (2) their revealed amounts could influence the provision of the good.

Unfortunately, the findings for SB_1, SB_2, and SB_3 rest on just one set of experiments and on the correctness of our assumption that Bohm's VIb treatment captures the true WTP amount.[28] Clearly more experimental research on overpledging is warranted.

Incentive Compatibility in Political Markets

Thus far we have considered incentive compatibility when public goods have been valued in the context of the private goods market model. Another body of research has investigated the incentive structures of various types of political markets and voting systems and the degree to which they would motivate rational individuals to reveal their true preferences. This research is of special relevance if a political market such as a referendum is used as the model for CV studies. Theorists have long been interested in how voting rules are related to the optimal provision of collective goods. Studies have found that almost all voting systems suffer from the

[28] For a somewhat different interpretation of Bohm's findings, see Vernon Smith (1979). Smith argues that single-response studies such as Bohm's will be less sensitive to the treatment variable than studies (such as those conducted by Smith and others) that use iterative procedures, because subjects in single-response studies are less familiar with what is being asked of them than subjects who have experienced the procedures repeatedly. Smith also appears to accept Bohm's judgment that the VIb results are invalid. With regard to the single-response versus iterative issue, we believe that many choices people are forced to make in real life (in a referendum, for example) are more appropriately modeled as one-time, unique decisions rather than as repeated decisions (such as those made by consumers who purchase bread or shampoo). Coppinger, Smith, and Titus (1980), for instance, found that it often took six trial runs of some of their auction mechanisms, with immediate feedback, before their subjects "learned" what type of behavior was advantageous for them.

possibility of coalition formation and from strategic voting in multistage balloting schemes (Zeckhauser, 1973; Sen, 1986).[29] Control of the agenda (for example, setting the quantity and tax price to be voted on in a referendum) and setting the conditions under which the vote will be used to make a decision (the plurality rule, for example) have been among the most frequently used elements of optimal provision voting systems.

However, this research has shown that there is one voting format which does motivate people to reveal their true preferences. It uses a single yes or no vote on a proposition to provide a fixed level of a public good for a specified tax price (possibly different for different voters). The best the voter can do is to vote yes if her willingness to pay is greater than the tax price and to vote no if her willingness to pay is equal to or less than the tax price (Zeckhauser, 1973).[30] Because of the one-time nature of the vote, the individual voter regards control of the quantity of the good offered and the tax price as exogenous, thus avoiding the issue of agenda formation.[31] Yet under such conditions, with a very large number of voters, it has been shown that there is no "rational" reason for an individual to vote, as her chance of influencing the final decision goes to zero as the number of voters increases to infinity.[32] Voting is assumed to be motivated by a sense of social and civic responsibility.[33]

Summary and Conclusions

We have examined that theoretical and empirical literature from several disciplines which bears on the assumption that surveys cannot satisfacto-

[29] Log-rolling and vote trading by politicians are examples of coalition formulation. A common example of a multistage vote scheme is the primary system in the United States, where the absence of a majority vote for any candidate in a primary requires the top two candidates to face each other in a runoff election. Strategic behavior at the first stage might be to vote for a candidate other than the one most preferred in the hope that the preferred candidate might thereby face a weaker opponent in a runoff election.

[30] Hoehn and Randall (1987) have used this result in the context of a single take-it-or-leave-it CV question (see chapter 7).

[31] If the vote on one referendum can influence the quantity of the good or the tax price in a future referendum, the incentive-compatibility property is lost (Gibbard, 1973; Satterwaite, 1975).

[32] Chamberlain and Rothschild (1981), considering the probability of casting a decisive vote and the convergence properties as the number of voters grows, show that if the number of voters is $2N + 1$, the probability of casting a decisive vote is on the order of $(1/N)$.

[33] Brennan and Buchanan (1984) argue that people may realize they have no chance of influencing the voting decision and thus may vote symbolically rather than voting their true preferences. We regard their arguments as another way of saying that people may behave in a more public-spirited manner in political markets than they do in private goods markets. Brennan and Buchanan are disturbed by the implications of this type of behavior for the use of voting outcomes to reveal economic preferences; we are not.

rily measure people's preferences for public goods, given strategic behavior. Our review of the theoretical literature showed that the free-riding behavior postulated by Samuelson is irrational outside of a naive, one-time prisoner's dilemma-game representation.[34] Where recurrent valuation or the possibility of it is postulated, free riding does not usually result in a stable Nash equilibrium, much less a dominant strategy; nor does it depict a realistic disequilibrium process. The incentive-compatible demand revelation literature demonstrates that formulation of the demand revelation problem is very complex. If side payments (individual taxes or subsidies) cannot be used, there is no way to guarantee truthful demand revelation unless the noncooperative behavior assumption is dropped or unless dishonest responses have positive costs. The ICDRD literature also suggests a number of features which tend to enhance truthful revelation of WTP amounts.

We then turned to recent theoretical arguments which identify incentives for cooperative behavior and suggest a plausible mechanism for the evolution of cooperative behavior from noncooperative behavior. For example, Ackerlof's recent incorporation of several fundamental psychological principles into a standard economic framework demonstrates that honesty in individuals yields positive long-run returns which are undermined by any lapse into dishonest behavior. The implication of these considerations is that unless there are very strong incentives, strategic behavior would be the exception rather than the rule.

We next examined the empirical evidence on the free-rider problem and the tendency toward cooperative or competitive behavior indicated by experimental economics. The results of these studies are consistent with the notion that strategic behavior is not inevitable in preference-revelation situations. Despite attempts by experimentalists to induce free riding, this form of behavior was the exception rather than the rule in all the experiments reviewed. Indeed, voluntary demand revelation performed as well as the more complex incentive-compatible demand revelation devices, and often better. The risk of potential group exclusion from provision of the good was identified as bringing the revealed WTP amount in the voluntary case near the true WTP amount, even when substantial immediate payment was required or anticipated. Two recent experiments (Schneider and Pommerehne, and Brubaker) provided strong support for the hypothesis that lying is not costless. Overall, the comparisons between true payment conditions and those which encouraged free riding suggest that under

[34] The two-person, zero-sum game theory (and the strategically equivalent constant-sum game) is often used as an alternative to the prisoner's dilemma formulation in modeling market behavior and military tactical problems. We note Shubik's warning in the *Handbook of Mathematical Economics* (1981) that this theory, too, is of limited value to the study of political economy because extremely few situations meet the conditions of pure opposition.

experimental conditions free riding accounts for a modest downward bias in the WTP amounts of about 10 to 30 percent. It was emphasized that these experimental findings come from a growing body of very consistent results obtained under a wide array of circumstances.

We also showed that the discrete-choice referendum model was incentive-compatible in the sense that a person could do no better than vote yes if her WTP amount for a good being valued by this approach was at least as large as the tax price, and to vote no if this was not the case. This finding offers the possibility of framing contingent valuation questions so that they possess theoretically ideal and truthful demand-revelation properties.

Overall, the theoretical and experimental evidence examined supports the view that strategic behavior is not nearly as severe a phenomenon in consumer decision making about public goods as many economists had feared. In the next chapter we turn to the evidence for the presence of strategic behavior in CV studies, and to how CV studies can minimize strategic responses.

7

Strategic Behavior and Contingent Valuation Studies

The picture of consumer behavior supported by the theoretical and empirical material presented in the preceding chapter contrasts sharply with the notion that consumers will inevitably lie when asked about their preferences for public goods. In this chapter we pursue the matter further by examining the role of strategic bias in contingent valuation studies themselves.

Expectations for Strategic Behavior in CV Studies

We begin by relating current CV practice to the six types of strategic behavior identified in table 6-2 and by exploring in more depth the motivational incentives to act strategically in the CV setting. Compared with the experiments described in chapter 6, in which we found only a low level of strategic behavior, here we find that strategic behavior in most properly designed CV surveys should be even lower.

Motivational States Induced by CV Scenarios

Of the six motivational types created by belief in different combinations of provision and payment conditions (table 6-2), four are not directly relevant to current CV practice, although it is possible to conceive of situations where some might be relevant. TP is not relevant because of the absence of markets. In valuing public goods, SB_3 is not relevant because it is difficult for CV studies to create the impression that respondents will have to pay the exact amount they offer. The assurance of anonymity given to respondents at the start of the CV interview, and the impracticality of an individualized collection system, render such a condition highly implausible. SB_4 and ME, as well as SB_3, are not relevant whether the amenity is a quasi-private or a public good, because CV studies avoid suggesting that the good will be provided whatever the respondent says, and there is no evidence that this contingency lacks credibility. This leaves two motivational states that are relevant to current CV practice—SB_2 and SB_1.

Strategic Behavior$_2$ (SB$_2$). Although atypical and undesirable, SB_2 is not unknown in CV studies. In the pay-fixed-amount condition, the respondent is told what everyone will have to pay before being asked how much the individual is willing to pay. In a study reported by Cronin (1982),[1] which valued different levels of water quality in the Potomac River, one subsample of local residents was told:

> Finally, let me inform you that the federal government will finance the cost of the pollution control. Your federal taxes will not vary more than a few pennies due to any decisions made on the size of the water pollution control effort in the Potomac.

This statement effectively assured respondents that their only payment obligation would be nominal, no matter how much they said they would be willing to pay. Under a fixed payment contingency, if the respondent values the good and believes provision of the good to be contingent on revealed preference, a strong motivation to strategic behavior$_2$ is produced. Instead of trying to influence the payment amount, as in the free-riding case, the respondent who values the good will be motivated to increase the likelihood of its provision by bidding an amount higher than his or her true WTP amount. In Cronin's study, Washington-area residents were free to bid more than they were willing to pay without fear that they would be held personally liable for the amount of their pledge.

Strategic Behavior$_1$ (SB$_1$). Given the inapplicability or impossibility of convincing respondents that they will have to pay the actual amount they offer, and the undesirability of telling them they will have to pay some fixed amount, the optimal practice (and the prevailing one) in CV studies is to use scenarios which imply a payment obligation but do not completely spell out the details. The wording "How much are you willing to pay in taxes and higher prices for an increase in X from . . . to . . ." is typical of many CV scenarios. It strongly implies a payment obligation (an implication that is reinforced when the context is a government-sponsored study and when a realistic payment vehicle is used), yet the relationship between the elicited amount and the actual payment structure to be enforced is otherwise unspecified. Other scenarios emphasize the payment consequences even more, but nevertheless are also vague about the relationship between stated and actual payments. For example, a second subsample in

[1] Cronin analyzed a data set collected for the Washington, D.C., Metropolitan Council of Governments. See also Davis (1980).

the Cronin study was told: "Finally let me emphasize that local households will have to pay the cost of the pollution control and that your response *will* influence the level of your local taxes" (Cronin, 1982; emphasis in the original). Scenarios such as these convey the idea, either implicitly or explicitly, that the amount of the respondents' eventual payments will depend on use of some arbitrary formula or process that spreads program costs among the citizens. They minimize, but do not eliminate, the motivations to engage in strategic behavior.

When it comes to eliciting amounts which are as close as possible to the respondent's true WTP amount, SB_1 is clearly superior to SB_4 because it is closer to the TP condition. The belief that provision of the good is assured can only weaken a person's motivation to devote the necessary thought and care to determining his or her true WTP.[2] The uncertainty of SB_1 regarding provision would be expected to work in the opposite and desirable direction. This is why SB_1 is almost always the type of motivational state specified in CV scenarios. In chapter 6 we noted that SB_1 is unique because in that state the incentives to behave strategically depend on the perceived uncertainty of the payment obligation,[3] and on assumptions about such factors as risk adverseness.

We believe the motivation for strategic behavior will be weak for most CV respondents because of the following considerations:

1. The informational requirements for strategic behavior are great.
2. CV surveys normally convey to respondents the impression that a large number of people are being interviewed. The perceived likelihood that overpledging by a *credible* amount will affect the outcome significantly, when it is averaged over a large number of people, is low.
3. Most of the payment vehicles customarily used in CV studies, such as taxes, higher prices, or higher electricity bills, appear to evoke strong budget constraints and other negative reactions. People do not lightly pledge to increase these kinds of payments, even in hypothetical situations.
4. Underpledging carries the risk that the public good will not be provided, as CV surveys normally present as a choice a situation in which provision is credibly uncertain.

[2] The SB_4 assured-provision condition can be shown to encourage free riding wherever there is any possible relationship between the revealed WTP and the actual eventual payment obligation.

[3] If the respondent's expected payment obligation is perceived to be less than the amount that he states, he will tend to overpledge; if his expected payment obligation is larger than the stated amount, he will tend to underpledge or free ride.

Each of these factors is consistent with the rational utility-maximizing economic consumer postulated by standard economic theory. The roles of other, extrarational, motivational factors also argue in favor of CV's ability to hold strategic behavior to an acceptable level. Before advancing to our heuristic model of strategic behavior motivation, we turn to a recent attempt to develop a rigorous model of the phenomenon.

The Hoehn and Randall CV Behavior Model

John Hoehn and Alan Randall have proposed a theoretical model of how respondents in contingent valuation interviews will react strategically to various features of survey instruments that are focused on the SB_1 condition.[4] On the basis of this model they reach several conclusions. One is that the WTP amounts an individual states in a CV survey will tend to undervalue true willingness to pay, while the WTA amounts given in a CV survey overstate true willingness to pay. Another is that the discrete-choice take-it-or-leave-it format is incentive-compatible: a respondent can do no better under it than to say yes if the payment obligation is less than his willingness to pay for the good. Finally, they conclude that the greater the attention a respondent gives to a CV scenario, the greater the convergence between WTP and WTA amounts toward his true values.

Hoehn and Randall ground these conclusions on a number of assumptions. In addition to those typically made in the incentive-compatibility literature, they assume that individuals are to some degree uncertain about how much satisfaction they would receive from a change in the level of the public good being valued in a CV study, that respondents are to some degree risk-adverse to paying more for the good than it is worth. Hoehn and Randall also treat the time and effort devoted to the valuation exercise as factors which constrain the preference-searching process. Further, they make assumptions that limit the possible ways in which respondents might resolve uncertainty over the payment obligation. Hoehn and Randall see respondents as either (1) believing they will have to pay the amount they state (this moves the incentives toward the TP or free-riding motivational states); (2) believing that the amount they have to pay is in some way proportional to the amount they state (which also tends to move the incentives toward the TP motivational state, especially when coupled with a belief that the project will be undertaken using a benefit-cost rule based on the amounts stated by all respondents); or (3) believing they will have to pay the cost on a per-capita basis (which leads to TP under any type of pluralistic voting implementation rule, and to some potentially strange behavior under a benefit-cost implementation rule). Use of a voting rule,

[4] See Hoehn (1983), Hoehn and Randall (1983, 1985a, 1987); see also Randall (1986b), and Hoehn and Sorg (1986).

which respondents tend to find most plausible, can obtain true preference revelation under certain conditions, but does so at the expense of forsaking Pareto optimality (Zeckhauser, 1973). Hoehn and Randall's assumptions of uncertainty, incomplete optimization, and risk aversion, however, tend to cause true preference revelation to diverge toward cautionary underpledging if preferences are elicited by the WTP (as opposed to the WTA) format.[5]

The Hoehn and Randall model is a useful, logically rigorous attempt to develop a theory of how the internal *economic* incentives motivate respondents in CV surveys. In its present form, however, the model suffers from several limitations. The first is generic to the incentive-compatibility literature in that the models in this literature ignore the other, less immediately economic factors that influence respondents' behavior in the survey situation. As Hoehn and Randall recognize in the most recent version of their model, the results they obtain are dependent on the assumption that respondents believe everything they are told in a survey—an assumption that is often false (see chapter 10). For example, if a respondent believes for some reason that he will escape having to pay the tax price he states (or some function of it), he may well give a WTP response much in excess of his true willingness to pay, a behavior which cannot be accounted for by the Hoehn and Randall model. Thus, a poor person renting an apartment in a public housing development is likely to have little disincentive to give a high WTP amount for an apartment-related amenity if he perceives it would be paid for by the federal government.

A second limitation, also acknowledged by Hoehn and Randall, is that their results are expressed in terms of expected values. Thus the *inequalities* Hoehn and Randall present—that WTP (as revealed in a CV survey) is less than or equal to true willingness to pay, and that revealed WTA will be greater than or equal to true WTA—may not be true for any particular CV survey, although they will be true on average. This property is inevitable as soon as one concedes that there is a random component to CV responses which is influenced by the survey instrument (see chapter 10). It is a property that makes the empirical testing of Hoehn and Randall's model difficult.

The Hoehn and Randall model leads to the conclusion that the internal economic incentives in a CV survey tend to make contingent valuation WTP amounts less than a perfect measure of a respondent's true WTP for

[5] Harstad and Marrese (1982), using an incentive-compatible revelation device in a laboratory setting, found strong evidence that subjects held suboptimal reservation prices (that is, their WTA offers were higher than those that would be in their best interest) and were satisficing (paying more than they had to). These findings support the Hoehn and Randall behavior model.

the good, a result with which we concur. We do not, however, believe that this behavioral model justifies statements to the effect that the WTP amounts obtained by CV surveys *always* underestimate true WTP, although such underestimation is likely to occur in many instances. Nor do we agree with Hoehn and Randall's conclusion that WTP amounts obtained in CV surveys through iterative yes-no elicitation methods (the iterative bidding game, for example) are superior to values obtained by noniterative elicitation methods. Hoehn and Randall reach this conclusion by extending their finding that a single take-it-or-leave-it offer is incentive-compatible to find in turn that a series of take-it-or-leave-it offers is also incentive-compatible; and by assuming that the iterative bidding game maximizes the time and effort respondents spend on thinking about the value (to themselves) of a specified quantity of a public good. One problem with Hoehn and Randall's argument is its implicit assumption that respondents treat the iterative discrete-choice situations as independent and exogenous. In our experience, respondents quickly realize the relationship between their answers to a take-it-or-leave-it question and follow-up questions.[6] Another problem is that in advocating iterated elicitation methods Hoehn and Randall fail to take into account the well-documented response-effect biases associated with these methods, such as starting point bias (see chapter 11).

A Heuristic CV Behavior Model

Generally speaking, CV studies place respondents in a hypothetical situation similar to that presented to participants in many of the experimental tests of strategic behavior described in chapter 6. In both cases individuals are told that they are one of a group of people being asked to value a nonmarketed good, that most people desire the good, and that it is not customarily available in a market. In both situations the possibility of direct group pressure is eliminated because people value the good without having contact with anyone other than the experimenter, and strategic calculations are made difficult because respondents are not given information about the size of the amounts stated by others before offering their own WTP amounts. Another similarity is the correspondence between CV scenarios and the group exclusion provision used by those experimental treatments that showed the lowest amount of free riding.[7]

Yet there are differences between contingent valuation studies and the experiments which bear on CV's potential for strategic behavior. For

[6] The take-it-or-leave-it offer immediately loses its incentive-compatibility property as soon as the agenda becomes manipulatable (that is, the future price or quantity offered is influenced by a respondent's previous answer).

[7] Group exclusion provisions, when used in CV surveys, are generally invoked by saying that the good will not be provided unless everyone is willing to pay for it.

example, subjects in experiments are often clearly informed about the conditions under which the good will be provided, and provision is guaranteed under these conditions. In Bohm's treatment II, respondents knew they would be able to see the television show for an amount proportional to their bid if the aggregate bids reached 500 kronor. This promise was credible because those promising the good (a university teacher and the Swedish national television company) were both reputable and in a position to provide the good. In CV studies, however, where the good can rarely be provided on the spot, the respondent must subjectively formulate the probability of provision of the public good, and the level of payment that ultimately would be required from him if it is provided. CV surveys also can, and generally do, credibly convey the condition that everyone will have to pay for the good if it is provided for in their taxes or in higher prices for consumer goods. This condition eliminates much of the incentive to free ride that is a prominent feature of experimental tests of strategic behavior.

In examining the implications of these and other differences between experimental and CV survey methods for motivating strategic responses, we begin by considering the likelihood of engaging in strategic behavior in the experimental setting as a function of the expected gain from engaging in strategic behavior, the expected cost of developing such a strategy, and any internal disincentives such as an inclination toward honest or cooperative behavior.[8] Advancing this argument heuristically rather than rigorously, we can formulate the likelihood of behaving strategically in the experimental setting as

$$L_S = f\,[EG,\ EC,\ ID], \qquad (7\text{-}1)$$

where L_S equals the likelihood of behaving strategically, EG equals the expected gain from such behavior, EC equals the expected cost of formulating the strategy, and ID equals internal disincentives to behave strategically.[9] In the CV setting we get

$$L_S = f\,[(P)EG,\ EC,\ ID], \qquad (7\text{-}2)$$

where P is a discount factor which is the respondent's subjectively formulated probability that the outcome of the CV survey will completely deter-

[8] Since the payoffs from actions are real in both the experimental setting and the real world, there is no difference in the monetary dimension between the two settings.

[9] Internal incentives to behave strategically are presumed to be absent in situations where there are no social norms against such behavior, as in gambling. All of the social sciences, except economics, assume there are social norms that encourage truthful responses to survey questions.

Table 7-1. Public Perceptions of Government Attention to the Views of Individual Citizens on Environmental Issues

Answer	Potential for strategic behavior	Percentage
A great deal	Highest	5
Some	Moderate	31
Little	Low	39
None at all	None	20
Other	Unknown	5

Source: Mitchell (1980b).

mine the provision and payment for the public good. Examination of each component on the right-hand side of this equation suggests that, other things being equal, a contingent valuation study will produce less strategic behavior than the comparable experimental treatment.

In the experimental setting, P is almost always equal to 1, whereas in the contingent valuation setting, owing to its lower consequence realism, $1 \geq P \geq 0$, a difference which reduces the incentive to act strategically in CV studies.[10] Since many of the respondent's internal calculations turn crucially on the value of P, is there any evidence available on the size of P? One piece of indirect evidence suggests that P is relatively low. The Council on Environmental Quality's 1980 *Public Opinion on Environmental Issues* (Mitchell, 1980b), whose introduction notes that the survey was done for the government, asked a national sample:

According to what you have heard or read or know from your own experience, how much attention do you think the Federal Government gives to the *views of individual* citizens like you on environmental issues? [A great deal, some, little, or none at all].

The percentage of those respondents who believed the federal government pays a "great deal" of attention to the views of individual citizens was very low, whereas a majority believed the government paid "little" attention or "none at all" (see table 7-1). Both findings are consistent with the arguments above. We note that the situation posed to the respondents in this question was very general. However, CV studies of broad public policies (such as national air and water policy) are similarly general. For local public goods the value of P is likely to be context-specific.

[10] We have implicitly assumed $\partial L_S / \partial P > 0$. It should be noted, however, that decreasing P increases the likelihood of random behavior. Further, a decrease in P which does not influence EC, EG, or ID is difficult to achieve. The easiest way to reduce P is to deemphasize the payment obligation, an approach that effectively decreases EG-EC so much that $\partial L_s / \partial P < 0$.

One of the key provisions of almost any contingent valuation scenario is that *everyone* will have to pay something if the good is provided. Thus, in addition to $1 \geq P$ in a CV survey, EG may be smaller even if P is taken to be 1 by the respondent. This characteristic essentially rules out the possibility of strong free riding by respondents in CV surveys, provided the scenario either implicitly or explicitly invokes the coercive power of the state to set entrance fees, issue regulations, regulate utility rates, or raise taxes. To some degree this feature of CV surveys mimics the matching provisions advocated by Guttman (1978) as a way of reducing strategic behavior. The assured "average-type" payment in the event of provision also tends to eliminate two of the motivations for strategic behavior: greed or the gain from free riding, since payment is enforced and not directly tied to the revealed amount; and fear that after paying one will not receive the good.[11] Our review of the experimental economics literature shows that almost every case of significant free riding has occurred when some subjects observed other subjects successfully free riding—a condition that cannot occur in a contingent valuation survey. It should be noted that EG is a function of the method used to elicit the WTP amount. At one extreme is the single (believed) take-it-or-leave-it offer which sets EG to zero. At the other, any method that directly elicits the respondent's maximum WTP amount allows EG to assume its largest possible value.

We turn now to the remaining variables in the equation. If the experimental treatment and CV survey have identical features, the expected gain from acting strategically (EG) will be the same for each, but both EC and ID will differ, and will differ in ways which should reduce incentives to act strategically in a CV survey. The cost of formulating a strategy (EC) is a particularly important factor in reducing the motivation to behave strategically in CV studies (Kurz, 1974; Brookshire and Eubanks, 1978). To formulate such a strategy, a respondent would have to know the sample size and the mean WTP amounts for all the respondents, including those already interviewed and all who remained to be interviewed. In contrast to CV studies, in experiments the sample size is sometimes revealed to participants, or can be fairly accurately inferred by the subjects. Some experimental treatments provide the subject with feedback on the responses of other members of the group. A respondent in a CV study may adopt some less information-intensive strategy, but this will almost always reduce expected net gain, $EG - EC$, below the achievable in the experimental setting. The internal disincentives (ID) to strategic behavior are also likely to be somewhat higher in CV studies than in experiments, particularly in those CV studies which use personal interviews. The presence of a trained, neutral, professional interviewer is usually assumed to invoke

[11] See Palfrey and Rosenthal (1984).

norms of honesty and cooperation that militate against dishonest behavior. Social pressures to appear neither cheap nor frivolous seem to be counterbalancing.[12]

The experimental studies conducted by Marwell and Ames (1981), Schneider and Pommerehne (1981), and Brubaker (1982), described earlier, demonstrate the importance of group exclusion for eliciting optimal WTP amounts. These researchers' scenarios specified that unless the participants produced a sufficient group WTP, no one would receive the good. Although most CV studies to date have not emphasized group exclusion, it is often implicit (and sometimes stated) in them that unless people are willing to pay enough for public goods, the goods will not be provided for anyone. Group exclusion is particularly important in CV studies based on the referendum model, with their collective decision-making framework and institutional means for providing the good.

Compared with the experiments, which show a low level of strategic behavior, we conclude from the evidence that strategic behavior in properly designed CV surveys should be even lower. The evidence is sufficiently extensive and consistent, we feel, to shift the burden of proof to the shoulders of those who would continue to challenge the validity of all CV studies on grounds of strategic behavior. This judgment should not be taken to imply that the CV analyst can totally dismiss the possibility of bias caused by strategic behavior. Although the prevailing payment provision in CV surveys, SB_1, is one of the least likely to elicit strategic behavior, some types of CV studies are more prone to strategic behavior than others. Strategic bias should take its place as one among other types of bias which may pose problems for improperly designed CV studies.

This completes our review of the evidence against the assumption made by some economists that the threat posed by strategic behavior on the part of respondents to CV surveys renders the method invalid. We now turn to the important question of what conditions promote strategic behavior in CV studies and to several related questions, including the likely impact of strategic behavior on CV benefit estimates if it does occur.

Controlling Strategic Bias in CV Studies

Conditions that Promote Strategic Behavior in CV Studies

The likelihood of any respondent engaging in strategic behavior (L_S) in a CV study lies along a line or continuum whose end-points are defined by L_S equals 0, when no discounted gain is expected from strategic behavior

[12] The presence of the interviewer may promote other biases, such as yea-saying, where the respondent answers yes to a question because he or she thinks it the socially appropriate response. These and other related biases are discussed in chapter 11.

$[(P)(EG) = 0]$; and by L_S equals 1, when an arbitrarily large discounted net gain from strategic behavior is expected. We hypothesize that respondents may be divided into four groups on the basis of their values for L_S, each with a differing potential for engaging in strategic behavior:

I. Those for whom $P = 0$ or $EG = 0$ or both. These respondents are located at the lower end-point.

II. Those for whom the initial comparison (presumed close to costless) of $[(P)(EG) > 0]$ and expected cost (including internal disincentives) indicates that net gain is negative or 0. These respondents will not engage in strategic behavior.

III. Those for whom the initial benefit-cost calculation indicates a possible net gain from strategic behavior, but whose revised estimate after determination of a strategy indicates a 0 or negative net gain. This group also will not behave strategically.

IV. Those for whom the initial benefit-cost calculation indicates a possible net gain and whose revised estimates, after determining a strategy, still perceive a positive net gain. This group will attempt to engage in strategic behavior.

For the reasons stated earlier, we believe that the first two groups contain the vast majority of the respondents to most CV studies.

There are situations, however, where a significant number of respondents might fall into group IV, unless special precautions are taken. If the good being valued is one from which respondents can be excluded by imposition of a user fee, such as access to a park, obtaining a hunting permit, or provision of a social service, the users or potential users of the good might be highly motivated to minimize the chance that a fee would be imposed, or, failing that, to minimize the size of the fee. If respondents believe the study will result in (high P) a fee being imposed on a good whose provision is not in jeopardy, then (1) the more specific the good being valued (a particular park, for example), (2) the greater the use of the good by the respondent, and (3) the lower the existing fee, the higher the $[P \cdot EG]$ for understating WTP. These conditions would dispose respondents to the free-riding condition SB_4 (see table 6-2). Conversely, if the probability that a fee will be imposed is perceived as low, and if the probability of government provision of the good out of general revenues is perceived as high, the same factors would predispose users to overpledge, as in SB_2.

In these examples, three factors appear to be particularly important in increasing the motivation to act strategically. The first two factors increase $[P \cdot EG]$; the third reduces ID. The first is the salience of the good to the respondent: specific goods currently used by respondents have an immediacy which can make them quite salient to these individuals. The second factor is the potential for being excluded from the good: if a respondent believes he or she may have to pay a user fee to gain access to the good, a payment obligation closely approximating the actual amount stated is created along with (where provision is assumed) a greater propensity for free riding. The third factor involves the assumed property right: if the good is currently available for a low fee or no fee, respondents may implicitly (or explicitly) claim a property right to its use and regard a perceived attempt to price it as illegitimate. This belief could nullify the norms which usually motivate CV respondents to truthful behavior.

Roberts, Thompson, and Pawlyk's (1985) study of the value of offshore oil and gas platforms for sport fishing illustrates the potential for strategic behavior and some of the strategies the researcher can use to minimize its effect. They employed a telephone CV survey to determine the value to recreational scuba divers of leaving abandoned offshore drilling rigs in place in the Gulf of Mexico. These rigs provide a habitat for an abundance of sea life. Recreational divers currently enjoy free access to this resource. The study's sample consisted of divers who had used the platforms in the year previous to the study. These respondents were asked to say how much they would be willing to pay for a hypothetical annual pass to the offshore structures, if such a pass were required. Recognizing the potential for the respondents to free ride, the researchers sought to lower the value of $[P \cdot EG]$ by assuring respondents that there "are no plans to charge for the use of the rigs."[13] They attempted to add credibility to this claim by explaining that they were "looking for a way in which we can place an economic value on the sport diving activity around Louisiana's offshore oil and gas structures." The delicate balancing of incentives to answer truthfully in scenario construction is well illustrated by this maneuver, since the attempt of these researchers to avoid the SB_3 type of free riding ran the risk of stimulating the overpledging of the SB_2 type. The fact that they got both very few $0 and very few high WTP amounts suggests that they were successful in avoiding strategic responses.

Each of the three types of survey administration—mail, telephone, and in-person interview—has characteristics that may encourage or discourage strategic behavior. Mail surveys are perhaps the most likely to encourage such behavior by decreasing *EC*, increasing *EG*, and decreasing *ID*. This is because the individual receiving the survey has time to formulate an optimal strategic response; has access to the entire questionnaire, which facilitates determining its purpose and ultimate use before answering the

[13] It could be argued that this procedure also increased *EC*, since introduction of this assurance made the optimal strategy somewhat less obvious.

WTP questions; and has only an impersonal communication channel to the researcher, which lessens the cost of directly lying to another individual.[14] A telephone survey, in contrast, is likely to discourage strategic behavior because its short length reduces the available time for strategizing. In addition, both the telephone survey and the in-person interview allow the researcher to control the flow of information and the sequence of questions to be answered. The contact with the respondent is more direct in the telephone survey than in the mail survey. This factor, which is likely to be particularly important in discouraging strategic behavior, is most present in the personal interview where the interviewer has been invited into the respondent's home.

Testing for Strategic Behavior

Three different types of field test for strategic behavior have been reported in the CV literature. None of them are able to definitively measure the presence of strategic behavior, however, because they cannot distinguish it from other possible types of response behavior. Nonetheless, the three types of test are so constructed that the failure to find strategic behavior is a good indicator of its absence. In all but one application of strategic behavior tests there have been negative findings, which supports the view that strategic behavior is not likely to present a problem in most contingent valuation studies. The one application for which a finding of strategic behavior was claimed is not particularly relevant to the way most CV scenarios are posed.

The first type of test provides respondents with the additional information they need to act strategically, and offers them the chance to change their WTP amounts accordingly. In the only application of this test, Rowe, d'Arge, and Brookshire (1980) first had their respondents reveal their initial WTP amounts for air visibility improvements in the United States Southwest; each respondent was then told that the other respondents had offered a specified mean bid. This specified mean bid was individually determined so that all the respondents, including those who may have acted strategically in giving their initial bids, could gain by revising their bids. Respondents were then encouraged to revise their stated WTP amounts in light of the new information. The drawback to this procedure as a test of strategic behavior is that the respondents could have used the mean WTP amount as a price and therefore as an indicator of the true value of the good (see the discussion on implied value cues in chapter 11). If they made this assumption, it would be rational for them to use the new

[14] The only CV studies we are familiar with whose responses may have been affected by strategic behavior were conducted by mail (Carson and Mitchell, 1983; Carson, Graham-Tomasi, Rund, and Wade, 1986).

information to honestly reappraise their initial WTP amounts.[15] Nevertheless, while revised bids might be caused by factors other than strategic behavior, the absence of revised bids is an excellent indicator of the absence of strategic bias. In the Rowe, d'Arge, and Brookshire study, out of 40 respondents only 1 submitted a revised bid when encouraged to do so. It is also noteworthy, in light of the experiments where economists were found to behave differently from noneconomists when given the opportunity to free ride, that the one person who did possibly act strategically was a randomly sampled economics professor.

The second type of test is a distributional one, first proposed and implemented by Brookshire, Ives, and Schulze (1976). They assumed that the distribution of true bids was normal, and made an intuitive argument that there are only two reasonable strategies available to a respondent who wants to move the survey's mean value closer to her own value for a good but does not know what the sample mean value will be without her response. If the respondent believes that the sample's mean WTP will be above her WTP amount, the appropriate strategy is to give the lowest possible WTP amount, such as $0. If she believes the sample's mean WTP amount will be below her value, the optimal strategy is to give the highest valuation she feels the researcher will believe. Thus, one indication of strategic behavior would be a larger than expected concentration of respondents at the low and high ends of the WTP distribution. Such concentrations would indicate strategic behavior even if other plausible distributions for willingness to pay, such as the log-normal, were chosen. Brookshire, Ives, and Schulze did not find such a distribution in their study.

The reliability of the distributional type of test is weakened by a number of plausible reasons, other than strategic behavior, why respondents might give low (zero) or high responses. Zero bids may represent honest responses by low-income respondents or by those who have negative attitudes toward the good in question. Some zero WTP amounts may also represent protest bids by respondents who refuse to play the game. Interpretation of responses at the upper end of a WTP distribution has fewer but similar problems, since high-income users of an environmental good who have strong positive attitudes toward the environment may truly be willing to

[15] Rowe, d'Arge, and Brookshire also exposed another subsample of their respondents to a supposed mean bid of other respondents. Subsample respondents were told the mean bid before giving their personal bids; this amount was described to each respondent as being in the $1.00 to $1.50 range. The authors report (1980:14) that on the average respondents who were given this information bid $1.70 a month less than those who did not receive the information. Presuming the samples are equivalent, this is a large difference. The authors interpret this finding as evidence for free riding. We find a price-setting explanation more plausible, although there is no definitive basis for choosing between the two explanations.

pay high amounts for it. In addition, some high or low amounts might represent meaningless bids by respondents who were unwilling or unable to give considered answers, but who were reluctant not to answer.

The available evidence supports the hypothesis that factors other than strategic behavior explain many of the extreme bids in contingent valuation surveys. Randall, Hoehn, and Tolley (1981) employ the concept of a core of usable bids to describe those responses remaining after the protest bids and the outliers—bids that are implausibly high or low given the respondent's income—are removed. The outliers, which usually constitute 5 to 10 percent of the sample's bids in CV surveys, will include many of the potentially strategic bids. Until recently the identification of outliers has generally been made on an ad hoc basis, but Desvousges, Smith, and McGivney (1983) have shown the value of using a more sophisticated systematic statistical approach proposed by Belsley, Kuh, and Welsh (1980). In chapter 10 we propose the use of the trimmed mean for this purpose. Analysis of the characteristics of respondents who give WTP amounts identified as outliers tend to show that these respondents have low incomes and low levels of education, and that they are disproportionately female and older. Such characteristics hardly describe those of whom strategic behavior is expected.

The third type of test for strategic behavior uses two equivalent subsamples, each of which is given a scenario that values the same good but provides different incentives for engaging in strategic behavior. This type of test was applied by Bohm (1984) in a study which measured the willingness of Swedish local governments to contribute to the basic costs of making detailed statistical housing information available to local planners. This study differs from others considered in this chapter in that the respondents were governmental units, and actual payments were required if the total WTP amount offered for the service by all the governments exceeded the fixed costs. Bohm divided 279 local governments into two groups, whose WTP amounts were solicited under conditions designed to provide incentives to strategically overvalue or undervalue the amenity. The governments in group 1 had a reason to understate their amounts, because they would be obligated to pay some percentage (up to 100 percent) of their stated WTP amounts (a form of SB_1) to the central government for the service. Units in group 2, on the other hand, had an incentive to overpledge because the most they would have to pay was a fixed amount (SB_2). Units bidding $0, or less than a set figure, would be excluded from receiving the service. Each local government was provided with complete information about the study, including explicit information about incentives to misrepresent the WTP amount, along with a moral exhortation which declared that "giving in to the incentive to misrepresent WTP would jeopardize the value of the investigation" (Bohm, 1984:144).

Participating units were also told that the amount offered by each local government responding would be publicized. The purpose of Bohm's approach was to bracket an upper and a lower bound of the true WTP for the service so that a clear-cut decision to provide the housing statistics could be made when the interval exceeded or fell short of the fixed costs. In this particular nonlaboratory and nonhypothetical application of the approach, the mean WTP amounts for each treatment were "quite close" and showed "little evidence of significant misrepresentation behavior" (Bohm, 1984:147).

The only other application of the third type of test was made by Cronin (1982), who claimed to have found strategic bias in his CV study of willingness to pay for water quality improvements in the Potomac River. One Cronin subsample in the Washington, D.C., area was told that payment would be in the form of an increase in local taxes, while the other subsample was told that the federal government would pay for the improvement at a cost of "at most a few pennies in your Federal taxes." Cronin found significant differences in the responses of the two groups, with the latter subsample willing to pay about 25 percent more than the former. His test is flawed, however, because valuing how much respondents want of a free good (a good with an effective zero price) is quite different from valuing how much they want of a good having a positive price (Kurz, 1974). When the possibility of a nontrivial payment is removed, a situation similar to the E_3 treatment of Schneider and Pommerehne and of Brubaker—where the good was guaranteed for any positive payment—is created, and yields similar results. Beyond that, a change in payment vehicles in Cronin's two survey treatments adds another source of variance, which cannot be ruled out as a cause of the difference he attributes to strategic behavior.

Summary and Conclusions

We have argued that strategic behavior should be treated as one of the many potential biases in CV studies. If it is present, would strategic bias be expected to over- or underestimate the respondents' true WTP, and by how much? In general, *if* strategic bias is present, slight overestimation is the expected result. As shown in chapter 6, the a priori prediction for SB_1—the optimal and most common payment motivation condition created in CV studies—is either under- or overpledging, depending on what the respondent believes about his expected payment obligation. For the reasons given there, even a rational, utility maximizing consumer would not have a strong motivation to under- or overpledge. The removal of outliers under certain conditions, advocated in chapter 12, will further reduce the effect of any strategic bias, especially in large samples.

In extending the analysis of strategic behavior to contingent valuation settings, this chapter began by reviewing the relationship of CV practice to the six payment/provision motivational states that lead to strategic behavior (according to standard economic theory). We found that SB_1, which of all six states corresponds most closely with conditions in most CV studies, produces weak motivation for strategic behavior. We next compared CV studies with the experimental settings described in chapter 6 to determine the extent to which experimental findings may be extrapolated to contingent valuation. We argued that CV studies and the experiments are comparable in many respects, and that where they differ the incidence of strategic bias should be even lower in CV surveys than in the free-riding experiments. This finding is a result of the inherent uncertainties in the payment and provision contexts of the typical CV setting, of the unavailability of the information required for optimal strategic behavior, and of guarantees against free riding by others. We argued that the conditions under which a mail survey is administered may facilitate strategic behavior if respondents are inclined to act in this manner, whereas in-person and telephone interviews tend to discourage strategic behavior.

Three types of tests for strategic behavior in CV surveys were examined. In each case, negative results were shown to be good indicators of the lack of strategic bias. The results of the only application of the first type of test (provision of information on the mean WTP amount) showed negligible strategic behavior. Among respondents in the second type of test (concentration of observations at extremes), which is routinely performed in CV studies that remove outliers, those giving outlier responses were shown to have demographic and attitudinal characteristics more suggestive of random or inconsistent behavior than of strategic behavior. It was also found that the mean WTP value of the outliers is often approximately equivalent to that of the remaining respondents, which suggests that undervaluers balance out overvaluers.[16] The third type of test (comparison of subsamples subjected to incentives to under- and overstate WTP) showed no evidence of strategic behavior in one application and evidence for strategic behavior in another. In the latter strategic behavior should have been expected since respondents were essentially asked how much of a free good they wanted—a question not relevant to most CV surveys.

Most of the arguments about strategic bias put forward in this chapter pertain only to WTP measures, and not to WTA measures. In most situations, the WTP question format appears to invoke the desired reaction in

[16] This result is consistent with Vernon Smith's (1980) observation that the majority of players are free riders who engage in a small amount of undervaluation, and who are counterbalanced by a minority of anti-free riders who engage in a larger amount of overvaluation, so that group mean and aggregate amounts are surprisingly accurate.

respondents—"What is the most I would be willing to pay rather than go without the good?" In contrast, posing the elicitation question from a WTA perspective seems to invite the strategic reaction, "What is the most I can get from selling?" instead of, "What is the least I would take in order to part with the good?" Only the answers to the latter question have meaning as a measure of economic welfare.

The evidence for willingness to pay, however, suggests that strategic bias is not a significant problem for CV studies under *most* conditions. Instead of being a fundamental, unavoidable threat to the CV method, strategic behavior is just one of many possible sources of bias which the designer of a CV study must take into account. There are CV designs, described in this chapter, which conceivably could stimulate a significant amount of strategic behavior unless mitigative measures are taken. Few CV studies to date have used these designs. The potential threat posed by respondents deliberately giving untruthful WTP values is likely to be much less serious than the possibility that they will give meaningless values—the subject of the next two chapters.

8

Can Respondents Answer Meaningfully?

In considering the evidence of the ability of surveys in general and contingent valuation surveys in particular to predict behavior, the basic issue is whether the necessarily hypothetical character of CV studies automatically renders their findings meaningless or, if not, under what conditions worthwhile CV data can be obtained. In this chapter, after a brief review of the critics' concerns on this issue and of the assumptions on which they are based, we discuss the nature of the market behavior that CV studies attempt to simulate, and then examine the findings from two areas of research that bear on the predictive ability of hypothetical markets: laboratory experiments comparing the effect on behavior of substituting actual monetary payments for hypothetical payments; and studies of the relationship between attitudes and behavior. In both sets of findings we discover grounds for optimism that CV findings can be meaningful.

Hypothetical Data

The researcher who wishes to conduct a valid CV study has to create a believable and meaningful set of questions which will simulate a market for the public good in question with sufficient plausibility that the respondents' answers may be taken as an accurate representation of how they would behave if confronted with an actual market for the good. Some critics are skeptical that hypothetical questions in an opinion survey can achieve this end.[1] Gary Fromm (Fromm, 1968:172) remarks, "It is well

[1] A similar debate about the validity of attitudinal data versus actual behavior occurs among transportation planners, who also want to predict behavior under conditions outside the range of previously observed behavior. Many transportation-demand modelers restrict themselves to observations of actual behavior ("manifest preference," in the field's vernacular); others, however, argue that appropriate attitudinal variables ("conceived preference") should also be incorporated in the models (Golob, Horowitz, and Wachs, 1979).

known that surveys that ask hypothetical questions rarely enjoy accurate responses." In a similar vein, Anthony Scott, an early critic of the method as applied to wild game resources, once declared, "Ask a hypothetical question and you get a hypothetical answer," adding, ". . . the results of this [CV] procedure have yielded extremely fanciful valuations" (Scott, 1965). More mildly, Freeman (1979b:104) and Feenberg and Mills (1980:92) argue that there are no apparent incentives in CV surveys to provide accurate responses.

Real Opinions or Nonopinions?

What is it about a "hypothetical answer" that makes it so problematical? One possibility is that the answers people give to hypothetical questions are likely to be off-the-cuff or careless responses which do not reflect their true taste preferences. This issue is a longstanding one among those who study the sources of public opinion. Twenty years ago Philip Converse dubbed such opinions "nonattitudes," which he defined (1974:654) as attitudes that are neither crystallized nor stable and therefore cannot be used as tools to predict other choice behaviors. He identified this phenomenon by examining the low correlations between respondents' expressed opinions on attitude questions when they were interviewed at successive points of time in panel surveys. In his original and much-cited paper on the topic, Converse (1964) examined what happened when, among other things, respondents in a national survey were explicitly invited to say they had no opinion on particular social issues "instead of laboring to concoct some kind of meaningless data point for us." At that time he concluded that nonattitudes characterize the views of a sizable proportion of the electorate on social issues. Converse's paper touched off considerable debate and stimulated a number of scholars to study the prevalence of nonattitudes.[2]

Converse subsequently tempered his views about the prevalence of nonattitudes (Converse, 1974:660), and other research suggests his worst fears were unfounded. George Bishop and his colleagues (Bishop, Hamilton, and McConahay, 1980), for example, have shown that college-educated respondents, at least, possess a stable conceptual framework for organizing their political attitudes that is inconsistent with the nonattitude hypothesis; and Tom Smith (1984) has shown that instrument error is an important contributing factor in apparent nonattitude responses. Another line of research also raises questions about Converse's assumption that

[2] See Converse (1964, 1970, 1974); for opposing views see Pierce and Rose (1974), Judd and Milburn (1980), Bishop, Hamilton, and McConahay (1980), and Brody (1986). For excellent reviews and evaluations of this literature, see Smith (1984) and Kinder and Sears (1985).

people are prone to concoct meaningless data points. In these studies, respondents are asked in the course of an otherwise typical survey whether they favor or oppose certain little-known (or even fictitious) programs, such as the Agricultural Trade Act of 1978. If people really are prone to express opinions in surveys no matter what they know, we would anticipate that most would try to answer such a question. In fact, most refuse to venture an opinion. In one version of a methodological experiment (Schuman and Presser, 1981:147ff) which asked whether respondents favored or opposed passage of the Agricultural Trade Act, 69 percent in a national survey volunteered the "don't know" response. In another version, which explicitly offered respondents the option of saying they did not have an opinion on this issue, 90 percent refused to give an opinion. Bishop, Tuchfarber, and Oldendick (1986) obtained similar results in parallel experiments which asked people's opinions about plausible but entirely fictitious issues.

Why do some people—a minority—go ahead and express opinions about fictitious or obscure programs? Bishop, Tuchfarber, and Oldendick (1986:248) hypothesize that "the less knowledgeable a person is about a given subject, the more easily he or she can be confused and pressured to give an opinion about it." But are the opinions given by the minority of respondents really nonopinions? Schuman and Presser (1981) do not think so. They found that those who gave opinions about the obscure programs named in their study did not randomly pick one of the offered response categories, as they would have if they were expressing a nonattitude. Instead, they seemed to construct answers to the posed question by first making "an educated (though often wrong) guess as to what the obscure acts represent." Once having settled on a constructed meaning for a program, respondents gave an opinion which appeared to be based on attitudes and beliefs relevant to the topic. In other words, the minority of people who expressed opinions in response to meaningless questions did so only after making them meaningful in their own terms.

We draw two conclusions from these studies. First, most respondents, particularly if they are given the chance to say they don't know, will stop short of giving meaningless answers. Second, when confronted by an unfamiliar, hypothetical situation, respondents tend to construct a meaning based on their previous experience and to arrive at an opinion which does indeed reflect their true taste preferences. The second conclusion is consistent with the observation of survey methodologists that many people find the opportunity to express their views to a sympathetic and nonjudgmental listener to be personally rewarding, especially when the purpose of the interview seems to be worthwhile (Sudman and Bradburn, 1982:5). In the context of the interviewer-respondent relationship people try to please the interviewer by giving as meaningful answers to the questions as possible.

The contingent valuation literature also offers evidence that the WTP amounts given in CV surveys are not random opinions. Beginning with Davis's original study (Knetsch and Davis, 1966), CV researchers have customarily reported the results of regression equations where respondents' WTP values are regressed on variables which theory predicts should be associated with this type of preference. Reasonably good fits are reported in enough cases (for example, R^2s of .20 to .50) to suggest that the CV methodology can measure people's values in a way that is consistent with their incomes and expressed taste preferences.

Can Hypothetical Studies Predict Behavior?

Even if it is granted that the opinions measured in well-designed CV surveys are meaningful in the sense of representing genuine views rather than random expressions of opinion, criticism of CV results for involving hypothetical answers can involve a second possibility—that the WTP amounts are too divorced from reality to predict behavior with the necessary degree of accuracy required for valid dollar benefits estimates. A recreationist may express a high WTP amount that correctly reflects his strong preference for a particular recreational good, but can we assume that the dollar amount he offers in the necessarily artificial world of the CV scenario will bear a sufficiently close relationship to the actual amounts he would pay if he were really confronted by a market in the good?

In 1980, Feenberg and Mills advised benefit measurement practitioners that "in the absence of empirical research that tests the accuracy of surveys predicting behavior, survey data should be avoided in benefit estimation to the extent possible" (1980:93). However, numerous tests of this kind have indeed been conducted, and they provide strong support for the ability of surveys to predict behavior.

These tests are of three types. The first is comprised of only a handful of studies involving expensive and arduous experiments in realistic field settings, where markets in public goods could be simulated using real money and compared with hypothetical CV markets (these studies are examined in chapter 9). The second type includes laboratory experiments that compare what happens when real dollars rather than hypothetical dollars are used as incentives in situations entailing choices. Tests of the third type examine the relationship between attitudes and behavior. These are numerous, and include studies by academic social psychologists as well as more practically oriented research by market research practitioners and others who measure people's buying and voting intentions in order to predict aggregate buying and voting patterns.

Laboratory Experiments Comparing Monetary and Nonmonetary Incentives

The notion that "'money talks' and real money 'speaks louder' than hypothetical money" (Bishop, Heberlein, and Kealy, 1983:629) is congenial to the economist's view of human motivation. In the few times it has been put to a rigorous test, however, hypothetical money has motivated research subjects equally as well as real money. Lichtenstein and Slovic (1973) demonstrated that the preference reversal phenomenon—where people's preferences between outcomes having the same expected utility changed when the outcome was presented as a loss instead of a gain, and vice versa—persisted when experiments were replicated in a Las Vegas casino, using real and binding cash bets. In another set of experiments, Grether and Plott (1979) also found that preference reversals were unaffected by the use of real payments. Knetsch and Sinden (1984) and Gregory (1982, 1986), in experiments to test whether the wide variations between WTP and WTA previously seen in CV surveys persisted when money payments and cash compensations were at stake, found that this was the case.

Attitude-Behavior Studies

Social psychologists have long been interested in the relationship between attitudes, as measured in surveys, and the behavior that the attitudes should predict. This interest was provoked by the startling (to social psychologists) conclusions of the famous LaPiere (1934) study. LaPiere spent several summers traveling cross country with a Chinese couple, during which they obtained service and accommodations in various restaurants and hotels. Six months after completion of their journeys, LaPiere sent a questionnaire to the same establishments asking if they would accept members of the "Chinese race" as guests. In contrast to the actual acceptance of the couple in all but one instance, 92 percent of the establishments replied that they would not accept the hypothetical Chinese. The LaPiere study was flawed in that it did not measure the attitudes of those people who had behavioral contact with the Chinese couple—LaPiere's letters were apparently directed to managers, while he and the Chinese couple were received by hostesses and waitresses—and did not specify that the Chinese were a well-dressed couple accompanied by a white university professor. Nevertheless, its dramatic findings still appeal to those who regard subjective measures as soft and unreliable guides to behavior.

Considerable research has been conducted on the attitude-behavior relationship in the fifty years since LaPiere's trip (Canary and Seibold, 1983). In reviewing its findings and assessing its implications for the CV method, we will pay particular attention to independently measured behavior instead of self-reported behavior, and to behavior in the real world instead of in laboratory settings.

According to Howard Schuman and Michael Johnson's (1976) review of the attitude-behavior literature, most studies that try to predict people's behavior from prior measures of their attitudes find a positive relationship, and a few are able to demonstrate an impressive predictive ability. For example, army trainees who said they were eager for combat were significantly more likely than others to perform well in combat several months later (Stouffer and coauthors, 1949). In another study, persons who said they support open housing were far more likely (70 percent) to sign an open housing petition three months later than those who had expressed opposition to open housing (22 percent) (Brannon and coauthors, 1973). A study of four elections showed that behavioral intention revealed in pre-election surveys correctly predicted the actual vote for 83 percent of the respondents who voted (Kelly and Mirer, 1974). Finally, a correlation of .55 was found between the intentions expressed by a sample of Michigan couples to have another child within the next two years and specific behavior toward having another child measured twelve months later (Vinokur-Kaplan, 1978). Schuman and Johnson conclude that for individuals the attitude-subsequent behavior correlations "are large enough to indicate that important causal forces are involved" (Schuman and Johnson, 1976:199). This cautious optimism is reinforced by the findings of transportation researchers who in surveys have compared transport-mode choice with actual behavior (Hensher, Stanley, and McLeod, 1975), and by Hill's conclusion in a more recent review of the attitude-behavior literature:

> A careful consideration of the research devoted to the interrelationship between attitudes and behaviors leads to greater optimism than that which characterized assessments of the area a decade ago. The debate over the existence or nonexistence of the interrelationship can be put to rest. Under a variety of conditions, attitudes have at least a modest utility in predicting behavior. (Hill, 1981:371)

The studies cited above predict the behavior of individuals from their attitudes. Another body of research involves *aggregate* attitude-behavior correlations, where an attitudinal measure for a sample is used to predict the subsequent behavior of the population from which the sample was taken. Studies have shown that while changes in individual intentions may shift over time owing to changed circumstances, they balance out in the aggregate as aggregation (over individuals, commodities, or time) tends to reduce the relative importance of neglected variables (Theil, 1971:181).[3] One study of women's intentions to have another child showed that while

[3] Theil provides an interesting discussion of how R^2 tends to decline in the estimation of consumer consumption equations as the data are disaggregated over individual and/or commodity groups. Whereas R^2s of .9 or more are common for highly aggregated time

only 41 percent had exactly the number of children they earlier said they intended to have, the actual average family size for the sample corresponded precisely with the average intended family size (Bumpass and Westoff [1969], cited by Ajzen and Peterson, 1986).

Since contingent valuation studies likewise attempt to obtain population estimates, and since these estimates are the primary goal of a CV study, studies of aggregate attitude-behavior correlations are of particular interest. There are some impressive examples of successful predictions of aggregate behavior in this literature. Properly conducted election polls, for example, are generally able to predict final election results with great accuracy.[4] According to Ladd and Ferree (1981), even the polls of the 1980 U.S. presidential election were no exception to this generalization because those polls taken immediately before the vote caught the last-minute shift which brought President Reagan to power.[5]

In the realm of economic behavior, impressive aggregate attitude-behavior correlations have also been demonstrated. Most widely known among economic behavior surveys are those that have been conducted each year since 1952 by the University of Michigan's Institute for Social Research (ISR). These are multiple surveys, using personal interviews or (more recently) telephone interviews of national samples to measure consumer sentiments and probe the psychology of economic behavior. The ISR Index of Consumer Sentiment (ICS) is an aggregate measure whose trend line reflects the changes in attitudes and buying expectations of the U.S. population. Consistently since the beginning of the ISR survey the ICS has declined substantially before the onset of every U.S. recession and advanced before periods of economic recovery (Curtin, 1982). An equivalent success rate in predicting GNP by aggregate consumer expectations is claimed by the Conference Board on the basis of its fifteen years' experience in conducting national surveys (Linden, 1982). In the Conference Board studies, which use a monthly mail survey, scales constructed from questions measuring consumer expectations about business condi-

---

series data, R^2s for equations based on the same individual-level data for specific commodities used to construct the aggregate time series may be .3 or lower. See also Cramer (1964), who proves that the R^2 estimated using the aggregate data must be greater than R^2s obtained using the underlying microdata.

[4] Properly conducted polls require professional interviewers, a scientifically drawn large sample, and carefully designed questions.

[5] Failures do occur in political polling, however, even when modern methods are used. In the 1982 congressional and state elections, the polls tended to overestimate front-runners' leads (Roper, 1983), particularly in Illinois (Kohut, 1983). Failures that occur when polling is used in surveys to measure benefits must be put in perspective, as what is regarded as a very large prediction error in a political candidate poll—5 to 10 percent—may be inconsequential for a CV survey.

tions, the availability of jobs, and personal economic fortunes over the following six months correlate closely with subsequent quarterly changes in the real gross national product.[6]

Both the ISR and Conference Board surveys also ask respondents about their intentions to buy particular kinds of large durable goods. Some surveys, however, apparently are able to predict demand for large consumer durables even to the extent of predicting sales of particular brands. In a recent article examining the factors which produce accurate and inaccurate polls, Roper (1982) provides several examples of successful buying-intention surveys. In one case periodic national surveys, conducted by his firm over a decade for a television set manufacturer, were able to predict net sales for the subsequent twelve months within 10 percent.[7]

Although we have identified studies that show reasonably substantial attitude-behavior correlations or the ability to predict consumer demand, many other studies fail to achieve the same level of success. This raises the question of which factors promote strong attitude-behavior relationships. Before answering that question it is useful to briefly outline a model of the attitude-behavior relationship.

A Model of the Attitude-Behavior Relationship

Fishbein and Ajzen and their associates have developed a widely accepted model of the attitude-behavior relationship. Figure 8-1 presents the version of their model that is relevant to those CV studies where a single behavior is to be predicted from one or more attitudinal measures. We will examine the model's several components and theoretical assumptions, using as an example of a single specified behavior the buying of national freshwater quality in a contingent market.[8]

[6] Questions have been raised by economic forecasters about the usefulness of consumer expectation surveys (Adams and Klein, 1972; Shapiro, 1972). Their criticism, which is often misunderstood by economists, is not that survey data fail to predict aggregate economic behavior reasonably well when compared to widely used macro models, but that they do not appear to contain much information which is independent of the major objective economic indicators. If consumers form their expectations in some rational, well-informed manner, one would expect this to be the case. Praet and Vuchelen (1984), in a recent article based on survey data for the European Community, argue that consumer surveys do contain original information that is helpful in short-term economic forecasting in European countries.

[7] After applying a correction factor for consistent underestimation owing to unintended consumer purchases. Such "calibration" is common in marketing surveys designed to predict purchases. If a systematic divergence between actual and CV survey behavior existed and could be quantified, calibration of CV results could be undertaken. For a discussion of calibration from a statistical viewpoint, see Scheffe (1973).

[8] The following discussion is based on Fishbein and Ajzen (1975) and Hill (1981). See Foxall (1984) for a discussion of the Fishbein and Ajzen model from the perspective of marketing and economic psychology.

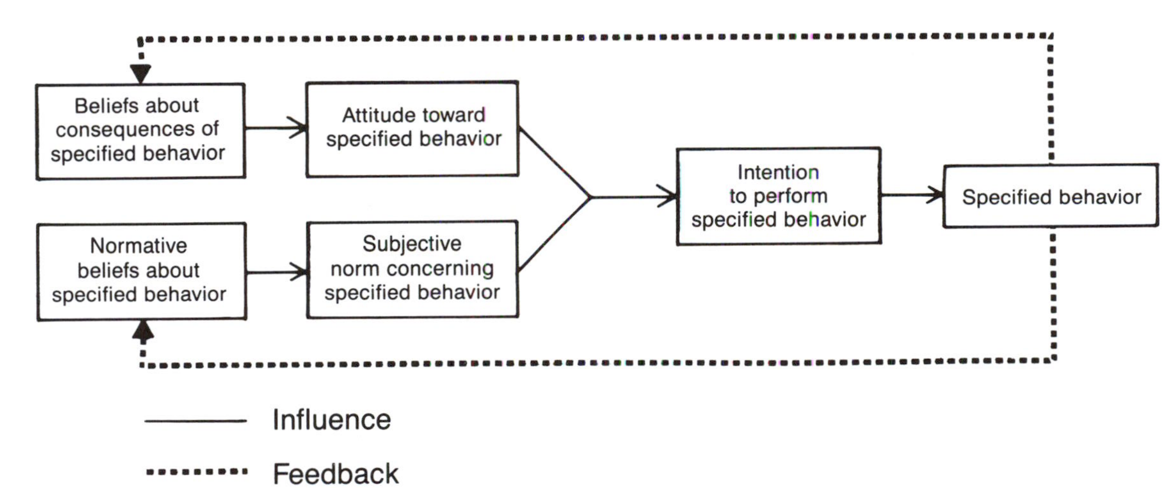

Figure 8-1. Fishbein-Ajzen model for predicting specific intentions and behaviors

The model's first component is beliefs. A belief is a probability judgment that links some object or concept to some attribute. According to Fishbein and Ajzen (1975), a person learns or forms a number of beliefs about an object on the basis of observation or information received from various sources. "The totality of a person's beliefs serves as the informational base that ultimately determines his attitudes, intentions, and behaviors" (Fishbein and Ajzen, 1975:4) For any particular attitude or behavior only a relatively small number of beliefs serve as determinants at any given time (Fishbein and Ajzen, 1975:219). Beliefs of two types are relevant for predicting behavioral intentions: (1) beliefs about the consequences of performing the behavior in question—a person may believe that government water-pollution control programs are ineffectual or that raising the quality of fresh water will improve the local drinking water; and (2) beliefs about how other people think one should behave (normative beliefs)—thus a person may believe that while friends (teachers, environmental groups, Boy Scouts) favor his paying for water pollution control, political associates (corporations, a political party, family) do not. The person may or may not be motivated to comply with any given reference group.

Attitude is the second component of the model. An attitude is a bipolar evaluative judgment of an object. It is essentially a subjective judgment based on some scale such as good-bad, strong-weak, risky-secure. In our example, a person who feels favorable toward improved water quality may also feel strongly that air quality should be improved and may regard accepting lower taxes for forgoing a cleanup in hazardous waste dumps as immoral. Fishbein and Ajzen consider attitudes toward an object to be a function of the person's beliefs about that object. A subjective norm is the totality of the normative beliefs held by the person and of that person's motivation to comply with them. It represents the balance of how relevant others expect the person to act with regard to the object. In our example, the norm might be to pay a "lot" or to pay a "reasonable amount" for water quality.

The last of the model's components, an intention, is an individual's probability judgment that she will undertake some action. This is what contingent valuation studies measure. Behavioral intentions in the Fishbein-Ajzen model are a function of both attitudes and subjective norms. To understand the importance of viewing intentions as a joint function of these two dimensions, consider the role both might have played in the 1950s if a liberal white southerner, who lived in the deep South and had nonliberal friends and associates, had been asked about his intention to offer hospitality to freedom riders. As a liberal, this person would have viewed the offer of hospitality as a good thing; as a white participant in southern society, he would have been influenced to some extent by the prevailing racist norms for how whites should act toward blacks, even if he did not share these norms.

The Fishbein-Ajzen model incorporates a feedback mechanism, reflecting the fact that the components exist in a process of dynamic adjustment. Attitudes, beliefs, norms, and expectations are influenced by behavioral experiences. Someone who paid little heed to water quality might modify his views about the subject after an extended canoe trip, for example. Because of this process, predicting behavior under novel conditions (the purchase of flowers at a convenience store) or predicting novel forms of behavior (the purchase of a kind of flower never before available) is more difficult than predicting customary behavior.

It might be thought that since CV studies measure behavioral intentions, and since behavioral intentions are the immediate determinant of behavior, the other components of the Fishbein-Ajzen model are irrelevant for CV studies. This is not the case, for several reasons. First, knowledge about the model's components can help the researcher specify the most appropriate context in a scenario. The relationship between behavioral expectations and behavior is contextual (Foxall, 1984) in the sense that whether or not the person ultimately behaves as she says she intends to depends on intervening experience, the presence or the views of relevant others, and the physical setting. To the extent that the behavioral context which leads to the most relevant decision in the CV context can be simulated, the validity of the study will be improved. Suppose research found that many respondents are influenced in valuing water quality improvements by the state of the economy. If the researcher wanted to predict support for water quality improvements even when the economy is growing slowly, she might include in the scenario a statement asking the respondent to make this hypothetical assumption. Such a device would serve two purposes: it would standardize one of the respondent's assumptions about the context in which the decision is being made, and it might improve the plausibility of the WTP estimate, since the decision maker would be assured that the estimate was not based on an overly optimistic context.

The second reason why all of the model's components are relevant to CV studies is that in order to properly specify a CV scenario, the researcher must identify those determinants of the behavioral intention which might cause bias or measurement error if not taken into account. For example, if respondents in a CV study intended to measure only air visibility benefits believe that visibility and the health effects of air pollution are closely correlated, the WTP amounts given are in risk of including health-benefit values as well. Again, if some respondents hold a strongly negative attitude toward the particular payment vehicle used in a CV study, or toward the notion of accepting payments for a degradation of environmental quality, the use of these elements in a scenario could bias willingness to pay for a good (Ajzen and Peterson, 1986).

A third reason why the determinants of behavior deserve attention has to do with validity.[9] One way of assessing the credibility of a CV finding is to see how strongly it is related to theoretically relevant attitudinal determinants. Bishop and Heberlein (1986) recommend this approach for establishing the credibility of contingent valuation WTP values for the existence benefits of acid rain deposition reductions, and provide an example of how theory can be used to develop a sophisticated explanatory model. Since existence benefits involve altruistic behavioral intentions toward the environment, Bishop and Heberlein draw on previous research on environmental altruism to identify factors that activate altruistic behavior. They argue that bid equations which were able to demonstrate significant positive relationships between CV existence values and measures of these (and other relevant) concepts would provide grounds for "arguing that legitimate economic values are being established at least to a rough approximation."

It is clearly desirable for the designer of a CV scenario to understand as fully as possible what determines the expressed values for the good under consideration. Before undertaking full-scale field testing of a CV instrument, directed discussion in small groups (focus groups), small preliminary surveys, and other qualitative exploratory research techniques may be used to obtain information on people's beliefs about the advantages and disadvantages of providing the amenity, what they associate with the amenity, and other relevant questions. The Fishbein-Ajzen model, which represents the state of the art in attitude-behavior theory, can serve as an important conceptual aid in this process.[10]

Determinants of Strong Attitude-Behavior Relationships

Fishbein and Ajzen's work yields three hypotheses—here labelled H1, H2, and H3—about three factors that enhance the ability of questionnaire items to predict behaviors. Properly designed CV studies can readily maximize two of these factors—correspondence and proximity. The third factor—familiarity—is more problematic.

[9] More specifically, with "theoretical validity," a concept discussed in chapter 9.

[10] Ajzen and Peterson (1986) provide an application of the Fishbein-Ajzen model to the CV method. Other models of the attitude-behavior relationship are reviewed by Davidson and Morrison (1982).

Correspondence. *H1. The greater the correspondence between the behavior and the attitude/intention measures, the more likely the latter will predict the former.*

Ajzen and Fishbein (1977) define correpondence as identity among (1) the action, (2) the target at which the action is directed, (3) the context in which the action is performed, and (4) the time when the action is performed. The measurement of a man's love for his wife is likely to be a poor predictor of whether he will give his spouse flowers (since there are many other ways to express love), whereas the measurement of that man's attitude toward the behavior of giving flowers should have much greater predictive power. Likewise, the behavior of spending money for public goods is best predicted by questions which specifically invoke this action.

Ajzen and Fishbein (1977) tested the correspondence hypothesis by comparing the reported attitude-behavior relationships in 109 attitude-behavior studies in which the target and action elements differed in degree of correspondence. They found the hypothesis strongly supported: every one of the 26 studies which had high correspondence, and used what Ajzen and Fishbein regarded as "appropriate" measures, showed attitude-behavior Pearson correlation coefficients of .4 or greater. In contrast, two-thirds of the studies where correspondence was partial had significant but "low" (<.4) correlations, and virtually every one of the 27 studies judged to have low correspondence had a statistically insignificant attitude-behavior relationship.

Contingent valuation survey questions differ from ordinary survey questions about public good preferences in two respects, both of which enhance the correspondence between the CV questions and the behavioral outcome they seek to predict. First, CV surveys introduce payment obligations, which is not done by many other surveys. The effect that questions involving payment obligations can have is shown in a Swedish study where respondents were asked whether or not they would like the Swedish government to increase aid to less-developed nations. Later in the *same* questionnaire the respondents were asked whether or not they would like this increase to take place "even if taxes would be raised in proportion." Half the supporters of increased aid vanished when the question was phrased the second way, leaving only 20 percent willing to pay for increased aid (Bohm, 1979:146). The shifts in opinion evoked by the wording changes in this study are understandable, because we would expect a higher demand for free goods. The Swedes who favored foreign aid in answering the first question consisted of two types of people: (1) those who favored it in the abstract but not when the cost implications were made explicit; and (2) those who both favored it in the abstract and were willing to pay for it. The later question induced those of type (1) to relinquish their support by introducing the contingency of cost. The correspondence hypothesis suggests that the second wording would more accurately predict the vote on a referendum to raise taxes for increased foreign aid.

The second way in which CV studies meet the correspondence criteria is to induce respondents to participate in a detailed hypothetical market in the good. This distinguishes the CV method from those ordinary surveys which ask respondents whether they would pay X dollars to obtain some relatively nonspecific public good. In one example of the latter approach, the Federal Energy Administration (FEA) asked a national sample, "Would you be willing to pay an additional $20 per year on your electric bill in order to cut down air pollution caused by power plants?" (Federal Energy Administration, 1977). In another study, done for the Environmental Protection Agency (EPA) in 1973, respondents were asked what they currently paid each year for their sewer service, after which the interviewer posed the following question:

> Some water pollution comes from incomplete treatment of sewer wastes. Water improvement requires additional treatment which would increase water and sewer bills or local taxes. How much *more* would you be willing to pay per year to provide extra treatment of sewage to clean up this form of water pollution? (Viladus, 1973:93)

Dorfman (1977) and Jackson (1983) have recently shown that data from such questions can be of some limited use for benefit-cost analysis. However, even though questions of this kind may provide useful information about the intensity of respondents' commitments to air or water pollution control, they are not ideal because they do not meet the correspondence criterion for predicting behavior. In the FEA and EPA instruments the markets are imperfectly described, and the good is only vaguely described as a "cut down" in air pollution caused by power plants, in the first instance, and as a "clean up" of water pollution from incomplete treatment of sewer wastes in the second. Jackson, in his paper, describes the good only as a "real clean-up in your natural environment." In none of these studies do questions mention where the cutdowns or cleanups would occur, or what amount of change in air or water quality would ensue from these actions. Also left unclear is whether respondents should answer for themselves or for their households.[11]

[11] Dorfman (1977) and Jackson (1983) note these and other problems with the National Wildlife Federation survey (Harris and Associates, 1969) that they used, but point out the unsuitability of other data sources for looking at national willingness to pay for pollution control.

Proximity. *H2. The fewer [the] intervening stages between a component in the Fishbein-Ajzen model and behavior, the greater the predictive power of that component.*

According to this hypothesis, if we have measures of behavioral intention, attitudes, and beliefs which are equally correspondent to a behavior we wish to predict, the strongest predictor will be the measure of intention, followed by the measure of attitudes. Several studies of environmentally related behaviors show just such a pattern. Heberlein and Black (1976) studied people's gas-buying behavior at a time (1973, before the Arab oil embargo) when the only motivation to buy lead-free gas was an altruistic desire to reduce the amount of potentially harmful ambient lead in the air. They found attitude-behavior correlations increased from .12 to .59 when they changed the attitude they measured in predicting the purchase of lead-free gas from a general concern about environmental problems to the degree of personal obligation the individual felt to buy this type of gasoline.[12] Weigel, Vernon, and Tognacci (1974) correlated subsequent behavioral commitment to the Sierra Club with three attitude scales whose specificity ranged from low (attitudes toward the environment in general) to high (expressions of interest in participating in Sierra Club activities). The pattern of correlations between their attitude scales and the behavioral participation measures replicated Heberlein and Black's findings, with a correlation of .60 for the behavioral intention scale-behavior relationship.

A search of the marketing research literature located only a handful of studies relevant to the proximity hypothesis. Juster (1966), in one of the few published studies to examine the relationship between purchase intention and purchase behavior, found a strong relationship between a probability-of-purchase scale for automobile purchases and actual purchase a year later, and a "moderately strong" relationship between probable and actual purchases of household durables. Half of those saying they were certain or practically certain to buy a car did so in the next six months. Those at the lower end of the probability scale tended to underestimate their actual purchase behavior.

The type of market research most analogous to CV studies is concept testing, a technique used when a firm wishes to find out if a potential market exists for a new product that would be significantly different from existing products and which has not yet been produced, even in prototype. Respondents are exposed to detailed descriptions of the product in a survey research setting, and asked if they would be likely to purchase such

[12] Strictly speaking, the obligation measure is not a measure of behavioral intentions, but it is close enough to warrant mention here.

a product to try it out, if it were available in a local store.[13] Unfortunately, the findings of most concept tests are not reported in academic journals, for proprietary reasons. However, Moore (1982) has reviewed the available literature on the topic, in addition to interviewing thirteen leading corporate practitioners about their companys' experiences with concept testing. According to Moore, companies that have compared intent-to-purchase indicators in concept tests with the subsequent trial purchases of the final product find that such indicators can predict the trial rate within 20 percent, apparently an acceptable level of accuracy for this type of research. "If the test predicts a trial rate of 50% the actual rate will fall between 40% and 60% about 80% of the time" (Moore, 1982:285). One corporate practitioner described to Moore an intent-to-purchase question which explained 80 percent of the variance in trial rates in a series of predictive tests. Other practitioners told Moore about instances in which the predictive power of the concept testing results were so good as to be "uncanny" or "scary."

The success of many political polls in predicting election outcomes also bears on proximity. Political pollsters measure many attributes of candidates in their polls, but they would not attempt to predict an election without posing a voting intention question. CV studies that use the referendum model and pose the elicitation question in the form of a voting intention maximize the proximity factor. Other contingent valuation studies also clearly meet the proximity criterion, as CV elicitation questions measure a behavioral intention to pay for an amenity.

Familiarity. The behaviors considered in the studies reviewed by Fishbein and Ajzen—practice of birth control, voting, buying gas, use of laundry detergents, breast feeding, destruction of school property, signing petitions, among others—are probably familiar to most people. The feedback process in the Fishbein-Ajzen model, according to which the experience of a behavior affects beliefs and, through beliefs, attitudes toward the behavior and intentions regarding it, suggests a third hypothesis:

H3. The more familiar the behavior, the more likely the respondent's attitude and/or behavioral intention will predict that behavior.

Russell Fazio (1981, 1982) and his collaborators tested this hypothesis in a variety of experimental settings by manipulating the degree to which subjects have direct experience with a behavior. Typically, subjects were introduced to a set of intellectual puzzles, either by being provided the opportunity to work example puzzles or by being shown examples by the

[13] If a concept is deemed worthwhile, it typically next undergoes a product test in which respondents evaluate prototypes; then follows selective market testing.

experimenter, and asked to express their interest in each type of puzzle on an attitudinal scale. Later they were placed in a room where they could voluntarily play with any of the different types of puzzle. Correlations between expressed interest and actual play were higher for those subjects who had the chance to work example puzzles. In other experiments, Fazio and his collaborators (1982) show that attitude-object relations are stronger and more accessible, as measured by the speed by which subjects respond to a situation evoking the attitude, when formed by direct experience.

The market research literature also reveals an awareness of and concern about the potential unreliability of predicting product use from responses based on a purely conceptual exposure to novel products. According to Tauber (1973), concept testing may discourage potentially successful major innovations because, as Moore (1982:290) puts it, "consumer attitudes, upon first exposure to discontinuous innovations, are not good predictors of what [consumer] actions will be after a prolonged exposure." Unfortunately, we have been unable to find any tests of Tauber's hypothesis in the market research literature.

The familiarity factor lies at the heart of the difficulties posed by the hypothetical nature of the CV method. In contrast to the other two criteria for a satisfactory attitude-behavior relationship, a greater or lesser degree of unfamiliarity characterizes many contingent valuation studies. Often one or more of the basic components of a CV scenario—the description of the good, the payment vehicle, or the elicitation procedure—is unfamiliar to a respondent. The very process of expressing a dollar value for a good described in a hypothetical scenario is also an uncommon experience for most people. The use of the referendum model in CV surveys is particularly helpful in overcoming the unfamiliarity of expressing preferences for public goods, as many people find it credible to be asked to value an unfamiliar and even a complex amenity through such a mechanism.

Summary and Conclusions

This chapter introduced the important issue of whether the hypothetical character of CV instruments renders them useless for measuring meaningful values. What is the likelihood that answers to a CV questionnaire would predict the relevant expression of preferences for public goods in a real-life setting? In addressing this question, we reviewed two types of evidence in the noncontingent valuation literature which bear on the issue of predicting behavior from hypothetical answers. The first was the findings of laboratory and field experiments that compare the results obtained by treatments using a hypothetical payment structure with those involving a nontrivial payment in real dollars. These studies revealed similar patterns of behavior whether a hypothetical or a real payment was involved.

The second, and more extensive, body of evidence concerned the relationship between attitudes and behavior. Much of the more than $1 billion a year spent on economic market research is for surveys designed to predict consumer behavior (Tull and Hawkins, 1984), and the percentage of political campaign funds devoted to polling increases each election year.

We also asked what factors are associated with successful behavioral predictions. To answer this question, we drew upon the Fishbein and Ajzen model to make distinctions among beliefs, attitudes, and behavioral intentions. Of the three determinants of strong attitude-behavior relationships—correspondence, proximity, and familiarity—CV studies, which attempt to measure behavioral intentions to pay for specified amenity levels, appear to meet the correspondence and proximity criteria. The key problem facing the designer of a CV study, we saw, is the novelty of valuing a public good, given the respondents' varying degrees of familiarity with the good being valued and how they currently pay for its provision. We noted that use of the referendum format is likely to help respondents cope with the novelty of the valuation choice.

In the next chapter we turn to the evidence for the validity of hypothetical questions as given in studies which have attempted to directly compare the values measured in CV studies with those measured in simulated markets.

9

Hypothetical Values and Contingent Valuation Studies

In chapter 8 we asked whether respondents could provide meaningful answers to CV scenarios given the hypothetical character of the markets, their complexity, and their novelty. Although we concluded that meaningfulness is a much more serious problem for the methodology than the threat of strategic behavior, our reading of the attitude-behavior literature suggested that surveys have the potential to obtain valid data on people's willingness to pay for well-defined public goods. How well do surveys realize this potential? In order to answer that question we must first clarify a subject of some confusion in the contingent valuation literature: the assessment of validity. After addressing that task in the first part of this chapter, we will examine the results of tests of the CV method's validity.

In considering the assessment of validity, we are guided by two assumptions. First, not all tests of validity are of equal importance. We will devote special attention to experiments which have compared hypothetical and real-payment markets for the same good, on the premise that the real-payment or "simulated" markets serve as a particularly useful criterion against which to judge the accuracy of hypothetical scenarios. Second, no single test is definitive. Both the nature of public goods and the multidimensional character of validity conspire against the results of any test "proving" that the method is a valid way to measure benefits. Moreover, even if a particular CV study is found to be invalid, the reason may lie in its poor design rather than in the method itself. Our objective in this chapter is to examine the overall evidence of validity.

Types of Validity

The validity of a measure is the degree to which it measures the theoretical construct under investigation. This construct is, in the nature of things, unobservable; all we can do is to obtain imperfect measures of that entity. In the CV context the theoretical construct is the maximum amount of money the respondents would actually pay for the public good if the appropriate market for that public good existed. In the hands of methodologists, validity is a multidimensional concept. The three types of validity distinguished by the American Psychological Association (1974)—content, criterion-related, and construct—usefully summarize the prevailing measurement approaches. Each offers a different strategy for assessing the measure-construct relationship, and each is applicable to contingent valuation studies in one way or another.

Content validity (sometimes called face validity) involves the issue of whether the measure adequately covers the construct's domain.[1] It differs from the other validity types in that it can only be assessed by a subjective judgment based on an examination of the instrument (usually, the wording of a question). Thus, the content validity of a fifty-item scale designed to measure people's historical knowledge might be evaluated by a panel of historians who would be asked how well these items cover the domain. To the extent that any of the items were inaccurate or that important historical periods were not adequately represented among the items, the panel presumably would question the scale's content validity. In a similar fashion, CV questionnaires may be assessed to see if they ask the right questions in an appropriate manner.

Criterion validity is concerned with whether the measure of the construct is related to other measures which may be regarded as criteria. It is far cheaper to use surveys to estimate crime victimization rates than to use police records for this purpose; but are self-reports an accurate measure? Self-reports have been validated by checking them against official reports (the criterion). The results have been generally favorable (see Turner, 1972), although certain kinds of events, such as assaults, are more likely to be underreported than less serious episodes, such as larcenies.[2] Suitable criteria are not always available to validate measures, as is the case with WTP amounts for public goods. However, several important studies have created markets for quasi-private goods—access to hunting preserves, for

[1] The following discussion draws on Bohrnstedt (1983).

[2] The difficulty of establishing criteria validity, even when an apparently clear-cut criterion is available, is illustrated by Miller and Groves' (1985) recent reverse record check of victimization reporting, which shows wide differences in validity depending on the procedures used to make the match.

example—in order to compare the resulting prices with values obtained from a hypothetical CV market for the same good.

Construct validity involves the degree to which the measure relates to other measures as predicted by theory (Carmines and Zeller, 1979). One form of construct validity, convergent validity, asks whether the measure is correlated with other measures of the same theoretical construct. For example, one way of estimating the total number of people employed in the United States is to use self-reports about employment status as measured in census surveys; an alternative measure of the same construct would be the number of people reported to be employed by businesses in census business surveys. Neither measure is sufficiently close to the construct to serve as a criterion for the other. Although these two independent measures produce somewhat different estimates, the fact that their changes over time correlate highly suggests they are measuring the same underlying construct.[3] The other form of construct validity, theoretical validity, asks whether the measure is related to measures of other constructs in a manner predicted by theory. Economic theory, for example, predicts that a measure of the quantity of a good that is purchased should be inversely correlated with the observed price of the good. Failure to find such a relationship raises questions about the validity of one or both of the measures.

Ever since the development of the contingent valuation method, its practitioners have sought to overcome economists' fundamental prejudices against subjective survey data by demonstrating the validity of CV-based benefit estimates. At first these efforts tended to confuse criterion validity with convergent validity by using comparable travel-cost or hedonic price estimates as if they were criteria for judging the accuracy of a CV measure, instead of being alternative measures of the same underlying construct. Further confusion was caused by failure to distinguish between the effects of reliability, or random error, and invalidity, or systematic error, a case in point being the identification of a type of error called "hypothetical bias," whereas hypotheticality is associated with random rather than systematic error. In recent years the discussion of these issues has become more sophisticated as better research designs and a greater variety of approaches have been used to assess the validity of CV estimates.

We turn now to a more detailed review of these types of validity and explore their implications for the contingent valuation method. Although no single validity study can be definitive, we will devote considerable attention to the most recent of Bishop and Heberlein's impressive series of

[3] In contrast to criterion validity, where one variable acts as a benchmark against which the other is evaluated, the correlation in this case validates both measures.

field experiments, which compare the results of hypothetical markets for hunting opportunities with the criterion of simulated markets for the same amenity.

Content Validity

As used by psychologists, content validity depends on the extent to which an empirical measurement adequately reflects a specific domain of content (Carmines and Zeller, 1979). In contingent valuation studies, the relevant domain is the structure of the market and the description of the amenity. To the extent that a CV questionnaire presents these in a fashion which does not accord with the relevant theory or with the character of the amenity, the study cannot measure what it is intended to. For this reason it is advisable for CV researchers to circulate their draft questionnaires among colleagues for comments before using them in the field, and always to make copies of them available when releasing their findings. Journal editors should request copies of questionnaires on which articles about CV are based, and include the questionnaires with the articles when sending them out for review. Those who wish to evaluate a published CV study or to use the findings of any CV study are advised to study the questionnaire on which the findings are based.

The kinds of questions appropriate to an assessment of a scenario's content validity include: Does the description of the good and how it is to be paid for appear to be unambiguous? Is it likely to be meaningful to the respondents? Is there anything in the scenario that might suggest to some respondents that the good would not be paid for? Are the property right and the market for the good defined in such a way that the respondents will accept the WTP format as plausible? Does the scenario appear to force reluctant respondents to come up with WTP amounts? Although the answers to these questions are necessarily subjective and open to debate,[4] this type of assessment should be undertaken whenever CV-based estimates are used for policy purposes.

Criterion Validity

Criterion validity has the greatest potential for offering a definitive test of a measure's WTP validity. Unfortunately, to assess this type of validity it is necessary to have in hand a criterion which is unequivocally closer to the theoretical construct than the measure whose validity is being assessed. A criterion of central importance to CV studies is actual market prices. Even though market prices are rarely available for public goods, several

[4] See, for example, Mitchell and Carson (1985) and Greenley, Walsh, and Young (1985).

researchers have conducted experiments in which the outcomes of hypothetical CV markets in recreational goods (such as hunting opportunities in special preserves) were compared with outcomes for identical markets in which these goods were bought or sold. This type of market, which can be created only for quasi-public goods from which people can be excluded, is called a simulated market because it is constructed by the researcher and does not occur in real life. In such hypothetical-simulated market (HSM) experiments the amounts respondents pay in the simulated market treatment are an appropriate criterion for the validity of the parallel hypothetical market.

Hypothetical-Simulated Market (HSM) Experiments

The HSM studies reviewed here valued quasi-private amenities which were relatively familiar to the respondents, and are therefore crucial negative tests of the CV method's validity: if people cannot accurately value goods of this kind in hypothetical markets, it is unlikely they would be able to improve their performance when asked to value public goods. A positive finding, on the other hand, is a necessary but not a sufficient condition for assuming the validity of hypothetical markets in public goods. People may be able to hypothetically value a hunting permit, but fail miserably when asked to say how much they would be willing to pay for improvements in the quality of river water.

It is of more than special interest that the first two HSM studies in the literature—Bohm's (1972) study of willingness to pay to see a television program in Stockholm, and Bishop and Heberlein's (1979, 1980) study of willingness to pay to hunt geese in central Wisconsin—appeared to cast serious doubt on the criterion validity of hypothetical markets. There are grounds, however, for questioning each study's conclusion that the hypothetical WTP measures are invalid. The analysis of the studies that follows is consistent with the findings of Bishop and Heberlein's more recent experiments.

Bohm's Stockholm television-viewing experiment. The earliest HSM experiment was Bohm's (1972) study of willingness to pay for the opportunity to see a television program, described in detail in chapters 6 and 7. The portions of the experiment relevant to this chapter are his treatments V and VIa (see table 6-2), particularly the latter. Neither involved actual payments. The subjects in treatment V were told that taxpayers at large would pay for the program. Those in treatment VI were the only subjects offered two forms of preference elicitation. Members of group VIa were told:

We would now like to know how much you would find it worth to watch this program, approximately in the same way as when you make up your mind what a visit to the theater or cinema is worth to you before deciding whether or not to pay the admission fee. In other words, we ask you to *try* to estimate in money what you think it is worth to you to watch this program here; more precisely the highest admission fee you would be willing to pay, *if* you had been asked to pay an admission fee for watching the program here. (Bohm, 1972:129)

The conditions under which the good would be produced and the individual would pay for it if it were produced were purposely left undefined. In these features treatment VIa was much closer to the usual CV format, which implies payment obligations, than was treatment V, which informed its subjects that they would not have to pay to see the show under any condition.

Bohm assessed the validity of the two hypothetical treatments by comparing them with the criterion of the WTP amounts obtained in treatments I through IV, which were simulated markets in the sense that they required actual payments from the subjects under varying conditions. In treatment V, the average WTP amount of 8.8 kroner was similar to the amounts given in the simulated markets (treatments I to IV). Bohm judged the bid of 10.2 kroner in treatment VIa to be higher, leading him to comment that "the results [of VIa] are of course compatible with the general view that, when no payments and/or formal decisions are involved, people respond in an 'irresponsible' fashion" (Bohm, 1972:125). In other words, when the realistic prospect of a payment obligation is removed (as it is in CV studies), people will not give considered bids.

There are two grounds for questioning Bohm's conclusion. First, the differences on which it is based—WTP amounts from treatments I–IV as criteria for treatment VIa—are not large enough to justify his strong statement. Specifically, out of four comparisons, the mean bid of group VIa was significntly higher than only one of the other criterion groups (group III). Moreover, inspection of the data provided in Bohm's article on the experiment shows that only group VIa had an outlier (of 50 kronor, where the median bid was 10), which raised its mean bid considerably.[5]

Second, Bohm's conclusion does not take into account what may be the best criterion of all, the results of treatment VIb. After first obtaining the WTP amounts by using the treatment described earlier, Bohm next asked group VI to take part in a real auction in which the ten highest bidders would be allowed to pay the amount they bid to preview the show (treatment VIb). Using the HSM market, VIb, as a criterion for the hypothetical

[5] Removal of the outlier reduces the mean payment of group VIa from 10.19 to 9.45 kroner and diminishes the difference between group VIa and group III so that it is barely significant. Bohm's groups were small, which raises the question of the power of his statistical tests to detect differences; see the discussion of this topic in appendix C. The variance of Bohm's data is such that his tests are only sensitive to differences of 30 percent or more.

WTP market[6], VIa, we find that in the two markets Bohm's subjects offered quite similar mean payments. In the simulated market (VIb) where real money passed hands the mean amount was 10.3 kroner, whereas in the preceding hypothetical market where people were asked to act as if they would make a payment the amount was 9.45 kroner.[7]

Our analysis of Bohm's study does not support his conclusion that people will act "irresponsibly" when given hypothetical payment conditions. The various comparisons suggest, if anything, that the hypothetical and actual payments tendered by his subjects are similar.

Bishop and Heberlein's Horicon goose-hunting experiment. This experiment was designed to compare different methods of valuing an early-season permit to bag a maximum of one goose in the Horicon zone of central Wisconsin. Bishop and Heberlein sent mail questionnaires to hunters who had applied for and received a free early-season goose-hunting permit from the Wisconsin Department of Natural Resources. In the simulated market treatment ($N = 221$), they offered to buy (WTA_s) the hunters' goose permits by sending them a negotiable check for a specified dollar amount. The offers consisted of one of nine different dollar amounts, ranging from $1 to a maximum of $200. The hunters were told they could keep the checks if they surrendered their goose permits by mailing them to the researchers. The separate sample selected for hypothetical treatments was asked, among other questions, whether they would be willing to pay a specified amount (WTP_h) for a (hypothetical) goose-hunting permit, and whether they would be willing to accept (WTA_h) a specified amount for such a permit. The distribution of the amounts offered in this treatment ($N = 332$) paralleled that of the simulated market, but no actual cash offer was made or requested. Bishop and Heberlein calculated the percentage of tendered offers which were accepted at each offer point. For example, in the WTA_s market, 90 percent of the 19 respondents who were offered $50 checks for their permits accepted it; this compares to acceptance by 40 percent of those ($N = 30$) who were offered a hypothetical $50 ($WTP_h$) for their permits. In each case the maximum ($200) amount was treated as a truncation point. Thus a box with probabilities (P) on the

[6] The test would have been more convincing if the possibility of interaction between the two treatments had been eliminated by the use of separate samples. People's bids in the auction may have been influenced by the amounts they previously gave in the hypothetical treatment.

[7] The latter amount excludes the outlier. The difference between the two amounts is not statistically significant at the .05 level. An examination of the distributions for the two treatments gives some indication of what happened when the condition shifted from hypothetical to real; there appear to have been varying reactions, with some people shifting down to the lowest amounts and others shifting up to the highest amounts from treatment VIa to treatment VIb.

vertical axis (0,1) and dollar amounts on the horizontal axis ($0, $200) was defined.[8] Using logit regression to map a curve through the (*P*,$) points, Bishop and Heberlein found a total consumer surplus of $63 for the actual cash offered (WTA_s), $101 for hypothetical willingness to see (WTA_h), and $21 for hypothetical willingness to pay (WTP_h). On the grounds that the simulated WTA_s value of $63 constituted a valid criterion for both the WTA_h and WTP_h treatments, Bishop and Heberlein concluded that both the hypothetical measures were biased, with WTA_h overestimating and WTP_h underestimating consumer surplus (1979:929). "The results show that contingent values could easily be in error by 50 percent or more" (Bishop, Heberlein, and Kealy, 1983:620).[9]

Although some have interpreted Bishop and Heberlein's Horicon findings as fundamentally supportive of the CV method's validity (Randall, Hoehn, and Brookshire, 1983), we do not share this view. The finding that concerns us here is the apparent discrepancy between the hypothetical WTP results of $21 and the behavioral criterion (WTA_s) of $63, since the WTP format is the format of choice for most CV studies. A difference of this magnitude would pose serious questions for the CV method's validity if it could be attributed only to the use of hypothetical questions and not to other factors. There are three grounds for believing that other factors may be involved in this discrepancy: (1) the WTA estimates are very sensitive to the decisions Bishop and Heberlein made regarding where to truncate the distribution and how to treat those who refused to accept an offer for their hunting permits; (2) several possible sources of bias inherent in the design of their experiment could have affected the WTP_h estimate; and (3) a WTA simulated market is not a good criterion for a hypothetical WTP market.[10]

The first point involves a problem generic to the single take-it-or-leave-it elicitation technique. In estimating the amounts, Bishop and Heberlein assumed that *everyone* who refused to accept a WTP or WTA offer was playing the game and refused because the amount was too high for them to be willing to buy or too low for them to sell. However, it is quite possible that one or two recipients of the $200 offer may not have cashed

[8] In the case of the WTP_h question, the maximum amount would have been reached only if 100 percent of the respondents had accepted the purchase prices proposed to them at each price level, including the $200 amount. Under Bishop and Heberlein's estimation procedure, the maximum value for WTA_s and WTA_h would have been reached even if only those respondents who were offered $200 accepted the offered purchase price.

[9] The Horicon experiments also included a travel cost study. The travel cost estimates of $9 to $32 (Bishop, Heberlein, and Kealy, 1983) that were derived also suffered in comparison with the simulated market value.

[10] See Carson and Mitchell (1983) for a more complete presentation of these three points, and Bishop and Heberlein (1986) for their views on the first point.

their checks because they doubted the validity of the offer, or because they threw the envelope away unopened, or for other reasons. To the extent that any of those who refused to buy or sell their permits in any of the markets are dropped from the analysis as nonplayers, the estimated consumer surplus is reduced.

Bishop and Heberlein also assumed that the correct truncation point for their logit analysis was the highest amount offered to any respondent ($200). The logic behind such an assumption appears faulty. Consider a situation in which 90 percent of those offered $20 for their permits ($WTA_h$) accepted the offer, while 90 percent of those offered $50 ($WTA_h$) also accepted. A finding of this kind would suggest that permits have a price of $20 and that as many permits can be bought at $20 as at $50. If we define a WTA amount in the standard fashion as the minimum compensation a purchaser would be willing to accept to give up a goose-hunting permit, the horizontal axis should stop at $20. If we extend the axis beyond $20, we add to the area above the curve (which represents the consumer surplus) and thereby increase the estimated value of the permits without purchasing an additional percentage of them. The implication of altering these methodological assumptions is considerable for Bishop and Heberlein's findings. We have shown (Carson and Mitchell, 1983) that different, but plausible, assumptions about nonplayer rates and appropriate truncation points yield the following alternative estimates for their study: WTA_h, $51–20; WTA_s, $43–28; and WTP_h, $14–10.[11]

This reestimation replicates, albeit at a much reduced dollar level, the large percentage difference Bishop and Heberlein found between simulated WTA and hypothetical WTP amounts. Does such inequality support their original conclusion that WTP_h is biased downward? We believe the evidence from their experiment is inconclusive on this point because they did not control for several potentially important biases, and because the WTA_s measure is a questionable criterion for a hypothetical WTP measure.[12]

One source of potential bias is that their simulated WTA market may have measured a different, and more valuable, commodity than the hypothetical WTP market. Since the simulated market required the surrender of the permit if the offer to purchase was accepted, hunters necessarily valued both the right to hunt a goose *and* the cost of changing whatever

[11] In contrast to the WTP_h case, the logit technique correctly estimates $WTA_{s/h}$ in all cases. It is the differing assumptions about the percentage of subjects playing and the truncation point that account for the differences in Bishop and Heberlein's and our estimates.

[12] On the basis of Willig's (1976) work on price changes, Bishop and Heberlein had concluded that true WTP and WTA should be close. Hanemann's (1986a) recent work on quantity changes (see chapter 2) suggests, on the contrary, that we might expect a sizable but finite difference between true WTP and WTA, since Horicon geese are likely to have a fairly small substitution elasticity, σ_0, with other water fowl hunting activities.

arrangements they had made to go goose hunting that season. As the hunters were asked to sell their permits shortly before the two-week season, changing these commitments cannot be assumed to have been costless. As the wording of the WTP_h question does not invoke this cost, the two markets are not strictly comparable.[13]

It is also possible that Bishop and Heberlein's hypothetical WTP amount was biased lower because that particular market contains unusually high (for contingent valuation studies) incentives for strategic behavior. The permits are presently given without charge. Hunters could easily be charged for the permits because, unlike true public goods, they are excludable. Since hunters commonly pay for other types of permits in Wisconsin, it is plausible that some or all of those giving WTP_h amounts might have anticipated the imposition of a fee, and might therefore have attempted to influence the future price of such a permit by expressing amounts lower than their true values.[14] Yet another possible source of downward bias for the WTP_h amounts stems from the payment vehicle used in Bishop and Heberlein's study. Since paying for hunting licences is a familiar activity for these hunters, the prevailing price of other permits they must purchase in the course of a hunting season may have served as a benchmark in their calculation of what they would be willing to pay for a goose-hunting permit. Thus, instead of revealing a maximum amount, some or all of the respondents in the WTP_h market might have expressed what they regarded as a "fair" or "reasonable" price to pay for such a permit.

Even if these biases could be eliminated from consideration, there remains the more serious question of whether the substitution of a WTP simulated market criterion would have led to a different conclusion given the strong evidence that consumers respond very differently to WTA and WTP markets. Bishop and Heberlein were well aware of this problem and originally intended to include a WTP_s market in their study design, but were unable to get permission from the relevant authorities to sell goose-hunting permits.

In our view, the evidence in the Horicon HSM experiment for the invalidity of the hypothetical WTP measure is less conclusive than first appeared. Bishop and Heberlein were well aware of the exploratory character of their study, explicitly noting at the conclusion of their original paper that "we must be careful at this point to avoid sweeping conclusions based on a single experiment" (Bishop and Heberlein, 1979:929). In a

[13] Bishop and Heberlein do invoke the cost of breaking commitments in the wording of the hypothetical WTA question, however.

[14] The identification of the researchers with the state university, and awareness that they had the cooperation of the Wisconsin Department of Natural Resources (from whom they received the names of the goose-permit holders) lends credibility to such a conjecture.

subsequent series of ingenious experiments, they further explored the relationship between hypothetical and simulated market payments. This time they were able to conduct a simulated WTP market. Compared with the Horicon findings, their Sandhill studies, to which we now turn, lead to a very different conclusion about the validity of WTP_h measures in valuing quasi-private recreational goods.

The Sandhill experiments. The study site for these important experiments was the Sandhill Wildlife Demonstration Area in Wood County, Wisconsin. The Wisconsin Department of Natural Resources (DNR) uses this extensive, 12-square-mile site, which is ringed with a deer-proof fence, to conduct deer management experiments. To maintain the deer population within habitat bounds, the DNR usually allows a limited number of people to hunt specified types of deer for a single day each year. This opportunity is so highly valued by Wisconsin hunters that a lottery system is necessary to allocate the annual free Sandhill deer-hunting permits. In 1983, 6,000 hunters applied for the 150 permits offered that year.

Bishop and Heberlein obtained the DNR's permission to use the lists of Sandhill permit applicants and permit recipients for two rounds of HSM experiments, which they conducted during the 1983 (Bishop, Heberlein, Welsh, and Baumgartner, 1984) and 1984 (Bishop and Heberlein, 1985; Heberlein and Bishop, 1986) seasons. These experiments involved the same type of simulated-hypothetical market comparisons employed in the Horicon study, only this time Bishop and Heberlein were able to sell permits as well as buy them. In both years the WTP experiments were conducted by mailing questionnaires to random samples chosen from the unsuccessful applicant pool.[15] Where the treatment included a bidding game format, brief follow-up telephone interviews were used to conduct the iterative game.

Because they were granted only 4 permits to sell in 1983, Bishop and Heberlein used auction markets for the WTP treatments in this round of experiments. Respondents were told that only the four highest bidders would win the permits. In all, four different auction formats were used in 1983, each with a simulated and a hypothetical WTP treatment. The total sample of 600 hunters was divided equally among the eight treatments. In the simulated market treatments the amounts the hunters chose to bid constituted binding contracts to buy a permit if the bid was successful. Respondents of the hypothetical treatments received a similar offer form on which they were asked to put down the highest whole-dollar amount

[15] In addition to the WTP treatments, each round of experiments included WTA versions, which we will not consider here.

of cash they would pay if the offer were real. The different types of auctions used in the 1983 experiments are described in table 9-1.

In designing the 1984 experiments, Bishop and Heberlein took into account the possibility that the auction mechanisms they had used the previous year might have caused respondents to behave differently than they would have in normal markets. As they observed in a preliminary report to the DNR,

> In auctions, people must think about their own values, as well as the probable bids of all other participants in the auction. In normal type markets, people need to decide whether their own personal valuation of the object is higher or lower than the market price. The purpose of the 1984 study is to see if people's behavior (both real and contingent) is different when they only have to consider their own valuation. (Bishop and Heberlein, 1985)

Armed with 75 permits in 1984, Bishop and Heberlein were able to conduct a simulated WTP market in which they offered 75 hunters, chosen randomly from the list of unsuccessful applicants for Sandhill permits, the opportunity to buy a permit for a specified price. The prices, randomly assigned to the hunters, varied between $18 and $512. Seventeen hunters chose to buy the offered permit for prices which ranged between $18 and $109. In the parallel hypothetical market, another 75 respondents were presented with a specified price and asked if they would be willing to pay that amount for a permit if the offer were real. The prices offered to this group were identical to those offered in the simulated market. A second hypothetical WTP market was also created, identical to the first except that a follow-up phone interview was used to conduct a bidding game whose starting point was the price initially offered to each of the 75 respondents assigned to the first hypothetical treatment.[16]

In the summary of the findings of the 1983 and 1984 Sandhill WTP experiments given in table 9-1, it can be seen that when real money was involved, the four 1983 auctions elicited very similar mean values of about $23. The hypothetical values in these auctions were somewhat higher, but only two of the four comparisons are statistically different at the .10 level. The 1983 Sandhill study showed much closer correspondence between simulated and hypothetical market prices than did the earlier Horicon study. The most important results, however, come from the 1984 experiment, which replaced the auction framework (with its potential for distorting effects) with the superior take-it-or-leave-it framework. Comparison E in table 9-1 shows a very close relationship between the mean

[16] In all these experiments, respondents were also sent a detailed follow-up questionnaire to obtain information about their attitudes, hunting experience, reactions to the experiment, and personal characteristics. The use of well-designed questionnaires and a $5.00 payment for a completed questionnaire achieved response rates of 90 percent or better.

Table 9-1. Willingness-to-Pay Results of the Sandhill Experiments

	Estimated bid		
1983 Experiments[a]	Cash (*N*)	Hypothetical (*N*)	Difference significant at .10
A. *Sealed Bid Auction*			
Submitted by mail	$24 (68)	$32 (71)	No
B. *Bidding Game 1*			
Initial sealed bid solicited by mail; allowed to change bid in subsequent telephone interview	19 (65)	43 (62)	Yes
C. *Bidding Game 2*			
Initial response to mail offer of fixed amount chosen at random between $1 and $500; this amount was used as a starting point in a subsequent telephone interview	24 (68)	43 (69)	No
D. *Fifth Price*			
Sealed bid solicited under condition that four highest bids would receive a permit by paying the amount offered by the fifth highest bidder	25 (69)	42 (70)	Yes
1984 Experiments[b]			
E. *Single Price Offer*			
Asked if willing to pay specified amount randomly chosen from $18 to $512 range	31 (70)	25 (62)	No
F. *Single Price Offer Plus Bidding Game*			
Same procedure as in E; this amount was used as a starting point in a subsequent telephone interview	No data	35	No

[a]Bishop and Heberlein (1986).
[b]Bishop and Heberlein preliminary estimates, personal communication to the authors, April 1986.

amounts of $31 for the simulated and $25 for the hypothetical markets. It is also interesting that the hypothetical value of $25 is closer to the real payment (WTP_s) amounts elicited in the 1983 experiments than is the 1984 real payment.[17]

We draw two conclusions from the Sandhill findings. First, as suggested by the criterion of simulated market prices, they demonstrate that hypothetical markets can value goods such as Bishop and Heberlein's hunting permits with considerable accuracy. This is an important finding because it establishes a necessary condition for the validity of CV studies. But as we pointed out earlier, this finding is only weak (if valuable) evidence that CV studies can accurately value quantities of goods like air- and water-quality improvements, which are much less well-defined in respondents' minds than Sandhill permits are to Wisconsin's avid and well-informed hunters. Second, the Sandhill data do not support the idea that the hypothetical character of CV studies leads people to give consistently higher or lower values than they would if they had to pay cash for the good being valued. The hypothetical offers are sometimes higher and sometime lower than the cash offers.

The Strawberry experiment. The criterion used in the final HSM study to be reviewed here is a purely private good. It, too, confirmed the ability of hypothetical markets to produce valid estimates of commodity values. In 1984, Dickie, Fisher, and Gerking (1987) conducted a field experiment to compare the results of hypothetical and simulated markets in a nondurable and familiar commodity—pints of strawberries. Interviewers appeared at the doors of 144 randomly selected households in Laramie, Wisconsin, displayed their wares before the household member identified as the person who regularly shopped for groceries, and informed him or her of the selling price, which varied between $0.60 to $1.60 per pint. In the simulated market the strawberries were sold for money if the respondent desired to make the purchase at the price offered; in the hypothetical market the respondents were told that the interviewer was conducting a market research study and then asked: "Suppose that a pint of strawberries can be purchased for $____. How many pints would you buy?" A comparison of demand curves failed to disprove the null hypothesis that the two treatments are statistically identical. After appropriate adjustments for outliers and interview effects were made, there was less than a 1 percent difference between the average hypothetical and actual prices. For this type of well-defined consumer good, CV estimates probably accurately predicted people's real market behavior.

[17] We do not discuss the findings of the various iterative bidding game treatments used in the Sandhill experiments, as this elicitation technique introduces the possibility of compliance bias by pressuring the respondent to bid higher than she really wants to go.

Hypothetical-Real Market Comparison

The HSM experiments are applicable only to hypothetical consumer market models which value semi-private goods. Referenda outcomes are a possible criterion for referendum-based hypothetical markets in public goods. Carson, Hanemann, and Mitchell (1986) conducted a study of this type by comparing the findings of a CV-like study of how California voters intended to vote on a proposition with the actual voting behavior of Californians on that proposition in a subsequent election.

The Field Institute's California Poll periodically conducts telephone surveys during California election campaigns which ask random samples of voters numerous questions about the issues and their voting intentions. In early October 1984, Carson, Hanemann, and Mitchell arranged to include in one of the institute's polls a few questions which asked the 1,022 registered voters about their voting intentions on Proposition 25. This measure, the Clean Water Bond Law of 1984, authorized a 20-year bond issue of $325 million, largely for the purpose of constructing sewage treatment plants. It was one of sixteen measures that would appear on the November 1984 ballot. Before the election, but after the survey, every registered voter received the California voters' pamphlet containing detailed pro and con arguments on the proposal.

This study measured respondent voting intentions in two different ways. In the first, which most concerns us here since it best represents a hypothetical version of the actual referendum, the interviewer read the brief description of the measure that would appear on the November ballot, then asked the respondent how he would vote if the election were held the next day. The second involved a set of questions eliciting the voters' intended vote under different price conditions. A demand curve for the water-quality improvement program was constructed from these data, which ranged from 89 percent approval at a price of $1 per year per household to 48 percent approval at a $50 annual household price.

The particular finding which provides evidence for the study's criterion validity is the comparison between the predicted vote as measured in the voting intention question and the actual vote on Proposition 25 a month later. Sixty-three percent of the respondents said they would vote for the measure, 13 percent said they would vote against it, and 25 percent said they were unsure how they would vote. Splitting the undecideds 50/50, 60/40, and 70/30 against/for (studies show that those who are undecided tend to vote more for the status quo than for change), gives predicted passing votes of 75 to 70 percent. The actual vote in favor was 73 percent, well within the 95 percent confidence interval for percentages of this size for the sample size used. In this instance, the hypothetical referendum received strong validation. That finding, which is replicated in other studies of referendum voter behavior conducted by political scientists (Magleby, 1984), suggests that a well-conducted CV study using a referendum format is likely to be a valid representation of voters' actual behavior in referenda.[18]

Construct Validity

Convergent Validity

Convergent validity, as noted earlier, concerns the correspondence between a measure and other measures of the same theoretical construct. To the extent that a correlation exists (that is, the measures converge) the validity of each measure is confirmed. This is in sharp contrast to criterion validity, where the criterion measure (such as Bishop and Heberlein's WTP_s) provides a standard against which the measure of interest can be judged. In convergent validity neither of the measures is assumed to be a truer measure of the construct than the other.

We emphasize the distinction between criterion and convergent validity because there has been some confusion on the point in the past. As a result of the credibility accorded by economists to the Observed/Indirect (travel cost and hedonic price) benefit measurement techniques described in chapter 3, considerable effort has been devoted to comparing the values for an amenity reached by one of these methods with the values obtained by the CV method for the same amenity. There has been a tendency to treat measures obtained by the Observed/Indirect methods as if they were criterion variables (Knetsch and Davis, 1966; Thayer, 1981; Brookshire, Thayer, Schulze, and d'Arge, 1982). In fact, for the reasons given in chapter 3, there are no grounds for considering travel cost- or hedonic price-based measures to be more accurate than CV-based measures.[19] These factors have led CV researchers to recognize that a correspondence between Observed/Indirect and CV-based estimates does not prove the accuracy of survey measures (Cummings, Brookshire, and Schulze, 1986; Cummings, Schulze, Gerking, and Brookshire, 1986:275; Smith, Desvousges, and Fisher, 1986), but rather contributes to the credibility of both.

[18] Fischel (1979) administered surveys to measure whether citizens in several New England towns favored or opposed letting a pulp mill, which would provide jobs for local residents, locate in their town. The findings of these surveys, whose wording closely resembled a CV scenario, were close to the outcomes of town meeting votes in those communities that voted on the issue.

[19] See also Smith, Desvousges, and Fisher (1986) for a comparison of the problems posed by CV and participation/travel cost analysis. Vaughan and coauthors (1985) conducted simulation studies which showed that it is often impossible to recover an individual's true WTP amount given the level of aggregation and price proxies typical of participation models.

What do the various studies which compare CV measures with travel cost and hedonic price measures show in the way of convergent validity? Overall, the correspondence between survey measures and the behavior-based measures is reasonably close. Before describing these findings, however, it must be noted that the benefits measured by the two types of methods are not strictly comparable. Travel cost analysis, for example, is an ex-post welfare measure which tends to exclude existence values, whereas contingent valuation is an ex-ante measure and includes these values. In one study valuing a recreation site (Thayer, 1981), contingent valuation and site-substitution measures were compared. The CV measure concerned what the sample was willing to pay to preserve the site, and included both recreational and existence values. The site-substitution technique measured the extra cost of going to an alternative site, and thus did not include intrinsic values. In general, hedonic price studies of air visibility benefits based on property values make the questionable assumption that all the benefits of air quality are capitalized into residential housing prices. This assumption ignores the possibility that some or all of these benefits may be incorporated into wage rates or into the prices of other goods and services.

The study *State-of-the-Art Assessment* (see Cummings, Brookshire, and Schulze, 1986) scrutinizes eight different convergent validation efforts, including four that used travel cost estimates, one that used a site-substitution estimate, and three that used estimates based on hedonic price measures.[20] A number of additional studies have compared benefits estimated by the CV method with benefits for the same good measured by an Observed/Indirect measurement technique: Blomquist (1984) used hedonic pricing estimates of the value of a view of Lake Michigan in Chicago; Loehman (1984) made estimates for air visibility in San Francisco based on a hedonic price analysis of property values; Hoehn and Randall (1985b) estimated a demand curve for visibility from the observation deck of the Hancock tower in Chicago[21]; Mitchell and Carson (1984) used Vaughan and Russell's (1982) estimate, based on a participation-travel cost model, of the benefits of moving the minimum level of national water

[20] The travel cost studies and their commodities are: Knetsch and Davis (1966) on recreation days; Bishop and Heberlein (1979) on hunting permits; Desvousges, Smith, and McGivney (1983) on water quality improvements; and Sellar, Stoll, and Chavas (1983) on boat permits. The site-substitution study is Thayer's (1981) study of a recreation site. The hedonic price studies include Brookshire, Thayer, Tschirhart, and Schulze (1982) on air quality improvements; Cummings, Schulze, Gerking, and Brookshire (1986) on municipal infrastructure; and Brookshire (1985) on earthquake information. The Cummings, Schulze, Gerking, and Brookshire study used wages, while the other two hedonic price studies used property values.

[21] See also Tolley and Randall (1985).

quality from boatable to fishable; Walsh, Sanders, and Loomis (1985) employed travel cost estimates for river use; Sorg and coauthors (1985) used travel cost estimates for cold- and warm-water fishing; Sorg and Nelson (1986) examined travel cost values for elk hunting; and Walsh (1986) used travel cost values for forest use.

Two of these studies are of particular interest. The Observed/Indirect measure used by Hoehn and Randall (1985b) was of especially high quality. They avoided many of the assumptions found in standard participation travel-cost and hedonic-price techniques because the demand curve they were able to estimate was very close to the direct demand curve for visibility. The Walsh, Sanders, and Loomis study is unique in limiting its travel cost-contingent valuation comparison to the stated-use portion of the total CV value. This is a more correct comparison because the total CV value also includes sizable existence and preservation components.

The results of these studies, taken as a whole, indicate reasonably strong convergent validity for CV. According to Cummings, Brookshire, and Schulze (1986), the comparisons they reviewed are within 60 percent of each other, and many are much closer. This is true of the additional studies as well,[22] including the Hoehn and Randall study, whose comparison measure is closer to criterion status than any of the others. Hoehn and Randall did not find a significant difference between their CV and demand curve estimates, and their CV measure had a much smaller standard error than their Observed/Indirect measure.

Theoretical Validity

Theoretical validity involves assessing the degree to which the findings of a study are consistent with theoretical expectations. Here the interest is focused on the determinants of a WTP amount rather than, as is the case with convergent validity, with the fit between two separate but equal measures of the same construct. Theoretical validity is most commonly measured by regressing some form of the WTP amount on a group of independent variables believed to be theoretical determinants of people's willingness to pay for the good being valued. The size and sign of the estimated coefficients are then examined and judged to be consistent or inconsistent with theory.[23] Another way to test theoretical relationships is

[22] We have not been able to review several other studies which compare CV results with those obtained by other techniques (for example, N. Smith, 1980; Duffield, 1984). Blomquist's (1984) review of statistical value-of-life estimates, obtained by CV and other techniques including hedonic wage equations, finds that CV estimates tend to be at the lower end, but not outside, the range of those obtained by the other methods.

[23] This type of procedure must be distinguished from a regression based on the "kitchen sink" approach, where every variable that "works" is entered into the equation to obtain the highest R^2 possible. Such equations are not satisfactory evidence for theoretical validity, although high R^2s obtained by such procedures are evidence for reliability.

to compare mean WTP values of different conditions for which theory suggests different values. For example, theory would suggest that larger risk reductions should be worth more than smaller ones and that shorter delays in providing an amenity are more valuable than longer delays in providing the same amenity.

Although the use of theoretically based regression equations in CV studies goes back to the method's earliest days, the publication of the findings of such equations in CV reports and elsewhere has become commonplace only in recent years. Knetsch and Davis (1966) used regression analysis to assess the theoretical validity of Davis's pioneering CV study of the value of a forest recreation area in northern Maine. Each of the variables in an equation, consisting of household income, years of experience by the household in visiting the area, and the length of stay in the area, were significantly related to the WTP measure. Together they explained a high percentage of the variance. On this basis, Knetsch and Davis (1966:135) concluded that "the economic consistency and rationality of the responses appeared to be high."

A wide range of amenities has been valued in the other studies that report regression equation evidence for the theoretical validity of their findings. These studies include Cicchetti and Smith's (1976b), Gramlich's (1977), Mitchell and Carson's (1981), and Carson and Mitchell's (1986) studies of water quality benefits; Walsh, Miller, and Gilliam's (1983) study of congestion in skiing areas and willingness to pay for expansion of skiing capacity; Walsh and Gilliam's (1982) study of the benefits of wilderness expansion; Majid, Sinden, and Randall's (1983) study of the benefits of proposed new parks in Australia; Tolley and Randall's (1985) study of air visibility benefits in the eastern United States; and Mitchell and Carson's (1986b) study of the benefits of reducing drinking-water risks. The latter three studies are notable for the attention they pay to the derivation and measurement of a formal theoretical model. The fact that the values elicited in these studies are related to the respondents' preferences and experiences in the predicted manner is important evidence for their theoretical validity. For small ticket amenities, however, regressions are often not reported, or, when they are, sometimes show that the theoretically suggestive predictors are statistically insignificant. This raises the question of whether people might be giving the equivalent of donations in these instances, or, alternatively, whether their WTP responses are biased by such factors as the expected level of an entrance fee payment vehicle (Mitchell and Carson, 1986d).

Whenever contingent valuation studies are designed with the intent of gathering data to be used for policy purposes, it is highly desirable that they take into account the need to produce theoretically based regression equations or comparable evidence of their theoretical validity, and that these be presented as a standard part of every CV study report.

Summary and Conclusions

In this chapter we have described the three major types of validity and reviewed the evidence for the validity of CV studies. The validity of a measure is the degree to which it measures the theoretical construct—for example, what respondents would actually pay for a public good if a market for it existed. Content validity is assessed by making a qualitative judgment about the degree to which a CV instrument and the analysis procedures used in conjunction with it cover the appropriate domain.

Criterion validity is assessed by comparing the findings of a CV study with a criterion variable, which is usually an important form of behavior external to the CV measuring instrument. The variable is believed to be a better approximation of the underlying construct, and therefore the closer the CV measure is to the variable's measure, the greater its validity. We identified two types of criteria relevant to CV studies. The first is the value of semi-private goods as measured in simulated markets. The findings of a series of important experiments which compared simulated market and hypothetical CV values for the same goods were reviewed. For goods which are well understood by respondents (hunting permits, admission to see a TV show), the correspondence between hypothetical and simulated markets was shown to be quite strong. This finding is consistent with the laboratory experimental literature on preference revelation, which shows, according to Schoemaker (1982:553–554), "*no evidence* that suboptimal laboratory behavior improves when committing subjects financially to their decisions." Because the simulated market-hypothetical market studies are based on a consumer market model, their findings are not directly relevant to the use of CV studies to value genuinely public goods. The second type of criteria is the vote in state referenda. We illustrated this criterion by describing one CV study which successfully predicted the vote on a water-pollution bond issue in California.

In the third type of validity, construct validity, a measure is validated by comparing it with other measures that theory suggests it should be related to. The convergent form of construct validity is assessed in CV studies by examining the correspondence between a CV estimate for an amenity's value with the findings of alternate ways of valuing the amenity. A positive correspondence is interpreted as validating both measures, since neither can be presumed to be a superior measure of the underlying construct. The various studies in the CV literature that compare CV estimates with those derived by Observed/Indirect methods fall into this category despite the tendency of economists to place more faith in Observed/Indirect measures. The large number of comparison studies show a reasonably high level of convergent validity, which is noteworthy given the various incompatibilities between the domains covered by CV measures on the one hand and Observed/Indirect measures on the other, with which CV measures have been compared.

Theoretical validity, the other form of construct validity, is assessed in CV studies primarily by regressing some form of the WTP amount on a group of independent variables believed to be theoretical determinants of the WTP amount. The outcome of interest is not the R^2—which is evidence for reliability rather than validity—but the size and sign of the estimated coefficients and whether they are consistent or inconsistent with theory. A number of CV studies have demonstrated this type of validity.

The findings of the various validity assessments reviewed in this chapter are generally favorable to the CV method's potential for measuring valid WTP amounts. The evidence presented from simulated-hypothetical comparisons is particularly impressive testimony to the ability of surveys to predict spending behavior involving quasi-private goods.

None of these conclusions say anything about the validity of a particular CV study, of course. The multidimensional character of validity and the absence of a clear-cut criterion against which to compare CV values for public goods means that the validity of individual studies cannot be established in a definitive fashion. Each contingent valuation study should be able to survive scrutiny of its scenario and its estimation procedures for content validity, and should provide evidence for its theoretical validity in the form of either theoretically based regression equations or experimental findings based on split-sample comparisons. In the next chapter we discuss strategies to improve the reliability and validity of CV studies.

10

Enhancing Reliability

Most narrowly defined, reliability refers to the extent to which the variance of the WTP amounts given by respondents in a contingent valuation survey is due to random sources, or "noise." If we have the results of a CV survey, the variance in the WTP amounts given for the good is the result of three principal factors. The first is the "true" underlying variation in willingness to pay for the good in the population of integer. Even if the values for the good were measured by a perfect instrument applied to a perfect sample, some people will be willing to pay more than others for the good. The second factor is the instrument: its concepts, its wording, its method of presentation. Imagine that the CV survey is administered at two different times, a few weeks apart, to exactly the same sample under identical conditions; and assume that when the survey is administered the second time the respondents have no memory of the WTP amounts they gave the first time (however unlikely), and that their preferences for the good in question have not changed in the interval. In this classic test-retest situation, the difference between the responses at time one and time two will reflect the true variance plus some noise introduced by the instrument's imperfections. The third source of variance in the mean WTP amount concerns the fact that we have only interviewed a sample of the population. Our sample of respondents is surrogate for a much larger population—perhaps 200,000, or even 240,000,000. In order to accurately represent the population, the sample was selected by rigorously following accepted random sampling procedures. If we had a perfect instrument and selected ten independent samples using identical random sampling procedures, the variance among the ten estimated willingness-to-pay amounts would reflect the variance introduced by our sampling procedures.

Efforts to enhance the reliability of a CV study's WTP amounts are of vital importance, because the more random the respondents' WTP amounts, the greater the chance that a study's mean WTP amount could be very different from the true WTP amount for the good. Some variance is inevitable in contingent valuation estimates, of course. But CV studies using relatively small samples or having scenarios that respondents find unclear or unrealistic could obtain estimates which differ from a true mean by large amounts. Studies having these characteristics are all too often published in good journals. As a basis for making benefit-cost determinations, their findings can easily be meaningless.

Another reason why the designer of a CV study needs to be concerned about reliability is that an instrument which leaves respondents confused, uncertain, or unbelieving is likely to tempt them to rely for values on cues in the instrument, rather than prompting them to search for their preferences. A survey instrument which provokes these responses is likely to cause systematic error or bias instead of promoting random guesses at what the good is worth.

We begin this chapter with a brief discussion of how the reliability of a CV study can be assessed, then go on to discuss how reliability can be enhanced in the design of the survey instrument and the sample. Although a clearly worded and realistic scenario helps improve the reliability of CV estimates, realism is sometimes difficult to achieve without inducing bias—a problem explored in detail and with particular reference to the nature of the good, the payment vehicle, and the elicitation method. We then turn to how the reliability of sample statistics from CV surveys can be enhanced by the use of robust statistical procedures and large sample sizes.

Assessing Reliability

Techniques are available to estimate the variance contributed by the instrument and by the nature of the sample. Unfortunately, these procedures are very expensive to carry out properly, and for that reason they are not performed routinely in sample surveys.

Variance due to the instrument can be assessed by test-retest methods which are well established in the psychometric and experimental design literature (Bohrnstedt, 1983, for example), where they are reported most often in studies having laboratory subjects or students. These methods are difficult and costly to use in surveys of general populations because the respondents have to be located again and convinced to repeat the survey. In one of the few CV test-retest studies on a general population known to us in the literature, Jones-Lee, Hammerton, and Philips (1985), in a national transportation-risk survey, reinterviewed a subsample of those interviewed a month earlier. The responses to the CV risk-reduction valuation question that was repeated showed no significant difference between the original and recall responses.[1] In another study, Loehman and De

[1] The distribution of the differences was relatively symmetrical, although the standard deviation was fairly large (Jones-Lee, Hammerton, and Philips, 1985:66). See also Hammerton, Jones-Lee, and Abbott (1982).

(1982) interviewed a group of about 45 college students twice, using the same CV instrument. The correlation between the original WTP amounts given and those given by the same students three weeks later was very high (r = .86). Heberlein (1986) obtained a substantial correlation between WTP amounts hunters gave for a deer-hunting opportunity one year and those they had given for the same opportunity a year earlier. In the absence of reinterviews, the burden on the researchers is to demonstrate that the individual willingness-to-pay amounts are not simply random responses. This can most easily be done by obtaining a respectable R^2 when regressing WTP on a set of independent variables, since the higher the R^2, the lower the random portion of the WTP response variance. If the independent variables used are those suggested by theory, regression analysis can also be used to demonstrate construct validity, as described in the preceding chapter.[2] Nevertheless, the reliability of a CV study which fails to show an R^2 of at least .15, using only a few key variables, is open to question. High R^2s for equations based on the kitchen sink approach, in which explanatory power is maximized by the indiscriminate use of as many variables as possible, are not particularly meaningful, especially for smaller samples where estimated R^2 and even $\bar{R}^2$ become very biased measures of the true variance.

High reliability, while desirable, does not convey any information about a study's validity or the absence of bias. Consider a study in which the WTP values clustered tightly around $100 for a particular environmental public good, in both the original survey and a follow-up survey of the same respondents three months later. If we knew that these values had been given in reply to a question which asked whether the respondents were willing to pay $100 for the good, we would suspect that the high level of reliability was obtained at the cost of severely biased responses.

Sampling variance can be estimated by use of a sampling design consisting of several large, independent subsamples (Cochran, 1977). Such an interpenetrating sample design allows the researcher to isolate the variance component that is due strictly to sampling by examining the variability of the sample statistics (for example, mean, median, and standard deviation) for each of the independent subsamples. The assessment of sampling variability through use of interpenetrating samples is directly analogous to the test-retest situation. In the first case the interview is assumed to be held constant while the samples vary; in the second, the sample is held constant by interviewing the same people at multiple times, while the interview situation varies.

[2] Knetsch and Davis (1966) provided early evidence for the reliability of CV measurements, as they were able to explain 59 percent of the variance in willingness to pay for use of a forest recreation area with a simple regression equation which included income, length of stay, and years of acquaintance with the area visited. This R^2 is considerably higher than normally found in contingent valuations of more abstract goods, or than is obtained from more heterogeneous populations than their Maine recreationists.

Estimating a study's reliability by the test-retest and the interpenetrating sampling approaches amounts to conducting multiple surveys—a prodigious investment of resources. More often than not, the same money would yield greater improvement in data quality if it were spent on increasing the sample size, improving the interviewers' training, conducting more elaborate pretests, and running split-sample experiments to test for key instrument effects.

The Survey Instrument and Reliability

We now turn to the survey instrument as a source of variance. Earlier we have argued that in CV studies it is important that the scenarios posed appear realistic to respondents. Realism in a CV scenario concerns the degree to which the valuation situation is plausible and meaningful to the respondent in the way intended by the researcher. Rowe and Chestnut (1982:70) concisely describe the characteristics of a good contingent valuation scenario:

> [It] must be informative; clearly understood; realistic by relying upon established patterns of behavior and legal institutions; have uniform application to all respondents; and, hopefully, leave the respondent with a feeling that the situation and his responses are not only credible but important.

The closer a CV scenario comes to meeting this description, the more reliable the WTP amounts collected are likely to be.

A number of factors contribute to making a scenario realistic. An obvious one is the degree to which respondents are familiar with the key scenario elements before the interview. These elements are the good, the method by which the good will be provided, the levels of its provision, the elicitation framework, and the payment vehicle. Familiarity may be acquired by experience or from indirect sources, such as television, reading, or hearing people talk. Direct prior experience is not essential for familiarity. Respondents who have never visited the Grand Canyon or seen a toxic waste dump may obtain considerable information about them from secondary sources. Nor does experience always result in familiarity. Although everybody experiences air visibility daily, some people are more aware of it than others. Because air visibility in Los Angeles is locally regarded as problematic, receives regular coverage in the news media, and is quite variable geographically, Los Angeles residents are more likely to be familiar with the concept of air visibility as an amenity than, say, residents of Chicago. Finally, familiarity with one scenario element, such as a method to reduce risk from toxic waste dumps, does not necessarily imply familiarity with the other scenario elements, such as different levels of mortality risk per 100,000 people.

Another factor affecting scenario realism is whether the key elements can be presented in the interview in such a way that respondents can easily grasp their meaning, however familiar or unfamiliar they may be. A respondent may not be familiar with smoke plumes from coal-fired power plants and their effect on views in the Grand Canyon, but simulated photographs can be used to portray the change in visibility. Illustrating changes in risks from hazardous waste dumps in relation to EPA risk standards, or in comparison with risks from other, more familiar, sources such as cigarettes, may enable respondents to grasp otherwise unfamiliar concepts.

A third factor influencing scenario realism is the degree to which the scenario appears plausible to the respondents. In an air visibility scenario an electric utility bill will be a more plausible payment vehicle than a sales tax, because the bill will have a more understandable connection to the cause of the visibility changes than the tax. To take another example, most respondents are likely to recognize that it costs money to improve freshwater quality, and that as taxpayers and consumers they ultimately will pay for that improvement. They are also likely to regard the status quo as the level of quality they are entitled to in exchange for their current payments; thus, being asked how much they would be willing to pay for improvement above the current level of provision is likely to make sense to them. However, if the elicitation framework for this improvement is altered to ask respondents how much they would be willing to accept (WTA) to forgo such change, the scenario is likely to seem implausible to respondents because (among other things) it seems to treat them as if they already owned the as-yet-unrealized improvements. Plausibility is also affected by the extent to which respondents believe the hypothetical improvements could be carried out. Respondents who do not believe nuclear power can be made safe will be incredulous if a scenario asks them how much they would pay for programs to reduce the risk from a given nuclear power plant to close to zero.

Also affecting the perceived realism of a scenario is the degree to which the scenario is compatible with prevailing moral assumptions. The use of a WTA format in a water-quality improvement scenario may not only appear implausible; respondents may also regard it as immoral to be "paid" to allow pollution to occur. It is possible to imagine scenarios that respondents would regard as both plausible and immoral; for example, a scenario for a proposed sale by the federal government of oil leases off the California shore, in which local respondents would be asked how much they would accept in reduced taxes for such a sale.

Scenarios that strike respondents as being unrealistic run the risk of eliciting one or more of three types of responses. The first and perhaps the least dangerous in terms of reliability—although highly undesirable for other reasons—are "don't know" answers. The other two—random guessing and responding to cues—occur when the respondent offers a WTP amount even though he cannot meaningfully connect the scenario to his preferences in the way desired by the researcher. Those who respond to cues rather than make a random guess resolve their uncertainty by basing their WTP amounts on some aspect of the scenario, such as the starting point in a bidding game, which supplies a cue to an appropriate value.

The Realism-Bias Relationship

It is important to note that there is no unique bias attributable to a scenario's lack of realism. This is why the concept of "hypothetical bias" (Schulze, d'Arge, and Brookshire, 1981; Moser and Dunning, 1986), which has been defined as "the potential error induced by not confronting any individual with the real situation" (Rowe, d'Arge, and Brookshire, 1980:6), is a misnomer. The only unique effect of a scenario's lack of realism is not bias, but random, directionless error (Thayer, 1981:32).[3] The researcher who wishes to make a scenario more realistic faces a tricky problem: on the one hand, an insufficiently realistic scenario will be vulnerable to bias; on the other, the elements which add realism to a scenario may themselves cause bias. Adding realism to a scenario may cause bias through an "information overload" effect whereby respondents ignore important information and focus on, and possibly misinterpret, unimportant information in determining their willingness to pay for the good in question.

From the researcher's point of view, CV scenarios contain two kinds of materials, those intended to be valuation-relevant and those intended to be valuation-neutral. The valuation-relevant elements are those the researcher wants the respondent to take into account in valuing the good—especially the description of the good and its provision. The remainder of the materials in the scenario are intended to convey a credible market for the good without affecting the WTP amount. The researcher does not usually want the respondent to be influenced by the particular elicitation method used in the study, much less by any of the numerous elements intended to be extraneous to the particular valuation situation. Thus, respondents ideally would be sensitive to the different levels of air visibility

[3] The effect of random error in CV studies may not be entirely without direction. Where the lower bound of a WTP distribution is constrained at zero, and there is no upper bound (as is the case with CV WTP distributions), increased variance resulting from random error will bias the mean WTP amounts upward from the true mean. This bias effect is likely to be most troublesome when the "true" mean willingness to pay for the good being valued is fairly close to zero and the distribution of the WTP amounts is characterized by a large variance parameter.

shown in a set of photographs of the same scene under different visibility conditions, and insensitive to the possible relationship between air visibility and sickness caused by air pollution, the particular characteristics of a payment card elicitation method, or the fact that the leaves are green in some of the photographs while in others they show fall colors.

Random guesses, to the extent that they occur, are pure noise and unrelated to bias or directional effects. However, the work on "non-opinion" behavior and decision-making heuristics discussed earlier suggests that uncertain respondents, when asked how much they are willing to pay for an amenity, are much more likely to rely on cues than to make guesses. In other words, instead of reacting to a scenario's valuation-neutral elements as intended, the respondent for whom the valuation situation is not very meaningful may treat them as valuation-relevant. In the absence of photographs that reinforce the concept of air visibility, for example, respondents may mistakenly regard the mechanism responsible for air visibility changes—air pollution—as valuation-relevant and include health benefits in the values given for the visibility effects.

While lack of realism increases a scenario's potential for bias, any attempt to increase a scenario's realism runs the risk that the reality-enhancing elements may be mistakenly regarded by respondents as valuation-relevant instead of valuation-neutral. The use of photographs to depict air visibility levels furnishes a good example of this problem. Photographs greatly enhance the realism of a visibility scenario because they vividly illustrate an amenity change—say, 20 mile visibility versus 30 mile visibility—which is virtually impossible to convey when only words are used. By focusing the respondent's attention on the amenity they also help to underscore the scenario's instruction to the respondent to disregard any possible health effects and to value only the visibility improvement. Unfortunately, air visibility photographs themselves contain numerous extraneous elements, such as the color of leaves, the angle of light, indications of the season, or the character of the setting. Although the researcher believes changes in these elements from one photograph to another are valuation-neutral, the respondent may unwittingly be influenced by them and give a WTP amount which is contingent, at least in part, on the particular views captured in the set of photographs used in the study. If shown a different set of views that illustrate identical visibility levels, the respondent might well give a different WTP amount.

Reducing Random Error

How can researchers enhance the reliability of their CV instruments without unduly biasing the findings? As noted previously, we do not yet have a sufficiently powerful model of response behavior in the survey interview to enable us to confidently predict when people will guess arbitrarily, when they will seize on cues, and when they will express their opinions.[4] In the absence of such a model, the following discussion is based on the assumption that procedures which increase a scenario's understandability, plausibility, and meaningfulness will enhance its reliability, even though these desirable qualities are often difficult to obtain without inducing bias in one way or another.

Careful use of various pretesting techniques to explore an instrument's weaknesses before taking it into the field is probably the single most effective way to enhance a study's reliability.[5] It is important to anticipate the mistakes that respondents may be prone to make and to provide opportunities during the interview for a respondent to learn about and correct any misunderstanding. For instance, there are two ways to ask about willingness to pay for different levels of a public good: the first is to ask for the marginal or incremental WTP amount for each change from a specific level of the good to the next higher level; the second is to ask for a total WTP amount from the base level of provision to each level of interest. On the basis of pretests, we have found that no matter which approach is used, a significant number of respondents will answer the WTP questions as if they were asked for the other value. It follows that higher-quality CV data can be obtained if the questionnaire is so designed that a respondent is shown his answers immediately after the sequence of WTP questions (summing, if the marginal method is used) and given the opportunity to change or reallocate his WTP amounts at that point. Pretesting is important because it is difficult to anticipate every eventuality

[4] Schuman and Presser (1981) offer some evidence that the stronger the opinion people hold on issues such as gun control or abortion, the less the random error in their responses (as measured in a panel study). This is an intuitively appealing finding because it seems plausible that the more people care about an issue, the better formed will be their opinion about it. However, even a generalization as simple as this one does not receive consistent empirical confirmation. Schuman and Presser point out that another study on the same topic (Judd and Krosnick, 1981) failed to replicate their original findings, and that in their own study people's resistance to the influence of cues from other aspects of the survey was unrelated to the strength of their opinions.

[5] Although the term "pretest" is sometimes used to refer to a small-scale test survey, we use it here to refer to the various research activities conducted in the course of developing a survey instrument. These include the use of focus groups, observed or tape-recorded interviews, the systematic debriefing of respondents, and minisurveys of 25 or more representative respondents. To the best of our knowledge, the survey research literature lacks an overall description of the diverse procedures employed by survey researchers in developing new survey instruments. Useful brief discussions of this topic are given in Sheatsley (1983) and in Converse and Presser (1986), and Hammerton, Jones-Lee, and Abbott (1982) provide an excellent description of how they conducted a pretest to determine the feasibility of using a survey to value safety risk reductions.

of this kind, especially in CV surveys which explore topic areas not previously studied in depth by researchers.

Reliability will also be enhanced if the researcher avoids putting too much pressure on uncertain respondents to answer the WTP questions. Schuman and Presser (1981), for instance, reduced the percentage of responses to questions meaningless to the respondent from 30 to 10 percent by adding the alternative "or do you not have an opinion on that issue" to the question. They also found the addition of this phrase helped the interviewers, who were otherwise sometimes reluctant to accept respondents' expressions of uncertainty at face value despite explicit instructions given to the interviewers to code such expressions as "don't knows." However, legitimating don't knows in this manner must be done with great care in CV surveys, for there is a delicate balance between making it too easy for respondents to say don't know when they would otherwise be able to come up with meaningful values and pressuring respondents to give values when they really cannot do so in a meaningful way.[6]

Another reliability-enhancing device is to offer the respondent more opportunity to think about the topic before posing the valuation questions. This may be done in the wording of individual questions. Questions that contain redundancies have been found to produce more accurate responses than shorter versions of the same questions (Cannell, Miller, and Oksenberg, 1981). It might also be done by asking questions, in the early part of the questionnaire, which pose tradeoffs between that good and other public and private goods. In implementing this approach care must be exercised to prevent the preliminary questions from highlighting the good being valued in such a way that respondents are led to systematically overvalue or undervalue it—a type of bias discussed in chapter 11 as "instrument context bias."

[6] Procedures to separate true zeros from protest zeros and to convert protest zeros to usable responses may be used to recapture respondents who essentially opted out of the valuation process. It is now a standard practice in CV surveys to ask respondents who gave zero WTP amounts why they did so, and to use their answers to this question to separate those who give protest zeros from those who really have no value for the amenity. Once those who protest in this fashion have been identified, it is possible to use predetermined follow-up questions to attempt to overcome their objections. If the pretest revealed that some respondents were reluctant to express WTP amounts because they believed the amenity should be paid for out of corporate profits, the follow-up could attempt to dissuade them from this view. Such follow-ups should not be ad hoc attempts by interviewers to convince respondents. Predetermined statements should be provided to all interviewers for anticipated questions and should be read verbatim in response to questions, without further elaboration. Otherwise, the comparability of the interviews is weakened. Any WTP amounts ultimately given by respondents who are persuaded to abandon a protest zero in this way are likely to yield better estimates than if their values were counted as missing and subsequently imputed using statistical techniques.

The Effect of Major Scenario Components on Reliability

Each of the major components of CV scenarios—the good and its provision levels, market type, payment vehicle, and elicitation method—affects reliability. We discuss each of these in turn below. By designing scenarios that meet the three criteria of understandability, plausibility, and meaningfulness, CV researchers can overcome the threat to reliability resulting from the hypothetical nature of a CV scenario. It should be kept in mind, however, that at times it may be desirable to sacrifice some reliability to avoid a larger threat of bias.

The good and its provision levels. Some goods and provision levels are easier for respondents to value in the CV framework than others, as we have seen. A high degree of familiarity may be assumed when semiprivate recreation goods, such as backwoods areas in Maine (Davis, 1964), lack of congestion at Colorado ski resorts (Walsh, Miller, and Gilliam, 1983), and offshore diving platforms (Roberts, Thompson, and Pawlyk, 1985), are valued by interviewing their users. In the Maine and Colorado studies just cited, the respondents were even interviewed on site, a practice followed in several similar studies (McConnell, 1977; Daubert and Young, 1981; and Thayer, 1981). Contingent valuation studies which valued air visibility in the Four Corners area in the Southwest (Randall, Ives, and Eastman, 1974; Rowe, d'Arge, and Brookshire, 1980) and Los Angeles (Brookshire, d'Arge, and Schulze, 1979) also benefited from having a familiar good.

Closer to the unfamiliar end of the continuum is the study of national water quality by Mitchell and Carson (1984). While the concepts of boatable, fishable, and swimmable water-quality levels were familiar to the respondents, the notion of a mostly nonlocal good—all freshwater water bodies in the United States—was a more difficult concept for them to comprehend. Perhaps the least familiar goods to have been valued by CV studies to date have been the prevention of a future change in the level of atmospheric CO_2 (d'Arge, Schulze, and Brookshire, 1980) and reductions in the health risks posed to local residents by nuclear power plants (Mulligan, 1978) and by toxic waste dumps (Smith, Desvousges, and Freeman, 1985). While respondents in the health-risk studies might have been familiar with the nature of the facilities posing the risks being valued, it is unlikely that they were familiar with the actual risk probabilities. In all three studies the causes or potential causes of the effects were particularly hard to grasp because they were complex and uncertain. In valuing such goods as nuclear power and toxic waste dumps, an additional factor is that the risk of death from these sources may be so emotionally charged that respondents find it difficult to accept the institutional arrangements presented in the scenario, and/or may substitute a higher level of risk for the level the researcher meant to convey.

Davis has placed such importance on familiarity that he suggested that only users of a good should be interviewed in CV studies, preferably while they were actually engaged in the consuming activity (Davis, 1964). Although some goods may be too removed from everyday life to be meaningfully valued by nonusers, the notion that only users can give meaningful WTP amounts is too extreme a position. Information is available to respondents from sources other than personal experience, and well-written scenarios and appropriate visual aids can assist respondents to understand the good and its provision levels. It is important that respondents who are unable to conceptualize the good sufficiently to be able to value it are provided with the opportunity to say they don't know or are too unsure about the good to give a WTP amount. Provided such WTP nonresponses are kept at a manageable level, there are techniques that can be used to correct for them (see chapter 12).

Type of market. The researcher has a choice between trying to imitate a private goods market or a political market. For valuing pure public goods, we feel that the political market model is usually the best choice, for reasons described in chapter 4. In terms of reliability, the referendum model greatly enhances a public good scenario's understandability, plausibility, and meaningfulness. For valuing quasi-private goods the appropriate market model is much more dependent on the particular good being valued and other aspects of the scenario, such as the payment vehicle and the possibility of exclusion from the amenity.

The payment vehicle. Although the good that researchers are commissioned to value is typically not under their control, the choice of payment vehicle usually is. The vehicles most often used in CV studies, such as utility bills, entrance fees, taxes, and higher prices, are likely to be familiar to most respondents.[7] What is novel, however, is the way these vehicles are used in CV studies. Respondents ordinarily think of their electric utility payments as a way to purchase electricity, rather than a way to buy increased air visibility. Likewise, while the concept of paying federal income taxes to obtain environmental protection (among other goods) may be familiar, the notion of voluntarily specifying how much in taxes a carefully described water-quality improvement is worth is not. The chosen payment vehicle, as already noted, should have a plausible connection with the amenity it is used to value.

The payment vehicle should be neutral with respect to the good unless the researcher intends to value a policy which is linked to a particular

[7] Those who rent apartments will be notably less familiar with property taxes, and, depending on rental arrangements, with utility fees. A "special fund" vehicle, designated for a particular purpose, will also be unfamiliar (Sutherland and Walsh, 1985). Use of such a vehicle requires pretesting to ascertain its credibility.

payment vehicle. In situations where respondents do not understand the scenario in the way intended by the researcher, a tradeoff between plausibility and understandability may be necessary to avoid misspecification. For example, despite its high level of familiarity and obvious connection with the good, the use of an entrance fee as a vehicle to value some aspects of a recreational site may be a poor choice because of the likelihood that it will encourage respondents to restrict their WTP amounts to the range associated with a "fair" or customary entrance fee (see the discussion of implied value cues as a source of bias in chapter 11). Similarly, if property taxes are used as a payment vehicle the researcher should be aware that negative feelings about such taxes may strongly influence the resulting WTP amounts. If the vehicle does influence the WTP amounts, it is the policy, rather than the public good independent of the payment mechanism, which is valued.

The elicitation method. Every hypothetical market described in a contingent valuation scenario must somehow elicit a price from the respondent. In chapter 3 we discussed the four different kinds of elicitation method—the open-ended question, the bidding game (with or without benchmarks), the payment card, and the take-it-or-leave-it methods. The criteria for choosing a particular elicitation method and payment format are threefold: the degree to which the method and the format are isomorphic with the reference institution, such as the referendum; the degree to which they simplify the decision of the respondent; and the degree to which they are free of bias. Selecting the optimal elicitation method for a given study usually involves tradeoffs among these criteria.

The choice of method may influence the reliability of the results (see chapter 4). As judged by interviewer evaluations and by the variance of the WTP amounts elicited by the different methods, the open-ended format seems to elicit less reliable results than the bidding game or payment card approaches, presumably because respondents are unfamiliar with the act of setting a price for a good. The payment card method appears to facilitate respondent valuations in a way that is likely to enhance reliability, especially in studies for costly programs that use the benchmark version of the payment card to indicate what respondents are paying for other types of public goods.[8] Evidence reviewed in the next chapter shows that

[8] In a field experiment conducted during the pretest stage of their study of national water quality benefits, Mitchell and Carson (1984) used payment cards with and without benchmarks. The benchmarks appeared to increase the reliability of the responses, as the standard error of the mean for the benchmark treatment was lower than that for the nonbenchmark treatment at the .10 significance level. This is noteworthy, given the low power of the test (the sample size was 100), which made it difficult to detect a statistically significant difference.

the bidding game technique is vulnerable to starting point bias. Heberlein (1986) also found it to be less reliable than the take-it-or-leave-it method when he reinterviewed the hunters who took part in the 1984 Sandhill study a year later (see chapter 9). He asked them to suppose that they had or did not have a permit for the 1985 Sandhill hunt (whether they had applied for one or not), and offered them the same opportunity to buy or sell the permit for the same price as a year earlier. Heberlein found that only 77 percent of those whose amounts were elicited by the bidding game format gave consistent answers, whereas 90 percent of those who received the take-it-or-leave-it treatment were consistent. The reliability coefficient for the latter treatment was .75, which compares with .45 for the bidding game respondents.

The high test-retest reliability shown by the take-it-or-leave-it elicitation method is not surprising, since the type of decision it requires from the respondent involves only whether the single amount suggested by the interviewer is above or below the amount the respondent is willing to pay for the amenity, and not an actual maximum WTP amount as required by the other methods. Viewed from another perspective, however, this type of discrete choice results in less reliable estimates because it yields less information from each respondent. Whereas the payment card or open-ended formats obtain a respondent's actual maximum WTP amount, the take-it-or-leave-it format only narrows down the range of a respondent's true value for the good. Thus, in order to obtain a mean WTP amount which is comparable in accuracy to, say, a payment card mean amount, a study using the take-it-or-leave-it format would require much larger sample sizes.[9]

The Reliability of CV Sample Estimates

No matter how realistic the scenario, the data collected from even the best contingent valuation survey instruments are useless unless sample statistics of sufficient quality for policy purposes can be obtained.[10] There are two primary ways in which the reliability of CV sample statistics can be enhanced: the first is through the use of sufficiently large sample sizes; the second is through the use of robust statistical techniques that guard against undue influence by outliers.

[9] A single follow-up amount added to the original take-it-or-leave-it amount offers the possibility of improving the accuracy for a given sample size (Carson, Hanemann, and Mitchell, 1986).

[10] This is true as well where the purpose of the CV study is experimental. Hypothesis-testing involving different treatments in CV surveys is discussed at length in appendix C.

Sample Size

To obtain an acceptable degree of precision in sample statistics, such as the mean WTP amount, contingent valuation studies require large sample sizes because of the large variance in the WTP responses.[11] This statement follows from the fact that the estimated standard error of the mean, SEM,

$$SEM = \frac{\hat{\sigma}}{\sqrt{n}}, \quad (10\text{-}1)$$

decreases as a function of the $\sqrt{n}$ for any estimated standard deviation of the WTP responses, $\hat{\sigma}$. The standard error of most other summary statistics, such as the median and the total willingness to pay, are also decreasing functions of the sample size.

Sample size selection procedures can make direct use of this fact if an estimate of the population variance, σ^2, is available and the researcher is interested in the absolute error. Standard statistical texts on sampling, such as Kish (1965), Cochran (1977), and Yates (1980), discuss how to choose sample size using simple random sampling or the widely used variants of single random sampling.[12] Typically, however, the researcher will be interested in the likely magnitude of the relative error (the percentage deviation from the true mean) rather than the absolute magnitude of the error. In this situation, the researcher needs to have a prior estimate of the coefficient of variation, V, where

$$V = \frac{\sigma}{\overline{TWTP}}. \quad (10\text{-}2)$$

A prior estimate of this quantity is often easier to guess in advance than σ itself, and tends to be much more stable than estimates of σ. The coeffi-

[11] Much of this variance is to be expected, and results from the diversity of opinion in large heterogeneous populations. Obviously, the researcher should expect a larger true variance in contingent valuation WTP amounts obtained from the general population than from homogeneous subgroups, such as fishermen or hunters. The variance in CV WTP responses resulting from the survey instrument itself only adds to the problems caused by a large natural variance.

[12] For personal interviews simple random sampling is seldom used, although the formulas of this type of sampling are often good approximations to the methods used commercially if the sample size is large. Commercial survey firms tend to use a combination of stratification and clustering techniques to draw what are often termed "area probability samples." Stratification tends to increase the efficiency of sample statistics, so that fewer observations are needed, while clustering tends to decrease efficiency, thus increasing the number of needed observations. If there is a lot of regional variation, the stratification effects may predominate. If the variation is between, say, income groups, clustering effects may predominate if the clusters are defined over small areas such as neighborhoods. Telephone surveys using random digit dialing, and mail surveys from complete listings, may closely approximate simple random sampling.

Table 10-1. Sample Sizes Needed (usable responses)

	Δ						
V, α	.05	.10	.15	.20	.25	.30	.50
$V = 1, \alpha = .10$	1,143	286	127	72	46	32	12
$V = 1, \alpha = .05$	1,537	385	171	97	62	43	16
$V = 1.5, \alpha = .10$	2,571	643	286	161	103	72	26
$V = 1.5, \alpha = .05$	3,458	865	385	217	139	97	36
$V = 2.0, \alpha = .10$	4,570	1,143	508	286	183	127	46
$V = 2.0, \alpha = .05$	6,147	1,537	683	385	246	171	62
$V = 2.5, \alpha = .10$	7,141	1,786	794	447	286	199	72
$V = 2.5, \alpha = .05$	9,604	2,401	1,608	601	385	267	97
$V = 3.0, \alpha = .10$	10,282	2,570	1,143	643	412	286	103
$V = 3.0, \alpha = .05$	13,830	3,458	1,537	865	554	385	139

V is the coefficient of variation, $\frac{\sigma}{\overline{\text{RWTP}}}$

Δ is the possible deviation as percentage of $\overline{\text{RWTP}}$

$\alpha = .05$ indicates that 95% of the time estimated $\overline{\text{WTP}}$ will be within Δ of $\overline{\text{TWTP}}$

$\alpha = .10$ indicates that 90% of the time estimated $\overline{\text{WTP}}$ will be within Δ of $\overline{\text{TWTP}}$

cients of variation from a number of contingent valuation studies that are given in appendix C may be helpful in making this estimate.[13] As a rough approximation, the necessary sample sample size, n_0, can be determined from the formula

$$\left[\frac{Z\hat{\sigma}}{\Delta\overline{\text{RWTP}}}\right] = \left[\frac{Z\hat{V}}{\Delta}\right]^2 \qquad (10\text{-}3)$$

where Δ is the percentage difference between the true willingness to pay ($\overline{\text{TWTP}}$) and the estimated $\overline{\text{RWTP}}$. The confidence intervals are of the form $t^{*}\Delta^{*}\overline{\text{RWTP}}$, where t is the Student's t-variate.[14] Essentially one is setting $\hat{\sigma} = \Delta\overline{\text{RWTP}}$ in setting the confidence intervals. The researcher should be careful to inflate this n_0 by the expected number of unusable WTP responses, such as nonrespondents and protest zeros.[15]

Table 10-1 presents the indicated sample sizes for different combinations of relative error (V), confidence levels $(1-\alpha)$, and the percentage difference between $\overline{\text{TWTP}}$ and $\overline{\text{RWTP}}$ that the researcher is willing to tolerate

[13] The estimated coefficients of variation V in contingent valuation studies almost always fall between .75 and 6. An initial estimate for V of at least 2 is advisable. Sample sizes between 200 and 2,500 are appropriate for this size V.

[14] Standard values for t are 1.96 (the 95 percent confidence interval) and 1.69 (the 90 percent confidence interval). Reasonable values for Δ lie between .05 and .3.

[15] The sample selection problems caused by these nonusable WTP responses are discussed in chapter 12.

(Δ). The sample sizes presume that simple random sampling was used to select the respondents. The table shows that a sample size of 683 usable WTP amounts would be necessary if a researcher anticipates a coefficient of variation (V) of 2.0 (a good initial estimate of V for contingent valuation studies),[16] is willing to accept a Δ of .15, and wants a two-sided 95 percent ($1-\alpha$) confidence level (t=1.96). If $\overline{\text{RWTP}}$ is 100, the 95 percent confidence interval for $\overline{\text{TWTP}}$ will be approximately [70, 130] in the example given above.

Outliers

Sufficiently large sample sizes cannot solve the problems presented by outliers, which tend to be a constant percentage of the sample regardless of its size.[17] Contingent valuation surveys are particularly vulnerable to outliers because WTP amounts, unlike ordinary scales, are unbounded at the upper end, and the most commonly used summary statistic in CV surveys—the mean WTP amount—is sensitive to outliers. It is easy to see that 1 respondent expressing a WTP of a million dollars will dominate the mean WTP amount for an amenity which is valued by 600 other respondents at $100 or less. Researchers will have no difficulty identifying a single outlier of this magnitude, or in justifying its removal from the data set. More difficult are the WTP amounts which, though questionable, cannot be dismissed out of hand. Take, for example, a respondent whose answers to other questions suggest only a moderate demand for a given amenity, but who expresses a WTP value which is the equivalent of 5 percent of his pretax income. A few outliers like this, if genuinely invalid, can significantly distort a benefits estimate.

What to do? One approach is to delete outliers on an ad hoc basis. This procedure suffers from the obvious drawback that it opens the researcher to criticism that he has engaged in selective deletion to achieve desired results.[18] A more defensible approach is to mitigate the effect of outliers through the use of robust statistical estimators.

Both the mean and the median are members of a family of estimators known as the α-trimmed mean. The median is the α-trimmed mean with α equal to .5, and the mean is the α-trimmed mean with α equal to zero.

[16] See table C-1 in appendix C.

[17] We use the term "outlier" here to include a wide variety of observations that are unlikely given the presumed distribution. Barnett and Lewis (1984) provide an extensive discussion of outliers and techniques for handling them.

[18] A careful examination of individual observations is highly advised, however. A few observations will fail simple "logical" checks that indicate data entry errors or improperly filled-out questionnaires. These values should be set aside and dealt with during the data imputation process, rather than treated as outliers.

Statisticians recommend α-levels of .05 to .25 for a robust estimator of the expected value of a distribution.[19]

The α-trimmed mean is easy to estimate. Order the sample observations, which we will denote by X_i from the smallest, X_1, to the largest, X_n.[20] The α-trimmed mean, $\overline{X}_\alpha$, is defined by

$$\overline{X}_\alpha = \frac{X_{(n\alpha)+1} + . . . + X_{(n-(n\alpha)}}{n - 2[n\alpha]} \tag{10-4}$$

for $0 \leq \alpha \leq 1/2$.

Bickel and Doksum (1977) show that an operational estimator of the variance of the α-trimmed mean, σ_α^2, is given by

$$\sigma_\alpha^2 = \frac{\sum_{i=[n\alpha]+1}^{n-[n\alpha]} (X_i - X_\alpha)^2 + [n\alpha] \cdot \left[(X_{[n\alpha]+1} - X_\alpha)^2 + (X_{n-[n\alpha]} - X_\alpha)^2\right]}{(1-2\alpha)^2 n}, \tag{10-5}$$

where $X_{[n\alpha]+1}$ and $X_{n-[n\alpha]}$ are the empirical order statistics.

If we assume that the X_i are from a symmetric unimodal distribution, then asymptotically the random variable

$$T = \frac{\sqrt{n}(\overline{X}_\alpha - \mu)}{\hat{\sigma}_\alpha}, \tag{10-6}$$

where μ is the population mean, has approximately a standard normal distribution (Z). Confidence intervals can be formed in the following manner:

$$\overline{X}_\alpha \pm \frac{Z(1 - 1/2\alpha)\hat{\sigma}_\alpha}{\sqrt{n}}. \tag{10-7}$$

Thus, one- and two-tailed t-tests can be formed so that comparisons with costs can be made for a benefit-cost analysis.[21]

[19] Stigler (1977) has examined a large number of experiments designed to measure physical constants (physical constants are most subject to normal measurement error), and has shown that a 10 percent trimmed mean would have resulted in a much more efficient and less biased estimate.

[20] Ordered in this way, the sample observations are known as empirical order statistics.

[21] See Bickel and Doksum (1977) for further details on these tests.

There are a number of other robust estimators (Mosteller and Tukey, 1977; Huber, 1981) which behave in a manner similar to the alpha-trimmed mean. Most of these methods can be extended to a regression framework.[22] Another approach used to remove the effect of outliers involves techniques based on the influence of an observation in the projection matrix (Belsley, Kuh, and Welsch, 1980; Barnett and Lewis, 1984). These techniques have been used in a contingent valuation context by Desvousges, Smith, and McGivney (1983). For the estimation of valuation functions, robust and resistant regression estimators should be considered (Krasker, Kuh, and Welsch, 1983).

Summary and Conclusions

We have seen that unacceptable variability in WTP amounts can be introduced through a wide range of sources. The most important, from a CV perspective, are those contributed by the survey instrument, the sampling variance, and the underlying variance of the "true" WTP amounts. The variance introduced by the instrument may be assessed by several techniques, including reinterviewing the same respondents after time has elapsed and comparing the variation in WTP amounts given by equivalent samples who received instruments that varied in one characteristic or another. A useful indicator of a CV survey's reliability is the presence of a respectable R^2 when the WTP amount is regressed on a set of theoretically relevant variables.

The reliability of the WTP amounts obtained by a given CV scenario is influenced by the degree to which respondents find it credible and realistic. Unfortunately, elements that enhance realism may pose the risk of bias. Keeping in mind the distinction between the valuation-relevant and the valuation-neutral elements of a scenario, those introduced to make a scenario more realistic should not be wrongly perceived by respondents as valuation-relevant. We have assessed here the realism-bias tradeoff for various types of goods, markets, payment vehicles, and elicitation methods.

We next turned to the separate source of variance contributed by statistical factors which affect the reliability of CV sample estimates. We emphasized that without large sample sizes, contingent valuation surveys cannot produce data of sufficient quality to be of use in policy analysis. Most CV studies reported in the literature use sample sizes far below those typically used by survey researchers (from 600 to 1,500 respondents) who

[22] For a review of the literature with an emphasis on distributional theory in complex situations, see Subramanian and Carson (1984), who also present simulations of how well various robust and nonrobust regression estimators perform in the presence of gross errors similar to those typically found in CV surveys.

wish to generalize to a population. Yet the coefficients of variation commonly found in CV studies suggest that they need to employ larger, not smaller, sample sizes than general surveys. We also highlighted the problem of outliers in CV studies, and advocated the use of robust estimators as a way to control the potential bias from this source, in preference to ad hoc deletions.

Reliability is closely related to bias. We saw that there is no unique bias produced by the hypothetical character of contingent valuation surveys, and that therefore it is meaningless to speak of hypothetical bias. Uncertainty induced by a poorly written, unrealistic scenario may lead respondents to make casual and thus unreliable responses, or it may make respondents more vulnerable to various instrument effects which tend to bias WTP amounts. A study whose WTP amounts are highly correlated with the starting points used in its bidding game format will yield estimates that are highly reliable, but so biased as to be invalid.

In the final section of the chapter we described techniques which help provide answers to two questions. The first (and deceptively simple) question is what size sample is needed to obtain sufficiently precise CV estimates.[23] CV surveys need relatively large samples because of the large variance of WTP responses. Table 10-1 provides guidance on the appropriate sample size for various combinations of variance, precision, and significance level. For the purpose of estimating benefits for policy purposes, table 10-1 implies that 600 usable responses is a minimum sample size. (Sponsors should realize that a CV survey is not worth undertaking if a large enough sample cannot be funded.) In answering the second question, what should be done with outliers, we recommended the alpha-trimmed mean as a substitute for the sample mean as a summary WTP statistic. Ad hoc procedures should be avoided whenever possible. If they are used to drop outliers, the justification for the decision process, and information about each outlier, should be provided in the study's technical report.

In the next chapter we turn to the potential biases in contingent valuation surveys.

[23] Williams (1978:225) argues that determining the optimal sample size for a study is one of the most difficult problems in applied statistics due to the presence of unknowns such as *V* and a lack of clear guidelines on acceptable precision.

11

Measurement Bias

In this chapter we outline the principal sources of bias in contingent valuation studies, the conditions that promote their occurrence, and approaches that may be used to minimize their effects. Our intent is to convey a sense of the variety and subtlety of systematic error and of the ways in which these biases can and should be taken into account in the design and execution of a CV study. Susceptibility to particular biases will vary greatly according to the nature of the amenity being valued.

The CV Bias Experiments

Those who pioneered the development of contingent valuation techniques in the 1970s were sensitive to the possibility that systematic error might affect their results. Economic theory dictated a concern for strategic behavior, and common sense suggested the possibility of bias in question wording or research procedures. In order to assess the effects of error and bias, researchers have conducted a large number of experiments, using experimental designs, to determine whether particular aspects of the instrument systematically influenced the WTP amounts. Most of these experiments have been conducted in the field rather than the laboratory. For example, a sales tax payment vehicle might have been used for half of a random sample of residents in a given location, while the other half would have been asked to value the same good by hypothetical increases in their utility bills. The respondents would have been randomly assigned to the two treatments, which would have been identical except for the scenario feature being tested. If the mean WTP amounts for the sales tax and the utility bill treatments were not statistically different at the 90 or 95 percent confidence level, it would have been concluded that the choice of payment vehicle had not biased the CV results. Some of these bias experiments, as we will call them, have been conducted in the course of efforts to obtain valid benefit estimates; others have been purely methodological in intent.

Bias experiments since the 1970s have tested variations in payment vehicles and payment cards, elicitation methods, the effects of using different starting points in bidding games, willingness to pay versus willingness to accept (compensation), different budget constraints, and variations in the information provided to respondents in the survey instrument.

The results of the early experiments were interpreted as showing that the contingent valuation method was almost uniformly robust regarding variations in starting points, payment vehicles, and the kinds of information provided about the amenity being valued (Schulze, d'Arge, and Brookshire, 1981). It now appears that some of these experiments were methodologically weak and that this conclusion was overly optimistic. Rowe and Chestnut (1983) point out some of the methodological problems involved in the early experiments and argue that while progress is being made in identifying and overcoming bias problems, "we are far from being out of the woods." An additional and serious problem with many CV bias experiments, both old and new, is that their sample sizes are too small to provide sufficient statistical power to reject the null hypothesis that there is no difference in treatment effects. These problems reduce the number of experiments whose findings we were able to draw upon for this chapter.

Since the statistical power problem is not well recognized in the contingent valuation literature, it bears further explanation. Consider a study having a total sample size of 75 people, in which 37 people received treatment A and gave a mean WTP amount of $3.91 (standard deviation = $5.65) for a good, while the 38 who received treatment B gave a mean WTP amount of $7.85 (standard deviation = $17.99) for the same good—a difference of almost 100 percent. Does the use of one treatment rather than the other influence the WTP amounts? On the basis of a *t*-test and a significance level of .05, the researcher rejected the hypothesis that the two means are different. However, the small sample size and large variances in his data result in a 95 percent confidence interval for the difference between the means of [−10.10, 2.22]. Thus the mean WTP for treatment B would have to have been $14.01, more than 250 percent greater than the results of A, before this test could show a significant difference in treatments.[1]

[1] Both type I errors (accepting alternative hypothesis H_1 when the null hypothesis H_0 is true) and type II errors (accepting H_0 when H_1 is true) should be taken into account in determining the sample sizes required to make a meaningful bias test. See appendix C for a detailed treatment of the power problem, and for tables showing the sample sizes needed for (1) different combinations of power and significance parameters, (2) acceptable differences in the percentages among subsample means, and (3) coefficients of variation. The choice of these features is to some degree a matter of judgment. For a given N, the power of a test declines as the size of the significance level for the test of significance increases, and rises as the standard deviation becomes smaller. The sample sizes required for tests of

Sources of Systematic Error in CV Studies

Over the years CV researchers have developed a list of potential biases, such as starting point bias, vehicle bias, and information bias, that most studies attempt to address in one fashion or another. This list is inadequate in several respects (Cummings, Brookshire, and Schulze, 1986): it is ad hoc, is insufficiently comprehensive, and in the case of information bias (effects caused by varying the information provided to the respondents in the scenario), it confuses legitimate effects with biases. In fact, it is to be expected that responses will vary according to the information provided, since the WTP amounts are intended to be contingent on the scenario. Perhaps the most troublesome problem is the concept of hypothetical bias. As noted in the preceding chapter, the hypothetical character of contingent valuation studies does not bias the result in any particular way, but represents instead a reliability problem. In this chapter we present, then, a framework—a typology—for bias in CV surveys that attempts to address these defects, first listing the assumptions behind our typology and the principal sources of bias in CV surveys.

Assumptions

Our framework, based on a set of assumptions about what motivates human behavior, encompasses a broader range of factors than economists normally recognize. We make the following assumptions about how people behave in an interview situation:

1. Shared meanings between people are problematic because meaning is subjective and contextual. This insight, which derives from the symbolic interaction framework in social psychology (Stryker and Statham, 1985), underlies our view that it cannot be taken for granted that what the researcher intends to convey by the research instrument's wording will be so understood by the respondent.

2. The public's values for public goods are an expression of choice based on a mixture of preference, analysis, and moral judgment (Lindblom, 1977; Etzioni, 1985). This assumption highlights the potential importance for the scenario of the societal context in which the respondent

---

significance at the traditional .05 level are quite large, hence we recommend the use of the .10 level as a compromise. For the example given above, a sample size of 210 would be required for a two-tailed *t*-test allowing a 10 percent chance of a type I or a type II error in determining whether a WTP amount obtained by a second treatment was 50 percent larger than the mean WTP amount of the first treatment. This is almost three times the original sample size of 75. Even larger samples would be required to detect differences of less than 50 percent.

makes her judgment. Factors such as the degree of controversy surrounding the provision of an amenity, the perceived views of important others on the topic, and the credibility of the sources of authority invoked (explicitly or implicitly) may complicate the task of designing a valid scenario.

3. People's tastes usually differ according to such factors as their socioeconomic status and previous experience (McKean and Keller, 1982). This assumption, documented in numerous social surveys, underlies the importance of probability sampling for CV surveys that intend to generalize to some larger group of people from those who were interviewed.

4. People (in our society) are motivated by norms of equity, fairness, and helpfulness (among others), as well as by a desire to maximize their personal advantage.[2] The tendency of some respondents to give nonopinions stems from their desire to fulfill the norm of helpfulness (by cooperating in the interview), and also from their desire to maximize their utility (by not appearing stupid in front of the interviewer).

5. Owing to human cognitive limitations, the effects of earlier experience, and a desire to minimize the costs of decision making, people at times simplify decisions by relying on preexisting knowledge structures (schemas) and resorting to systematic rules of thumb or judgmental heuristics when making choices under conditions of uncertainty. This assumption is based on the work of Simon (1957) and of Kahneman, Slovic, and Tversky (1982), among others.[3]

6. When people lack well-defined opinions on an issue, their judgments in response to survey questions are particularly sensitive to the demand characteristics of the interview (Bishop, Tuchfarber, and Oldendick, 1986), or to how the questions are phrased and responses are elicited (Fischhoff, Slovic, and Lichtenstein, 1980).

7. Every class of valued public goods is subject at some point to satiation or diminishing marginal utility. This is a standard assumption made by economists and bears on the issues of how to sequence WTP questions and aggregate WTP estimates across studies.

[2] The empirical basis for this generalization is documented in some detail in our discussions of existence benefits (chapter 3) and strategic behavior (chapters 6 and 7). Kahneman, Knetsch, and Thaler (1986) provide evidence for community standards of fairness that apply to price-, rent-, and wage-setting by firms.

[3] See Nisbett and Ross (1980) and Abelson and Levi (1985) for useful syntheses of the literature on this topic, and Schoemaker (1982) for an analysis of the implications of the literature for the expected utility model.

Principal Sources of Bias

Our typology is based on what we consider to be the four principal sources of potential systematic error in contingent valuation willingness-to-pay estimates. These are:

1. Use of a scenario that contains strong incentives for respondents to misrepresent their true WTP amounts.
2. Use of a scenario that contains strong incentives for respondents to improperly rely on elements of the scenario to help determine their WTP amounts.
3. Misspecification of the scenario by incorrectly describing some aspect of it (according to theory or the policy-relevant facts of the case), or, alternatively, by presenting a correct description in such a way that respondents misperceive it.
4. Improper sampling design or execution, and improper benefit aggregation. If adjustments are not made for sampling and aggregation errors, they can bias the aggregated benefit estimate.

In this chapter we will consider only those biases resulting from the first three sources; those resulting from the fourth are discussed in the next chapter.

A Bias Typology

The principal biases that appear at this time to be most germane to CV studies are outlined in table 11-1 and examined in detail in the pages that follow. Several have received considerable attention in the CV literature, and most of the rest are known to practitioners in the field. Our approach to these biases differs somewhat from earlier approaches in making a distinction between biases and misspecifications, in rejecting the concept of an "information bias" (for the reasons given in chapter 2), and in including biases caused by incorrect sampling or aggregation procedures.

Incentives to Misrepresent Responses

In the first major category of potential biases are those resulting from incentives to respondents to misrepresent their stated WTP amounts. These incentives, caused by one or more aspects of the interview situation, can lead to two types of behavior: (1) strategic behavior, which is a deliberate attempt to influence either the future payment or provision of the public good in question, and (2) compliance behavior, which attempts—consciously or unconsciously—to fulfill what the respondent perceives as the expectations of either the sponsor of the survey or the interviewer.

Table 11-1. Typology of Potential Response Effect Biases in CV Studies

1. *Incentives to Misrepresent Responses*

   Biases in this class occur when a respondent misrepresents his or her true willingness to pay (WTP).

   A. *Strategic Bias:* where a respondent gives a WTP amount that differs from his or her true WTP amount (conditional on the perceived information) in an attempt to influence the provision of the good and/or the respondent's level of payment for the good.

   B. *Compliance Bias*

      1. *Sponsor Bias:* where a respondent gives a WTP amount that differs from his or her true WTP amount in an attempt to comply with the presumed expectations of the sponsor (or assumed sponsor).
      2. *Interviewer Bias:* where a respondent gives a WTP amount that differs from his or her true WTP amount in an attempt to either please or gain status in the eyes of a particular interviewer.

2. *Implied Value Cues*

   These biases occur when elements of the contingent market are treated by respondents as providing information about the "correct" value for the good.

   A. *Starting Point Bias:* where the elicitation method or payment vehicle directly or indirectly introduces a potential WTP amount that influences the WTP amount given by a respondent. This bias may be accentuated by a tendency to yea-saying.

   B. *Range Bias:* where the elicitation method presents a range of potential WTP amounts that influences a respondent's WTP amount.

   C. *Relational Bias:* where the description of the good presents information about its relationship to other public or private commodities that influences a respondent's WTP amount.

   D. *Importance Bias:* where the act of being interviewed or some feature of the instrument suggests to the respondent that one or more levels of the amenity has value.

   E. *Position Bias:* where the position or order in which valuation questions for different levels of a good (or different goods) suggest to respondents how those levels should be valued.

3. *Scenario Misspecification*

   Biases in this category occur when a respondent does not respond to the correct contingent scenario. Except in A, in the outline that follows it is presumed that the *intended* scenario is correct and that the errors occur because the respondent does not understand the scenario as the researcher intends it to be understood.

A. *Theoretical Misspecification Bias:* where the scenario specified by the researcher is incorrect in terms of economic theory or the major policy elements.

B. *Amenity Misspecification Bias:* where the perceived good being valued differs from the intended good.

   1. *Symbolic:* where a respondent values a symbolic entity instead of the researcher's intended good.
   2. *Part-Whole:* where a respondent values a larger or a smaller entity than the researcher's intended good.
      a. *Geographical Part-Whole:* where a respondent values a good whose spatial attributes are larger or smaller than the spatial attributes of the researcher's intended good.
      b. *Benefit Part-Whole:* where a respondent includes a broader or a narrower range of benefits in valuing a good than intended by the researcher.
      c. *Policy-package Part-Whole:* where a respondent values a broader or a narrower policy package than the one intended by the researcher.
   3. *Metric:* where a respondent values the amenity on a different (and usually less precise) metric or scale than the one intended by the researcher.
   4. *Probability of Provision:* where a respondent values a good whose probability of provision differs from that intended by the researcher.

C. *Context Misspecification Bias:* where the perceived context of the market differs from the intended context.

   1. *Payment Vehicle:* where the payment vehicle is either misperceived or is itself valued in a way not intended by the researcher.
   2. *Property Right:* where the property right perceived for the good differs from that intended by the researcher.
   3. *Method of Provision:* where the intended method of provision is either misperceived or is itself valued in a way not intended by the researcher.
   4. *Budget Constraint:* where the perceived budget constraint differs from the budget constraint the researcher intended to invoke.
   5. *Elicitation Question:* where the perceived elicitation-question fails to convey a request for a firm commitment to pay the highest amount the respondent will realistically pay before preferring to do without the amenity. (In the discrete-choice framework, the commitment is to pay the specified amount.)
   6. *Instrument Context:* where the intended context or reference frame conveyed by the preliminary nonscenario material differs from that perceived by the respondent.
   7. *Question Order:* where a sequence of questions, which should not have an effect, does have an effect on a respondent's WTP amount.

Strategic bias. Strategic bias occurs when respondents deliberately shape their answers to influence the study's outcome in a way that serves their personal interest. We discuss strategic behavior at length in chapters 6 and 7, reaching the conclusion that this type of bias is only problematic under certain worst-case conditions. Although such conditions are generally rare in CV studies, and strategic bias is much less of an impediment to CV studies than many economists have assumed, the possibility of strategic bias must be taken seriously.[4] It is particularly important that contingent valuation instruments create a plausible payment obligation. Questionnaires that unnecessarily call attention to the hypothetical character of the valuation exercise, either directly by using wording such as "pretend" or "assume a hypothetical situation" when eliciting the WTP amount, or indirectly by the use of a highly abstract or implausible payment vehicle, run the risk of encouraging overbidding by respondents who want the good to be provided. This situation would be exacerbated by wording that emphasized the importance of the respondents' answers to policymakers.

Compliance bias. Social psychologists and survey researchers tend to believe that people generally are motivated to tell the truth to an interviewer, but are prone to shape their answers to please either the interviewer or the sponsor, especially when they do not have a strong or well-considered view on the survey topic (Schuman and Presser, 1981). This belief is plausible, but surprisingly poorly documented. A recent review of the complex social-desirability literature concludes that "perhaps the problem is not as overwhelming as it appears to be" (DeMaio, 1984:279). To the extent that it is a factor, this motivation accounts for the several forms of compliance bias. One form is *sponsor bias,* where respondents attempt to answer questions in ways they believe will meet the expectations of the sponsor of the study. The potential for sponsor bias is one of the reasons why surveys or polls conducted directly by interest groups are given little credence by reputable pollsters, politicians, and the media. Environmental groups who wish to demonstrate that the public is willing to pay for particular environmental improvements, or corporations who wish to demonstrate that the public does not, will find it necessary to engage a survey organization whose skill and neutrality are above reproach if the results of a CV survey they commission are to be taken as credible benefits estimates. To date, most contingent valuation surveys have been conducted under the auspices of universities or reputable survey or research firms.[5]

[4] Because of economists' widespread concern about strategic behavior in surveys, researchers reporting the findings of CV surveys are advised to explain why strategic behavior is not a threat to their findings.

[5] Wherever the possibility exists that the sponsor or organization conducting the survey has an identifiable interest in the outcome of the survey (as in the case of a government agency valuing an amenity that is perceived to be a part of its mission), special effort should be taken to legitimate the full range of responses, including $0 amounts.

If telephone or in-person interviews are used, the potential exists for *interviewer bias*. The bias would be present to the extent that respondents shape their answers in a way that they think will either please the interviewer or will increase their status in the interviewer's eyes. Because in CV surveys respondents are asked to undertake an unfamiliar task, some will look to the interviewer for confirmation that they are giving the "right" answer (Bishop, Tuchfarber, and Oldendick, 1986). An interviewer in a recent CV study of drinking water risk-reduction benefits (Mitchell and Carson, 1986c), conducted in southern Illinois, made the following note right after an interview:

> Respondent wanted a great deal of assurance she was giving "normal" responses. She asked what most other people had said [in answer to the valuation questions] and said things like, "That's a lot, isn't it?" [when the risk reduction levels to be valued were described to her] often during the interview.

Tests for interviewer effects on responses require a situation where the respondents are randomly assigned to the interviewers and each interviewer conducts a reasonably large number of interviews. Because it is usually difficult to randomly assign respondents to interviewers in in-person surveys, telephone interviews are employed for most interviewer bias experiments. Studies of error induced by interviewers in surveys have generally found that while the effects are relatively small when the interviewers are experienced (Groves and Magilavy, 1986), interviewers who are college students produce much larger response effects than other interviewers (Bradburn, 1983). We know of no interviewer bias experiments in the CV literature that randomly assigned respondents to interviewers, although Smith, Desvousges, and Freeman (1985) approximated this condition in their study of risk from hazardous waste, and Desvousges, Smith, and McGivney (1983) controlled for different respondent characteristics in their Monongahela water quality study. In both studies no evidence was found for interviewer bias.

These favorable findings were based on interviews conducted by highly trained professional interviewers at some of the country's survey research centers. The interviewing procedures used by such organizations are designed to make each interview as uniform in structure as possible in order to maximize the comparability of the results. A key requirement is that the questions must be asked exactly as they are printed in the questionnaire, with no exceptions. According to the Research Triangle Institute's *Field Interviewer's General Manual,* "A carefully worded and carefully asked question will elicit a fairly precise response. *Any deviation* from the exact wording of a question, whether deliberate or not, can easily change the task the respondent is asked to perform" (1979:23; italics in the original). Likewise, any explanations of the question offered the respondent must be restricted to those previously prepared and provided to all the interviewers for this eventuality; ad hoc explanations of what a question means, however well-intended, destroy comparability. Studies have shown that these goals are not consistently met, even with well-trained interviewers (Rustemeyer, 1977; Cannell, Lawson, and Hausser, 1975; see also, Turner and Martin, 1984).

Implied Value Cues

A significant effort is often required of respondents in a contingent valuation survey because of the length and complexity of the scenario and the need to give a dollar value for the good. This creates an incentive to adopt strategies to lighten the task. In our discussion of reliability, we noted that one such strategy is to casually pick a number without much reflection. Unfortunately, techniques intended to overcome this behavior, such as the bidding-game elicitation format, may provoke other respondent strategies that lead to biased responses.

According to Kahneman, Slovic, and Tversky (1982:14), people who are uncertain about a phenomenon are prone to make estimates by starting from an initial value which they adjust to yield the final answer. When it occurs, this strategy, called anchoring, leads to bias toward the initial value because the adjustments are typically insufficient (Slovic and Lichtenstein, 1971). In CV studies anchoring is most likely to occur when respondents fasten upon elements of the scenario that are not intended by the researcher to convey information about the value of the good and use them as cues to the good's approximate "correct value." The major sources of such cues in CV scenarios are: the elicitation techniques used by researchers to help respondents formulate their WTP values,[6] the method by which the respondent is asked to (hypothetically) pay for the good, and the materials used to describe the amenity being valued. Overall, elicitation techniques are the most important source of this kind of bias, although certain types of payment vehicles and descriptive devices, such as risk ladders, may also be troublesome.

Starting point bias. Starting point bias occurs when the respondent's WTP amount is influenced by a value introduced by the scenario. The bidding game and take-it-or-leave-it elicitation techniques pose the most obvious threat of this kind since they directly confront the respondent with a proposed amount that the respondent is asked to accept or reject. Confronted with a dollar figure in a situation where he is uncertain about an amenity's value, a respondent may regard the proposed amount as conveying an approximate value of the amenity's true value and anchor his WTP amount on the proposed amount. This propensity is likely to be exacerbated by yea-saying, the tendency of some respondents to agree with an

[6] See chapter 3 for descriptions of the major elicitation techniques.

interviewer's request regardless of their true views (Arndt and Crane, 1975; Couch and Keniston, 1960).[7] In the bidding game, even if a respondent rejects the initial bid, starting points well above the respondent's true WTP will tend to increase the revealed WTP, while starting points well below it will tend to decrease it—a pattern found by Roberts, Thompson, and Pawlyk (1985) in their study of the value of a marine recreational resource to skin divers. Although studies using the single-price-offer format obviously find various proportions of respondents who do refuse to pay amounts that are above their true willingness to pay for a good, an indeterminate percentage of those who do accept the price presumably do so only because their tendency to yea-say overcomes their unwillingness to accept that price.

A preponderance of the tests for starting point bias in CV studies shows that this bias occurs when the bidding game format is used (Cummings, Brookshire, and Schulze, 1986; Mitchell and Carson, 1985; Roberts, Thompson, and Pawlyk, 1985; Boyle, Bishop, and Walsh, 1985; Welle, 1985), and that starting point bias is often largely relative to the final WTP bids (Rowe, d'Arge, and Brookshire, 1980).[8] No generally valid method exists to adjust the findings obtained by the bidding game to compensate for the effect of the starting point.[9] For these reasons we do not concur with the judgment of the Water Resources Council's *Principles and Guidelines* (1983:74) that the bidding game is the preferred elicitation format for contingent valuation studies.

Range bias. The vulnerability of the bidding game format to starting point bias led Mitchell and Carson (1981, 1984) to develop the payment card elicitation technique. While the large number of amounts on the

[7] The problem of yea-saying may be seen as akin to the biometrician's problem of how to estimate the effect of a stimulus against a non-zero background. Hanemann (1984b, 1984c) and Carson and Mitchell (1983) have considered the yea-saying problem in a CV context. If a willingness-to-accept compensation format or very high initial starting points are used, nay-saying rather than yea-saying is likely to be observed.

[8] The only convincing test that shows the absence of starting point bias is Thayer (1981). Some of the tests claiming to show no starting point bias (for example, Brookshire, d'Arge, and Schulze, 1979) have no power to statistically detect the very large difference they observed; see appendix C for a discussion of this issue.

[9] The standard test for starting point bias (Thayer, 1981) has been to perform an analysis of variance or run a regression of the form

$$\text{WTP} = a + bS + e,$$

where WTP is an $n \times 1$ vector of revealed WTP, S is an $n \times 1$ vector of the starting points used, a and b are coefficients to be estimated, and e is a vector of error terms. Carson, Casterline, and Mitchell (1985) argue that more complicated formulations may have to be considered as possible reaction functions, since starting points above and below the respondent's true WTP are likely to yield different response patterns, which may be nonlinear.

typical payment card reduces, if it does not eliminate, the possibility of starting point bias, the presence of these amounts introduces the possibility of range bias, where the information presented on the payment card influences the respondent's WTP amount. This effect is to be expected if respondents consider the range of behaviors contained on the payment card as reflecting the researcher's knowledge of or expectations about the distribution of these behaviors, and use them as a frame of reference in estimating and evaluating their own behavior.[10]

Where the payment card technique is used, design configurations might bias WTP amounts in the following ways: (1) the maximum amount on the card may be lower than the maximum WTP of the respondent and thus artificially constrain his or her WTP amount; (2) the maximum amount may be taken to imply a reasonable upper bound and induce the respondent to give a higher amount than he or she would give if the maximum amount were lower; and (3) the amounts shown on the card may not include the amount the respondent is willing to pay and may therefore lead the respondent to choose a WTP amount that is either higher or lower than the preferred value.

Bias resulting from the first restriction is fairly easy to avoid by using a sufficiently high value as the upper limit on the payment card, although this may encourage the second form of range bias. Range bias of the third kind is more subtle and difficult to minimize. Although payment card formats encourage respondents to pick any number on the card or any number in between, in practice respondents almost always choose either a value from the list or a common number that is not on the list, such as $25 or $100. As a result of this propensity, the choice of values given on the payment card may lead to a form of implied value bias caused by the absence of a listed value in the region of interest to the respondent. For example, if the true WTP of a significant number of respondents falls into the gap between zero and the first positive amount on the payment card, these respondents are likely to express WTP values that are either higher or lower than their true values. The larger the gap in the area of interest,

[10] Schwartz and coauthors (1985) found a strong bias of this kind in their experiments on the effect of the range of the response categories on their subjects' self-reported television viewing. In one experiment, the first of six closed-ended answer categories was "up to 1/2 hour" in treatment A, and "up to 2 1/2 hours" in treatment B. In treatment A, the "more than 2 1/2 hours" category was the highest-offered category, whereas treatment B offered five categories above 2 1/2 hours, the highest of which was "4 1/2 hours." Compared with treatment A, twice as many subjects reported watching more than 2 1/2 hours of television in treatment B, whose format carried the implicit suggestion that this behavior was normative. These authors also show that the evaluations of the importance of television in their lives "as well as of the variety of their leisure-time activities were affected by the comparison standards suggested by the scales" (1985:394).

the greater the potential bias. If the card provides a large enough selection of amounts and the sample is sufficiently large, such rounding off should not cause much trouble in computing means or medians. The implications of rounding off for more sophisticated multivariate analysis are not well understood, although we believe they may be serious if most of the true WTP amounts fall between two consecutive amounts marked on the payment card.[11]

Relational bias. Relational bias occurs when the amenity being valued is linked to some other public good in such a manner that the other good (the "reference good") implies a value for the amenity being valued in a way unintended by the researcher. The reference good may be invoked by a specific scenario feature, like the benchmark amounts listed on a payment card, or the payment vehicle. Alternatively, certain types of amenities may invoke a reference good without any prompting by the scenario. In either case the respondent relies on the price implied by the reference good instead of trying to determine the maximum amount the good in question is really worth to him.

The use of benchmarks (that is, goods with a dollar amount attached to them) on payment cards is intended to remind respondents that they are already paying for many public goods through taxes and the prices they pay for regulated goods, and to provide a general idea of the magnitude of these payments. Because benchmarks pose an obvious threat of relational bias, only goods that are not directly related to the amenity being valued should be used as benchmarks. In a study of environmental benefits, for example, appropriate benchmarks would be such nonenvironmental goods as police and fire protection, the space program, and roads and highways. In conducting a test for relational bias in their national water quality studies, Mitchell and Carson (1981, 1984) systematically varied both the dollar levels for the (nonenvironmental) benchmarks on the payment cards and the number of benchmarks in an effort to see whether the amounts or the number of benchmarks influenced the respondents' WTP amounts. Although the hypothesis that the respondents based their values on the benchmark amounts was not supported in this case (Mitchell and Carson, 1981, 1984), the potential for relational bias in other situations should not be underestimated.[12] When payment cards with benchmarks

[11] The tendency to choose common numbers or the amounts actually listed on the payment card is likely to be more of a problem when the revealed WTP amounts for the good in question are concentrated in a fairly small interval, such as a range between $1 and $30, than when the good is a broad national program for which the spread of WTP amounts will be much greater.

[12] Randall, Hoehn, and Tolley (1981) show that payment cards that use "trivial" private goods, like toothpaste, as benchmarks tend to push WTP amounts upward in comparison with payment cards that use only public goods.

are used, care should be taken that the benchmarks are reasonably uncontroversial or unevocative. Displaying how much the average household is being taxed to support welfare payments, for example, might create a bias if many respondents framed their WTP amounts solely in terms of that controversial item rather than in the context of all the public activities identified on the card. And relational bias is not restricted to payment cards. Slovic, Fischhoff, and Lichtenstein (1980) show that similar phenomena occur when risk ladders display certain types of risk.

Another source of relational bias is respondent awareness of prices or costs for goods that resemble the one being valued. Government agencies often recover part of the cost of providing quasi-private goods by imposing charges, such as park entrance fees and hunting and fishing license fees, for their use. The size of these fees is determined by a political process rather than market demand. The presence of these "prices" poses a serious problem for CV surveys that attempt to value goods of this kind, as many respondents will spontaneously take the prevailing range of fees as representing a "reasonable" or "fair" value for the amenity, and will therefore restrict their WTP amounts to the range implied by these reference values. For example, when Sorg and Brookshire (1984) used an open-ended elicitation format to value an elk resource, more than half of the respondents simply gave the current cost of an elk-hunting license ($25). Harris, Driver, and McLaughlin (c. 1985) describe a study by Nachtman (1983), who found that of bus riders who gave a WTP amount for enjoying the recreational opportunities in the Maroon Bells area of central Colorado, 40 percent gave the same amount ($2.00) as the shuttle-bus fare they had paid to get into the area. Scenarios for quasi-private goods must overcome the respondents' propensity to anchor their WTP amounts on prevailing prices.[13] Certainly the use of entrance fees and license fees as payment vehicles should be avoided unless the researcher intends to make his findings contingent on these payment vehicles.[14]

Importance bias. Importance bias stems not from any individual component of a contingent valuation scenario, but from the experience of

[13] The use of a utility bill as the payment vehicle for air pollution amenities may also suffer from this problem if the true willingness to pay is much larger than the respondent's current utility bill. Here the current level of a respondent's utility bill implies a starting point, and limits the range of likely WTP amounts.

[14] Some have argued that relational bias can be corrected for by treating the initial WTP amount as a starting point for an iterative bidding process. This requires the strong assumptions that the initial amount will always be less than the respondent's true WTP amount, and that the iterative process will not bias the findings by putting pressure on the respondent to pay more than he really thinks the amenity is worth. In the absence of empirical evidence to the contrary, we are skeptical about both assumptions.

taking part in a CV study. Importance bias occurs when a respondent infers that one or more levels of the amenity must have value, because otherwise such an elaborate and expensive effort would not be made to get his opinion on the matter. This tendency, which may also be considered a form of compliance bias, was identified in the pretests for the study of drinking water risks mentioned earlier. During a session with a focus group in which a small number of local residents were asked to react to various possible scenario formats, the discussion leader described a particular technique that might be used to communicate the types of low-level risk which were to be valued in the study. The participants' immediate reaction was to advise against its use, in no uncertain terms, because "you won't get high values, if you give people information in that way." Since the leader previously had gone to considerable lengths to convince the participants that no particular outcome, either high or low, was expected or intended in the proposed study, this unexpected response was a useful reminder that respondents are susceptible to being influenced by the act of taking part in a CV study.

In order to minimize this effect, the scenario must be so designed that respondents who are not willing to pay anything for the amenity feel comfortable in giving that response. If this condition is not met and some respondents express values because they feel it is expected of them, the benefits will be overestimated. In the drinking water study, Mitchell and Carson (1986c) attempted to guard against importance bias in several ways. One of these was to include in the scenario, just before the actual elicitation of WTP amounts, a statement which informed respondents that some people vote yes and some people vote no to each of the hypothetical referenda they were going to be asked to value. The aim of this procedure was to legitimate zero values for those who felt the risk reduction was too small to be worth any money. The procedure appeared to be successful: many respondents did in fact give zero values for very small risk reductions, and the interviewers, who had been alerted to the importance bias phenomenon, reported that the vast majority of the respondents showed no signs that they believed the study was intended to obtain high WTP amounts.

Other features of the instrument can lead to importance bias if they exaggerate the desirability or undesirability of the amenity. Examples include the use of evocative symbols in artwork on a mail questionnaire (such as skulls and crossbones or pictures of pristine or grossly polluted streams); wording that inappropriately evokes catastrophe, national pride, the desirable qualities of an amenity (such as the towering cliffs and spectacular canyons of a wilderness area), or "villains" (such as mining interests or big corporations); or a presentation that is too one-sided and ignores important tradeoffs.

Position bias. This last type of implied value cue is closely related to importance bias. Here respondents who are aware that they are being asked a sequence of valuation questions regarding provision of different levels of goods take that sequence to imply something about the value of the levels or goods that is unintended.[15] They may, for example, assume that the last of a series of improvements they are asked to value is particularly worthwhile, or, if they are uncertain that any of the improvements are worth something to them, that the last is the only one worth valuing.

Scenario Misspecification

The researcher conducting a contingent valuation survey faces the task of obtaining relevant preferences from the respondent. Misspecification occurs when the respondent incorrectly (from the standpoint of theory or policy) perceives one or more aspects of the contingent market and the good to be valued. Misspecification may be theoretical or methodological, depending on the source from which it arises. *Theoretical misspecifications* result when the researcher describes a scenario that is incorrect from the standpoint of economic theory or the known facts of the situation. In this case, the respondents' values cannot reflect the appropriate contingencies even if the respondents understand the scenario perfectly. What can loosely be called *methodological misspecifications* result when the market described by the researcher is formally correct, but one or more elements are inadequately communicated so that the respondent does not perceive them in the way intended by the researcher. This problem is fundamental to survey research: studies have repeatedly showed the error of assuming that respondents necessarily understand even seemingly objective terms such as "unemployment," "victim," or "pollution" in the same way (Turner and Martin, 1984, 1:410ff).[16] Methodological misspecification is a serious threat to the reliability and validity of contingent valuation surveys and tends to be underestimated by researchers untrained in survey research techniques. As Sudman and Bradburn observe,

> The fact that seemingly small changes in wording can cause large differences in responses has been well known to survey practitioners since the early days of surveys. Yet, typically, the formulation of the questionnaire is thought to be the easiest part of the design of surveys—so that, all too often, little effort is expended on it. (1982:1)

The relationship between theoretical and methodological misspecification is diagrammed in figure 11-1.

[15] Such a sequence is often used in conjunction with some sort of scale card, such as a risk ladder.

[16] In a study of the degree to which respondents understand survey questions, Belson (1968) found that the best-understood questions were interpreted exactly as researchers had intended by about 50 percent of the respondents.

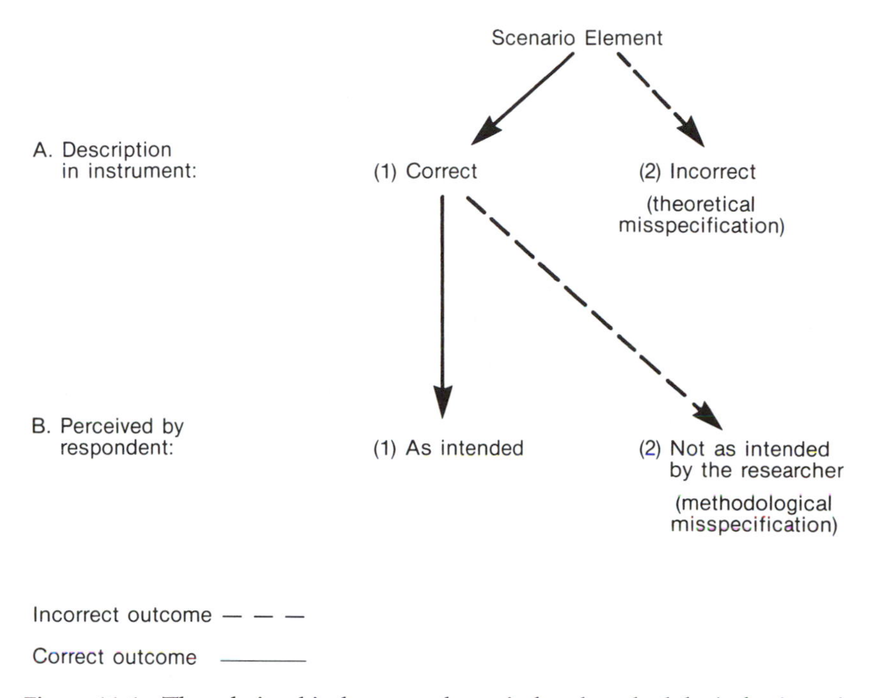

Figure 11-1. The relationship between theoretical and methodological misspecification

Bias will result if the respondent does not correctly perceive the scenario element as intended and if the misperception has a directional effect. For example, a researcher who uses a property tax payment vehicle may intend it to be a neutral form of payment, whereas the respondents, because they believe strongly that their property taxes are excessive, might react by expressing lower WTP amounts for the good than they otherwise would have been willing to pay. While there are potential biases associated with each of the misspecifications described in the typology below, they may just add random error to the WTP estimates or even be benign.

Biases caused by respondent misperceptions are among the most important and most problematic sources of error in CV surveys. These biases have the potential to dramatically affect a study's findings. As an example, consider a survey in which the researcher thinks she is asking the respondent to value a low-probability risk, whereas the respondent interprets the scenario's risk description as implying a much higher probability of risk. Even when respondents misperceive parts of a scenario that are less central than the level of the amenity—such as the payment vehicle or the implied budget constraint—a sizable bias can result. Misspecifications are problematic because scenarios offer many opportunities for them to occur, they are difficult to detect in the absence of an objective measure of the true

WTP amount, and the tendency of respondents to be uncertain about their WTP amounts makes them susceptible to extraneous influences of this type (Fischhoff, Slovic, and Lichtenstein, 1980).

At various points in this book we have emphasized that to the extent that contingent valuation surveys attempt to convey concepts or settings that are more complex or unfamiliar than those used in ordinary surveys, the method faces even greater communication difficulties than those faced by researchers using conventional attitude questions. There is, however, an important sense in which the contingent valuation method is *less* likely to suffer from misspecification than a study which uses ordinary survey questions. Ordinary survey questions presume that respondents can meaningfully report their attitudes and actions, abstracted from any situational context (Turner and Martin, 1984:417). While questions such as "Are we spending too much, too little, or about the right amount for water pollution control?" are simple and therefore relatively easy to communicate, their extreme generality does not take into account the well-documented situation-specific character of human decision making.[17] The result, as Fischhoff, Slovic, and Lichtenstein point out, is to increase "the probability that the elicitor and respondent will be talking about different things without solving the problem of inadvertent control. Indeed, one might even argue that impassive elicitation is the most manipulative of all. For it means that the entire questioning experience is conducted under the influence of unanalyzed predispositions and presumptions of the elicitor . . ." (1980:124). Contingent valuation scenarios, in contrast, leave less to the respondents to define as they see fit by making explicit many of the parameters of the valuation situation.

Most scenario misspecifications occur because the designer violates one or more of the informal question-wording guidelines developed by survey researchers (Payne, 1951; Bradburn and Sudman, 1979; Schuman and Presser, 1981; Dijkstra and van der Zouwen, 1982; Sudman and Bradburn, 1982). Misspecification can be minimized by developing the necessary theoretical economic framework for the particular problem before designing the questionnaire, and by an intensive program of questionnaire development, including the use of focus groups and careful pretesting of draft questionnaires.[18] After the interviewing has occurred, the researcher

[17] For an overview from the perspective of symbolic interactionism, see Stryker and Statham (1985); for specific studies, see Lazarsfeld, Berelson, and Gaudet (1948), and Crespi (1971).

[18] According to the Panel on Survey Measurement of Subjective Phenomena convened by the National Research Council, one general failing of survey researchers and their sponsors is that they devote too little attention and money to pilot studies and the methodological research necessary to improve the quality of their survey findings (Turner and Martin, 1984:315).

may identify possible misspecification biases by personally debriefing the interviewers, examining their written comments on the interview schedules, and analyzing certain types of response patterns. An awareness of the scenario misspecifications most likely to be found in CV surveys can greatly aid these endeavors.

Methodological misspecifications that are both likely to occur unless precautions are taken and are likely to bias WTP amounts if they do occur fall into two categories: those involving the actual amenity, and those involving the other aspects of the questionnaire that create the context in which the amenity valuation is made. In examining these misspecifications in detail, it must be kept in mind that the list of biases that follows is neither exhaustive nor the most relevant to every type of CV study.

Amenity misspecification bias. Particularly important are those misspecifications involving the amenity being valued. The description of the good in CV surveys typically contains several elements such as the time period within which the good is to be provided; the location, cause, and size of the change; and the nature of the amenity itself. A study of visibility benefits, for example, may describe the amenity as an increase in average annual visibility from 20 miles to 40 miles in the greater Los Angeles area due to a reduction in the emission of particulate matter from electric power plants. Since people tend not to have previously well-defined values for many of the goods valued in CV studies, there is considerable potential for them to ignore some or all of the details in a scenario, or to distort them by the unconscious use of judgmental heuristics. Such heuristics have been identified by psychologists studying the conditions under which people make errors in perception (Fischhoff, Slovic, and Lichtenstein, 1980; Kahneman, Slovic, and Tversky, 1982). The effect of these distortions may be to completely invalidate a contingent valuation study. Amenity misspecifications may be classified as *symbolic, part-whole, metric,* and *probability of provision* biases.

Symbolic bias occurs when respondents react to an amenity's general symbolic meaning instead of to the specific levels of provision described. It is at the heart of Kahneman's (1986) concern that respondents, when asked to value amenities with important nonuse values, have a propensity to respond to the symbol rather than to the substance. This would happen if respondents offered an answer to the question "What is more important, health or income?" before learning how much health and income were at stake. As evidence of "symbolic demand" Kahneman (1986) describes a telephone CV survey, conducted by himself and Jack Knetsch, in which virtually identical WTP values were expressed by independent samples for a water quality program to preserve fishing, despite the fact that for some samples the benefit was restricted to a single region in Ontario and for another sample the program applied to the entire province. While such an effect can be induced, it is not inevitable, as other, comparable studies which used lengthy in-person interviews to value local (Desvousges, Smith, and McGivney, 1983) and national (Mitchell and Carson, 1984) fresh-water quality improvements have found that the more geographically extensive water improvements were valued several times higher than the local improvements, as would be expected if respondents' values were contingent on the amount of water involved.

Although we are less pessimistic than Kahneman about the ability of respondents to meaningfully value nonuse amenities when they are presented in a carefully conceived and pretested instrument, his point that symbolic factors can interfere with the valuation process is well taken. It is a reminder that the interviewing procedure and the wording of the scenario have to overcome a natural inclination of respondents to think of a contingent valuation survey as an ordinary public opinion survey. Factors likely to promote this misconception are CV surveys that value small-ticket items, use simple scenarios, and employ impersonal survey methods. For example, the implied consequence to a middle-income person of saying that he would pay $5 a year to double the acreage of local parks may not be serious enough in itself to stimulate him to consider the real value to him of this particular level of park provision. Needless to say, a difference between $1 and $5 in the mean WTP amounts for a park program may have large consequences for policies regarding the park.[19] Simple scenarios, such as one which describes an amenity change in a couple of sentences, may be insufficient to communicate the atypical nature of the CV-type survey question. Similarly, it is likely to be more difficult for respondents to grasp the type of valuation desired and to focus on the valuation-relevant items when these are presented over the phone or in a mail survey.

Other factors of quite a different type may also provoke symbolic responses unless special precautions are taken. These include scenarios involving a controversial topic on which the respondent has strong feelings or a topic that the respondent finds emotionally threatening. An example of the first might be a CV study that tries to value nuclear power risks; respondents may be unable to transcend the temptation to use the study to record their opposition or support for the controversial technology. Topics that stimulate strong emotional feelings might make it difficult for respondents to focus on the valuation-relevant aspects of a scenario—for example, a CV study that sought to value a reduction in the risk of contracting AIDS from contaminated blood.

Part-whole biases are major amenity misspecifications, and are also a result of the tendency of respondents to respond to public goods as global

[19] See Mitchell and Carson (1986c) for further discussion of this issue.

symbols without paying sufficient attention to the specific description offered in a CV survey. Here, however, the misconception is more specific. The dimensions of a good that are particularly prone to this misperception are its geographic distribution, its benefit composition, and the package of policies of which it is a part. Consider a respondent who is asked how much she is willing to pay for water quality improvements in a local river basin. If she is unable to isolate that river basin in her mind from her state's or region's other rivers, the respondent may in fact value a larger range of waters than intended by the researcher. Such geographical part-whole bias accounts for Kahneman's results in the Ontario fishing survey.

Benefit part-whole bias occurs when respondents are unable to differentiate between benefit subcomponents or between the subcomponents and the value for all types of benefits. We have described in chapter 3 the reasons why we believe respondents are likely to find it difficult to arrive at a meaningful value for, say, the option value of a good that is distinct from other types of benefits accruing to that resource (such as present-use or existence values).[20]

The third area where the potential exists for respondents to confuse a global entity with the component the researcher wants to value is the policy package of which the amenity is a part. Normally, respondents in CV studies are offered the opportunity to value only a single type of policy. Unless precautions are taken, respondents may treat the proposed policy as symbolic of a larger policy package and, without realizing their error, assign to the proposed policy some of the values they have for related policies. In our study of national water benefits, we specifically reminded respondents that no matter what amounts they gave for water quality, they would still pay for air pollution control (the other major federal pollution control program). To determine whether our efforts to avoid policy-package part-whole bias were successful, we designed an experimental test as part of this study; the results suggested that this particular bias was not present (Mitchell and Carson, 1984).

Among general strategies to minimize part-whole biases are inclusion of a description of the larger entity in the scenario, with a warning not to confuse the larger entity with the amenity changes being valued; and making the description of the good more salient by the use of such descriptive devices as maps (Desvousges, Smith, and McGivney, 1983). Another strategy, which appears to be compatible with respondents' cognitive processes, is to have them first value the total resource, even if this is not the subject of the study, and then have them allocate their total WTP amounts

[20] See also Mitchell and Carson (1985) and Greenley, Walsh, and Young (1985) for discussions of possible misspecifications leading to geographical part-whole bias and benefit part-whole bias.

for the component of interest. For example, respondents would first value air visibility improvements in the United States as a whole before specifying how much of this amount they would allocate to New Mexico or Los Angeles. These and other strategies are discussed in chapter 12, where we address the part-whole issue in the context of aggregation and disaggregation.

Still another type of amenity bias to be considered is *metric bias,* which occurs when the respondent values the amenity according to a different metric than the one intended by the researcher. At issue is the degree of precision with which people can meaningfully distinguish among different levels of the good described in the scenario. Consider a risk scale where risk levels are given in annual deaths per 100,000, and where two equivalent subsamples of respondents are asked to value a series of risk reductions.[21] Group A's reductions are 1.0, 2.0, and 5.0 to .05 per 100,000, and group B's are 5.0, 10.0, and 15.0 to .05 per 100,000. The researcher intends the scale to be cardinal so he can map the WTP values for specific risk levels on a scale and then trace out a complete demand curve for risk reductions of various magnitudes. Respondents, however, may perceive only an ordinal metric composed of low, medium, and high risks, with little differentiation within the broad categories. Since group B is valuing larger risk reductions than group A, there would be evidence consistent with the absence of metric bias (and with the presence of a cardinal metric) if group B's mean WTP values were to be higher than group A's. If, despite the difference in risk reductions, the values are roughly the same, it would appear that the respondents do not differentiate among risk reductions at this fine a level, but are valuing low-level risk reductions in general.

Probability of provision bias occurs when the perceived probability that the good will be provided differs from the researcher's intended probability. Survey designers usually try to convey the idea that the indicated level of the public good being valued will definitely be provided if enough money is raised. If skeptical or cynical respondents discount this certainty, the good will be undervalued. For example, researchers who ask respondents to value reductions in risk levels such as those posed by hazardous waste sites or nuclear power plants face the difficult task of convincing respondents that these reductions will in fact occur if the government program being valued is implemented.

Context misspecification bias. The second major source of biases involving misspecification of the market scenario is the various elements in the questionnaire that provide the context for valuing the amenity. Six of the seven elements discussed below are aspects of the scenario itself; the other involves those parts of the questionnaire that precede the scenario.

[21] This example is similar to a test conducted by Mitchell and Carson (1986b).

The payment vehicle has been identified as a crucial element of the scenario, and one that may carry value in itself. Typical payment vehicles used in contingent valuation surveys include park entrance fees, utility bills, property taxes, sales taxes, special funds, and higher prices and taxes. A number of CV experiments have found that the public's willingness to pay for public goods is influenced by the type of payment vehicle used in the study (Rowe, d'Arge, and Brookshire, 1980; Greenley, Walsh, and Young, 1981; Brookshire, Randall, and Stoll, 1980). Again, no misspecification is involved, nor can bias be said to exist, if the researcher understands the relevant effects of the payment vehicles on the WTP amounts and knowingly accepts them. Of course, if the payment vehicle does influence the findings they cannot be generalized to other situations where the good would be provided by a different vehicle, or where the vehicle is less controversial. One recent practice among CV practitioners has been to use the relatively neutral vehicle of higher taxes and prices whenever appropriate, in order to avoid the possibility of *payment vehicle bias*.

Property right bias is caused by the respondents' ambiguity about whether they have a right to ("own") the public good in question or whether they consider it appropriate that they have to pay to acquire it. Two basic property-right reference frames have been used in CV surveys. One employs a purchase structure and asks people how much they are willing to pay for the amenity; the other employs a compensation framework and poses the valuation question in terms of the amount people are willing to accept to give up the amenity. Chapter 2 discusses the reasons why the WTP structure is preferred in CV surveys, and why the dollar amounts for comparable amenities differ greatly according to whether the WTP or the WTA property-right reference frame is used.

The particular method of provision (or the particular agent providing the public good) used in a CV scenario can also bias WTP amounts. This is *method of provision bias*. Provision of a good by public charities, like the Salvation Army, tends to evoke a higher WTP than its provision by some vague government authority or industry. However, some state governmental agencies, such as state fish and game commissions, also appear to elicit high WTP amounts. Conversely, provision by the federal government may bias WTP amounts downward if respondents are influenced by the perception that the federal government generally wastes money—a perception consistently documented in public opinion polls.[22]

It is perhaps not recognized as widely as it should be by public opinion researchers that a well-designed CV scenario significantly inhibits what might otherwise be a casual expression of approval, because it asks people

[22] Ways to minimize protest zeros given by those who regard government as wasteful are considered in the discussion of item non-response bias in chapter 12.

to express their preferences in dollars instead of in degrees of agreement or disagreement about a statement. This said, there is a potential for *budget constraint bias* in CV surveys if respondents, in reacting to the hypothetical market construct, do not take their ability to pay into account, as they would with any comparably sized purchase or a referendum issue on the ballot. Where the indicated WTP amounts are relatively nominal, the threat posed by budget constraint bias is small. However, when major and costly programs are valued, misspecification of the budget constraint could have a sizable influence on the benefit estimates.

The simplest case of budget constraint bias arises when the intended constraint is the household's income and the respondent mistakingly thinks the constraint is his personal income.[23] (The reverse can also occur if the desired budget constraint is personal income rather than household income.) More complex is budget constraint bias in studies that attempt to value a program that respondents are currently paying for in their taxes or in the price of goods (such as the national air or water pollution control programs). In order to obtain the full consumer's surplus for the current level of that good's provision, respondents have to realize that the amount they give is not meant to be an addition to their current payments, but the amount they would give if they were not making any payments at all for this purpose. If the scenario does not communicate this condition, the misspecification would tend to bias the WTP amounts downward. Another situation, encountered in using payment cards with benchmarks in our water quality study, is the tendency of respondents to want to reallocate their payments for other goods—notably the defense program—to the amenity being valued. If permitted, this would have introduced a decidedly inappropriate budget constraint and biased the WTP amounts for water quality upward. The avoidance of budget constraint bias requires careful pretesting to identify potential problems. Once this has been done, it is usually possible to redesign the instrument to minimize budget constraint bias.

We have noted earlier the crucial role played in contingent valuation studies by the elicitation question and the potential that the various elicitation formats have for implying a value. *Elicitation question bias* occurs when the elicitation question used by the researcher fails to successfully convey a request for (1) the highest amount the respondent will realistically pay before preferring to do without the amenity being valued, and (2) a firm commitment to pay that amount.[24]

[23] Budget constraint bias can also arise in this context when the respondent being interviewed really knows little about the household's income. The problem can be avoided by interviewing that person in the household who identifies himself or herself as the household head.

[24] In the discrete-choice framework, where a particular price is offered on a take-it-or-leave-it basis, the issue concerns whether the commitment is to pay the specified amount.

Some of the formats that have been used or proposed for use in CV studies are:

A. Would you be willing to pay ___ if it resulted in (improvement)?
B. Would you pay ___ for (improvement)?
C. What is the highest amount you would *be willing to pay* for (improvement) before you would feel you are spending more than it's really worth to you?
D. Would you personally *be prepared to pay* an extra ___ to (obtain improvement)?

The Water Resources Council's *Principles and Guidelines for Water and Related Land Resources Implementation Studies* recommends B instead of A on the grounds that respondents may interpret the latter as an appeal for voluntary contributions (1983:80). Formats C and D are less vulnerable to this criticism, but they possibly could be strengthened by removing the italicized words in each. No experiments have been conducted to test the effect, if any, of these particular wording changes, and it is difficult to determine a priori the direction of the potential bias. For amenities that are valued at higher than nominal amounts, the direction of the bias is likely to be downward, as respondents are likely to be averse to making too high a commitment to pay for the good.

Another factor promoting elicitation question bias is a tendency on the part of some respondents to believe that any amount they commit themselves to that happens to be greater than the amount actually required to accomplish the improvement will be wasted. This misunderstanding leads them to temper their WTP amounts by what they consider a "reasonable" upper limit, based on what they think the improvement should cost per household. We discovered this belief when conducting the pretests for our study of drinking water benefits. Respondents in the small Illinois town where the study was conducted were anxious to know what the actual costs of the proposed risk reductions would be so that they could avoid giving more than necessary and avoid the risk of having city hall use the money for other purposes.

Avoidance of elicitation question bias requires careful attention to the wording of the scenario, and exploratory studies aimed at determining the likelihood of this effect. In extreme cases, use of the single-price elicitation method, where the respondent is offered one dollar amount and asked whether or not she is willing to pay that amount for the amenity, may be required.

Instrument context bias occurs when the questionnaire materials that precede the CV scenario influence the WTP amounts in a directional fashion. This is an area of current interest to survey methodologists, whose experimental findings thus far suggest a tentative conclusion that this effect is not as widespread as previously thought. The basis for anticipating instrument context effects is a theory of how humans process information.

The theory holds (among other things) that when people are called upon to judge an event, they sample only the most easily accessible subset of the information available in their memory (Wyer and Srull, 1981). Kahneman, Slovic, and Tversky (1982) call this judgmental shortcut the availability heuristic; it is exemplified when an individual makes a judgment about the general risk of heart attacks primarily on the basis of recent heart attacks suffered by her acquaintances.

In experimental psychology the phenomenon of "priming" is caused by this context effect. Experiments have been conducted in which researchers manipulated the information subjects received shortly before they undertook a task to see if the information, which was not formally related to the task, primed or influenced their performance. In some cases a priming effect has been observed. In one set of experiments (Wyer and Srull, 1981:181–183), subjects were asked to rate a person on a series of traits (such as hostility, boredom, selfishness, and kindness) solely on the basis of a paragraph that described the events in one afternoon of the target person's life. The paragraph description was ambiguous with respect to hostility: it described such actions by the target as refusal to let a salesman enter the target's apartment, which might or might not be undertaken in a hostile manner. A seemingly unrelated exercise, conducted at the beginning of the experimental session, introduced hostility-related items to one group of subjects and kindness-related items to a second group. The subjects' subsequent ratings of the target person's personality were affected by these items in the direction predicted by a priming effect.

Survey methodologists have demonstrated similar effects. George Bishop (1985) has found that respondents were more likely to say that they "follow what's going on in government and public affairs" if they were asked a question about this immediately after a series of "rather difficult questions about their United States Congressman's record" than when the same question was asked before the questions about the congressman's record. Some methodologists (for example, Tourangeau and coauthors, 1985) have theorized that context effects like these are more likely to occur when people are asked about an issue that is obscure or requires a complex judgment.

In contingent valuation surveys respondents make complex judgments after a considerable amount of material has been introduced by the interviewer. Much of this material is contained in the scenario, of course, and is intended to influence the values by describing the conditions of the market. However, there is the possibility that context effects will be inadvertently induced by the questions that precede the scenario. In a typical CV study the interviewer usually begins the interview with questions about such topics as the respondent's experience with the amenity, his general attitudes toward its preservation, or even about his personal characteristics. Although these questions are not part of the scenario, and are usually not intended to influence the WTP amounts, they may in fact do so under certain conditions. For example, asking a series of questions about trust and confidence in public authorities immediately before a scenario might bring negative cognitions about officialdom to mind, and so lead respondents to express a lower value for an amenity in a scenario that assumes that public authorities will guarantee the amenity's provision, than if the questions about trust had not been asked.[25] Similarly, asking a series of questions about the desirability of environmental quality, without posing some of these questions in the form of economic tradeoffs (such as jobs versus environment), could exacerbate the potential for compliance bias by emphasizing the normative character of environmental improvements.

Clearly, the potential for context effects is present in contingent valuation surveys. How serious is this threat? Unfortunately, no studies of the phenomenon have been conducted in regard to CV surveys. The available evidence in related areas (Tourangeau and coauthors, 1985) suggests that although context effects can and do occur in attitude surveys, their occurrence is sporadic and unpredictable. They are not consistently related to surveys whose questions involve unfamiliar or complex subjects. For surveys in general, Tourangeau and his coauthors conclude that "in *most cases* questions about related issues can be placed close together in an interview without affecting the results and that, when there are context effects, *in most cases* they will be small" (emphasis in the original). This is a reassuring finding, although until more evidence is available, including experiments conducted in the contingent valuation context, CV researchers should be sensitive to possible context effects.

The order in which questions are asked sometimes, in unintended ways, influences response patterns in surveys (Schuman and Presser, 1981). *Question order bias* is potentially troublesome in CV surveys when values are elicited for several different amenities or different levels of an amenity. For some goods there is a natural sequence to the provision of the good that prescribes the order of questions, and respondents answering later questions are intended to take their answers to earlier questions into

[25] In the nomenclature of theories of cognitive processes (see Wyer and Srull, 1981), these preliminary questions would evoke a "schemata" or ensemble of feelings, beliefs, and images about a topic, often organized around a central idea or image (for example, distrust). Often a person is capable of viewing a given situation from the perspective of several diverse schemata. To the extent that schemata determine valuation, the evocation of one schemata (distrust) rather than another (working in the public interest), or both at the same time, may distort the outcome.

account. One example is where the provision of higher levels of an amenity requires incremental improvements in lower levels of provision, as is the case in raising the boatable level of water quality to swimmable levels.[26] In such a case question order bias is not a problem, since the respondents are valuing the correct sequence of policy changes. In other instances, the researcher wants the respondents to value each good independently, on its own terms, without reference to their WTP amounts for other goods also valued in the same survey. Here the question order may influence the amounts given for one or more of the goods that are being valued, despite the interviewer's instructions to the respondents to disregard previous answers. Random assignment of different question orders will detect this effect, if it is present.[27]

Summary and Conclusions

Instrument effects and respondent misperceptions are common problems in survey research. In this chapter we have reviewed a large number of potential threats posed by these sources to the validity of contingent valuation studies. In the absence of a fully developed measurement model in the survey research literature, we developed a working model in the form of a set of assumptions about respondent behavior in CV surveys and the sources of error that follow from these assumptions. Three source of error—incentives to misrepresent responses, implied value cues, and misspecification of the market scenario—were considered in detail. A fourth major source of error, improper sampling and aggregation, is discussed in chapter 12; it occurs in the survey sampling, design, or data-analysis stages.

Some, but not all, of the biases considered in this chapter have been previously identified in the contingent valuation literature. We intentionally excluded several types of bias, particularly hypothetical bias and information bias,[28] that have received considerable attention but which are not really meaningful categories of bias. Where possible, we drew upon the results of experiments conducted by CV researchers to detect the

[26] Survey researchers have a tendency to see question order effects where economists see natural marginal valuation and substitution effects. For this reason it is important to decide whether or not these potentially legitimate economic effects are an integral part of the scenario being valued.

[27] See Mitchell and Carson (1986b) for a test for question order effects. In that study no effect was detected.

[28] The question of what information should be conveyed in the limited time available in a CV interview is, however, an important and underappreciated issue. See Samples, Dixon, and Gower (1985) for one of the few studies that attempts to address this problem.

presence of one bias or another. We also called attention to the methodological problems that make some of these experiments less useful than they otherwise would be.[29] Our hope is that future bias experiments will be based on sample sizes that are large enough to allow meaningful conclusions to be drawn from their findings.[30]

In our treatment of each type of bias we offered suggestions for avoiding or minimizing the bias effects whenever possible. Whether or not a study is vulnerable to one or more of these biases depends on a number of factors, including the survey method used, the nature of the amenity, and the purpose of the study. The researcher must carefully assess the potential sources of bias to which his or her study may be vulnerable. To do so will often require intensive preliminary research of a qualitative nature in order to better understand how and why potential respondents will react to the various scenario elements, and the inclusion of one or more bias experiments in the study design to verify that the most likely sources of bias are not present in the study.

That the contingent valuation method is vulnerable to instrument effects and to miscommunication between what the interviewer says and what the respondent understands will come as no surprise to anyone familiar with survey research. The array of potential biases described here is a forceful reminder that the contingent valuation method cannot easily be applied in an off-the-shelf manner. The possibility of biased CV benefit estimates should not be taken to imply that bias is inevitable in CV studies, any more than the possibility of having biased coefficient estimates in an econometric model necessarily implies that any particular set of coefficients is biased. The more we understand about how people answer contingent valuation questions, the greater the likelihood that we can minimize extraneous sources of error and obtain better benefit estimates.

[29] Unfortunately, some reports of CV experiments fail to provide the basic information necessary to evaluate the findings.

[30] Appendix C presents information that may be used for this purpose.

12

Sampling and Aggregation Issues

Contingent valuation researchers usually want to generalize their findings in various ways. Typically, they use their data to estimate benefits for a much larger group of people than those they actually interviewed. Sometimes they use the WTP amounts from a particular CV study to infer values in other contexts or to value a broader or narrower set of policies than those specifically addressed in the original study. In this chapter we address the issues involved in sampling and aggregation.

We begin with a discussion of biases caused by sampling design and execution errors. For individual WTP responses to be usable in benefit-cost analysis, they must be extrapolated to the appropriate population. We identify and discuss ways to minimize or compensate for (where possible) the effects of four potential sources of sampling and nonresponse error. We then turn to the question of the temporal stability of CV findings, and address the issue of whether the findings from a CV study conducted at one point in time may be used to estimate the benefits of the same amenity at a later point in time. The next section of the text considers the validity of aggregating separate estimates for geographical entities or for different amenities. Under what circumstances can benefit estimates of air visibility improvements for one area be combined with the estimate of the same improvement for another area? We conclude the chapter with a discussion of the fallacy of motivational precision and its bearing on the use of CV surveys to measure separate values for benefit components.

Bias from Sampling Design and Execution Errors

Probability sampling procedures provide surveys with a straightforward way to generalize from the responses of a relatively small number of respondents to much larger populations.[1] These procedures are based on

[1] See Sudman (1976) and Cochran (1977) for two standard treatments.

the principle that each economic agent (such as an individual or a household) in the population of interest has a known probability of being selected.[2] Sampling issues have not received much attention in the contingent valuation literature until recently,[3] although they represent a substantial threat to the accuracy of aggregate WTP estimates. One reason for this neglect is that, until recently, CV researchers were preoccupied with overcoming measurement problems (for example, Cummings, Brookshire, and Schulze, 1986) and treated aggregation almost as an afterthought. Another factor is that economists have lacked training and experience in statistical survey techniques, as they have typically used existing survey data and have relied on such agencies as the Census Bureau and the Bureau of Labor Statistics and on survey organizations such as the University of Michigan's Institute for Social Research to take care of sampling and missing-value problems before releasing their data for public use.[4]

Determining who to interview for a CV study and how to locate and interview these people involves a series of decisions. First, the researcher must decide how to define the population of economic agents likely to be influenced by the change in the level of the public good. Do the agents include the residents of a particular town or other geographic area? Are they the users of the amenity? Are these agents to be individuals or households? Next, the researcher must decide how to identify or list this population. The list or method of generating such a list is known as a sampling frame. It is from this list that the actual sample is drawn. The third step is to attempt to obtain valid WTP responses from each of the economic agents chosen to be in the sample. Unfortunately, there will be a sizable number of respondents who fail, for some reason, to give valid WTP amounts. These "nonresponses" can lead to nonresponse bias or to sample selection bias, or both, unless corrective steps are taken. The eventual benefit estimates can become biased as a result of the sampling decisions and procedures at each or all of these stages. These issues are treated below in the context of discussions of the four types of potential sampling design and the execution biases listed in table 12-1.

[2] Where the main purpose of a study is to assess the effects of different scenario design features (such as the payment vehicle) on the WTP amounts, random assignment of the respondents or subjects to the different treatments, rather than random sampling, is necessary (see appendix C).

[3] See Desvousges, Smith, and McGivney (1983), Mitchell and Carson (1984), Bishop and Boyle (1985), Mills (1986), Moser and Dunning (1986), and Edwards and Anderson (1987).

[4] However, the procedures used by these agencies often have sizable implications for economic analysis which are only now becoming recognized. See, for example, Lillard, Smith, and Welsh (1986) and David and coauthors (1986).

Table 12-1. Potential Sampling and Inference Biases in CV Surveys

1. *Sample Design and Execution Biases*
   - A. *Population Choice Bias:* where the population chosen does not adequately correspond to the population to whom the benefits and/or costs of the provision of the public good will accrue.
   - B. *Sampling Frame Bias:* where the sampling frame used does not give every member of the population chosen a known and positive probability of being included in the sample.
   - C. *Sample Nonresponse Bias:* where the sample statistics calculated by using those elements from which a valid WTP response was obtained differ significantly from the population parameters on any observed characteristic related to willingness to pay; this may be due to unit or item nonresponse.
   - D. *Sample Selection Bias:* where the probability of obtaining a valid WTP response among sample elements having a particular set of observed characteristics is related to their value for the good.

2. *Inference Biases*
   - E. *Temporal Selection Bias:* where preferences elicited in a survey taken at an earlier time do not accurately represent preferences for the current time.
   - F. *Sequence Aggregation Bias*
     1. *Geographical Sequence Aggregation Bias:* where the WTP amounts for geographically separate amenities that are substitutes or complements are added together to value a policy package containing those amenities, despite the fact that the amenities were valued in an order (for example, independently) different from the appropriate sequence.
     2. *Multiple Public Goods Sequence Aggregation Bias:* where the WTP amounts for public goods that are substitutes or complements are added together to value a policy package containing those amenities, despite the fact that the amenities were valued in an order (for example, independently) different from the appropriate sequence.

Population Choice Bias

Population choice bias occurs when the researcher misidentifies the population whose values the study is intended to obtain. Populations may be defined in terms of the element (the individual recreator, for example), sampling unit (cars entering recreation areas),[5] locale (in two counties in northern California), and time (in July 1988). Choosing the correct population is simplest when the population who will pay for the good (or who would be presumed to pay according to a given payment vehicle, such as a local tax) coincides with the population who will benefit. The greater the divergence between those who pay and those who benefit, the more problematic choice of population becomes. Consider the case of the huge Four Corners power plant at Fruitland, New Mexico (Randall, Ives, and Eastman, 1974). Residents of the area and visitors who come to enjoy the scenery use the public good of air visibility without paying the cost of maintaining it. This payment obligation is borne by those in Los Angeles and elsewhere who purchase their electricity from the utility which owns the plant. Nevertheless, area residents and visitors may be a crucial population for a WTP study of the aesthetic benefits of local air visibility, since they directly experience the benefits.

Four relationships—A, B, C, and D—between paying for and using a public good are described in table 12-2. Relationship A is the optimal position for respondents in a CV survey, and where possible payment vehicles that move people toward A should be used. Depending on the situation being valued, respondents may fall into any one of the four relationship categories. Let us suppose that the researcher wishes to estimate the noise pollution control benefits for a village which is considering ordering quieter garbage trucks. If the cost would be paid out of village property taxes, the appropriate population would appear to be the residents (households) of the village. Because many of those who would benefit from the amenity are also village residents, this population is close to category A.

Choosing the appropriate population becomes more complex when such factors as absentee landlords and spillover effects are considered. A population composed of those who live in political jurisdictions responsible for public goods almost inevitably includes both users and nonusers. Those who reside in a city having a public school system, for instance, include the childless, people whose children are too young or old for public school,

[5] In this discussion, "unit" is often used when "element" is technically the correct term because we frequently specify households as the relevant economic agents. In this and many other (but not all) instances, the population unit and the population element will be equivalent.

Table 12-2. Relationship Between Paying for a Good and Benefiting from Its Provision

Benefit from the good		Pay for the good: Yes	Pay for the good: No
	Yes	A	B
	No	D	C

and those who send their children to private schools. In our village example, those in relationship B would include visitors to the town who would benefit from quiet garbage trucks even though they would not pay for them, since they are not subject to town property taxes. Category D would include deaf residents and absentee property owners. However, use of the town population as a sample would leave out some Ds. Presuming that property taxes are the source of the town's revenue, absentee landlords would not be represented in a sample of town residents; a population consisting of property tax payers would include them, but it would exclude renters.[6] People in relationship D may hold existence benefits for an amenity. It is conceivable that some absentee landlords who were former residents might gain utility from knowing that their friends who continue to live in the village have a quieter environment.

Turning to the case of the Four Corners air visibility amenity, Los Angeles residents who never recreate or intend to recreate in the Four Corners area may value the knowledge that the extraordinary air visibility there is unaffected by emissions from the plants which provide their electricity. Indeed—and here we come to relationship C—the preservation of air visibility in the Southwest might be worth something to residents of Ohio as well. Selection of a local or even a regional population for a CV study of this amenity would be inadequate if the researcher wished to include existence values in a national estimate of the benefits of high visibility in the Southwest.

After the entire population has been defined in terms of the relationships given in table 12-2, the researcher has to define the relevant economic agents. Because payments for most pure public goods are made at the household level (property or income taxes, for example), the household

[6] We recognize that renters eventually pay all or some of the taxes imposed on landlords.

rather than the individual will most often be the relevant economic agent. When this is the case, the appropriate sampling procedure is to allow any adult who claims to be a household head to be a spokesperson for the household—the current practice of the U.S. Census Bureau.[7] Where the amenities are quasi-private goods such as hunting permits, however, individuals rather than households are likely to be the appropriate economic agent.

Sampling Frame Bias

After the population of interest has been identified, the sampling frame must be defined. The frame may be an existing list of the sample units of interest, or, more commonly, a method of generating a list. If the population and the sampling frame diverge, *sampling frame bias* can occur. This type of bias makes it difficult, if not impossible, to accurately generalize the results of the study to the population initially defined by the researcher, even if there are no other problems in carrying out the survey.

The procedures for defining the sampling frame vary according to the survey method used—in-person, telephone, or mail.[8] The sampling frame for in-person surveys of people who live in a given area is normally based on a physical enumeration of geographically defined occupied dwelling units. For large areas, various stratification and clustering techniques have been developed to make enumeration costs manageable (Cochran, 1977). Nongeographically based populations often pose more difficult problems for in-person surveys. Suppose that those who use a beach or visit a park comprise the population of interest. A valid sampling frame should make it possible for the sample to represent the visitors according to the time of day they visit, the day of the week, the season of the year, and, possibly, how they use the facility. The sampling frame for telephone surveys can either be chosen from the numbers listed in phone books—which entails the very real problem of unlisted numbers, both voluntary and involuntary[9]—or, preferably, by using the random digit dialing method. This method, which selects numbers at random from the universe of usable numbers for the population of interest (see Frey, 1983:57–86), ensures that unlisted as well as listed numbers are included in the sample. Mail

[7] See Becker (1981) for the implications of using households as the unit of analysis.

[8] For nontechnical descriptions of sampling frame development procedures, see Sudman (1976) or Tull and Hawkins (1984).

[9] Approximately 95 to 96 percent of American households have telephones. Rich (1977) reports that the rate of unlisted numbers in urban areas soared 70 percent between 1964 and 1977. Groves and Kahn (1979) report an unlisted rate of 27 percent for their latest national sample. According to Frey (1983:62), "when you add new, but unpublished, listings to this figure, it is possible that at any one time, nearly 40 percent of all telephone subscribers could be omitted from the telephone directory."

survey sample frames are based on lists of potential sampling units. They face the problem of obtaining up-to-date addresses for every economic agent in the population of interest. This is often a difficult task for surveys of the general public because of the frequency with which people in our society change their residences.[10]

Nonresponse Problems and Biases

No matter what sampling plan and survey method are used in a contingent valuation survey, some level of nonresponse to the WTP questions is virtually inevitable, with the consequence that the number of those who give valid WTP amounts will be smaller than the number of originally chosen sample elements (Panel on Incomplete Data, 1983). There are two distinct ways in which a member of the sample can fail to respond to a WTP question. In the first, known as unit nonresponse (Kalton, 1983), the person (or household) fails to respond to the questionnaire. This occurs when people cannot be found at home by phone or in-person interviewers, when they refuse to be interviewed when asked to participate by the interviewer, or when those sampled in a mail survey fail to return the questionnaire.

In the second, called item nonresponse, a respondent answers some or most of the questions on a survey but fails to answer a particular question of interest, such as the WTP question. Although all surveys suffer from some level of item nonresponse, some CV surveys have unusually high item nonresponse rates for the WTP questions. With the exception of questions which ask for the respondent's income, item nonresponse rates exceeding 5 to 7 percent are relatively rare in ordinary surveys (Craig and McCann, 1978). In CV surveys, however, nonresponse rates of 20 to 30 percent for the WTP elicitation questions are not uncommon where (1) the sample is random and therefore includes people of all educational and age levels, (2) the scenario is complex, and (3) the object of valuation is an amenity (such as air visibility) which people are not accustomed to valuing in dollars. *Up to a certain point,* these higher levels of nonresponse to WTP questions are acceptable or even desirable. It is unrealistic to expect that 95 percent of a sample will be able and willing to expend the effort necessary to arrive at a well-considered WTP amount for certain types of amenities. Given the choice between having someone offer an unconsidered guess at an amount or having them say they don't know how much it is worth to them, the latter behavior is preferable, provided appropriate procedures to compensate for the resulting item nonresponse are used.

[10] There are likely to be fewer problems of this kind where the appropriate sampling frame consists of a current list of addresses held by a government agency—a list of the holders of fishing or hunting licenses, for example.

Item nonresponses on WTP questions fall into four general categories: don't knows, refusals, protest zeros, and responses that fail to meet an edit for minimal consistency. In a well-designed contingent valuation study, the don't know, refusal, and protest zero categories usually account for the bulk of the item nonresponses, and it is possible, through questionnaire design, to influence the distribution of nonresponses across these categories. Protest zeros are perhaps the most troublesome, as it is necessary to distinguish them from the bids given by respondents who prefer to forgo the good in question rather than to have to pay for it. Depending on the amenity being valued and the payment vehicle, protest zeros can constitute a substantial portion of the $0 bids. In the Monongahela study by Desvousges, Smith, and McGivney (1983), for example, approximately half of the zero bids were identified as protest zeros. The technique (developed by CV researchers) of asking respondents who gave zero bids why they did so is an indispensable aid in separating the latter group from those who reject the evaluation process itself for one reason or the other.

In the CV studies with which we are familiar, responses set to "missing" (often called outliers) during the consistency edit are typically from very low income respondents who gave WTP amounts representing an implausibly large percentage of their income, or from upper-income respondents who gave zero or very low WTP amounts even though their answers to other questions in the survey indicated a strong demand for the good. Various techniques can be used to perform the consistency edit, ranging from simple logical edits to statistical techniques based on different definitions of outliers (Barnett and Lewis, 1984). As the criteria for defining an observation as an outlier are to some degree judgmental, it is important that researchers explicitly describe (for each outlier, if necessary) the reasons why they reject WTP amounts as invalid.

Both unit and item nonresponses result in the loss of valid WTP amounts from those originally chosen for the sample. If 1,000 households were drawn by probability-based methods for a CV sample, and valid WTP amounts were obtained for only 800 of those households, the researcher would have to determine what effect the missing 200 households would have on the WTP estimate. Put another way, could the values of the 800 people in the realized sample (those for whom valid WTP amounts were available) accurately represent the values for the amenity held by the population from which the original 1,000-household sample was selected? If nonresponse in a CV survey was not associated with the magnitude of the WTP values held by the original sample, the failure to interview some respondents from the original sample would not cause bias (provided the sample size were reasonably large), although it would affect the reliability of the estimates. However, the factors determining the nonresponse rate are likely to be corrected with the magnitude of the WTP values. Researchers have found that nonresponse is often associated with lack of interest in the topic of the survey (Stephens and Hall, 1983), and it is likely that those who are less interested in an amenity would hold different values for it than their more interested counterparts. They have also found that response rates typically vary across population subgroups, such as lower-income people, and there is ample evidence that survey variables are often associated with the characteristics of these subgroups (Kalton, 1983:16).

To determine whether the nonresponse in a given study results in bias, two questions need to be asked, one relating to whether differential response rates from identifiable categories or groups of households exist, and the other to whether there are systematic differences between those within a particular group who did and did not answer the WTP questions. Bias will occur to the extent that these between- and within-group differential response rates exist *and are related to the value for the good.* A given CV study may suffer from a between-group sample nonresponse bias, a within-group sample selection bias, or both.[11] It should be clear that the failure to observe a characteristic related to WTP (income, for example) can change a sample nonresponse bias into a sample selection bias, and that obtaining a previously unobserved characteristic can change a sample selection bias into a nonresponse bias. To be more explicit, let

$$\text{WTP} = f(X,\beta) + U, \tag{12-1}$$

where $f(X\beta)$ is a regression function based on X—a matrix of predictor variables—and U is a vector of error terms. Sample nonresponse bias occurs when the sample distribution of Xs differs significantly from the joint population distribution of Xs, and sample selection bias occurs when the sample distribution of U differs significantly from the population distribution of U.

The flow chart in figure 12-1 identifies the conditions that result in one or both of these biases by posing a sequence of questions and answers. Where every element in the original sample is included in the realized sample of valid WTP responses (see question 1), no bias related to nonresponse will occur, and the higher the response rate the lower the potential bias. Where nonresponse is present in some form, the second question serves to identify whether or not sample nonresponse bias is present. It asks whether nonresponse rates vary *across* definable population subgroups which hold different WTP values for the amenity being valued. If, for

[11] The term "nonresponse bias," as used in the survey research literature, often refers to both the between- and within-group biases.

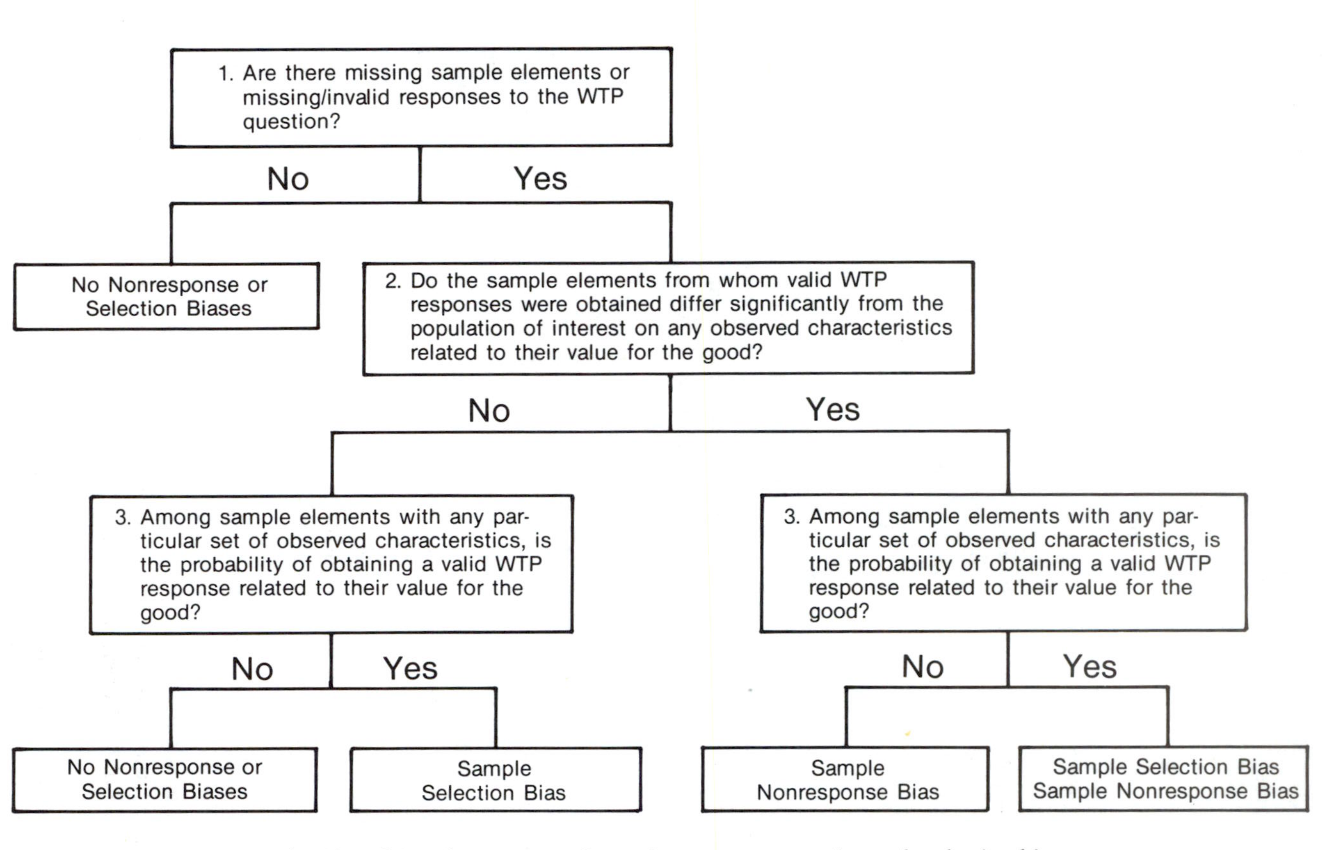

Figure 12-1. Decision tree for identifying the presence of sample nonresponse and sample selection biases

example, low-income people are underrepresented in a realized sample relative to their proportion in a population, and if these subgroups hold lower WTP values than those in higher income subgroups, then the WTP estimate would be biased upward.

Whether or not sample nonresponse bias is present, it is also necessary to ask whether, *within* any population subgroup, the probability of obtaining a valid WTP response from a sample element is related to that element's true WTP amount for the good. If the answer to this third question is yes, sample selection bias is present. It would be present, for example, if in a given sample those with low incomes who did offer WTP amounts were more interested in the amenity and valued it more highly than those with low incomes who were unwilling to make the effort to answer the WTP questions.

Sample nonresponse bias and sample selection bias are discussed in more detail below, along with the available correction procedures.

Sample Nonresponse Bias

Sample nonresponse bias, the between-group source of bias, is a persistent problem in CV surveys because different categories of respondents, such as those with greater or lesser education or incomes, tend to have different nonresponse rates, and these categories of people also tend to hold different values for many public goods. In-person surveys often underrepresent males and young blacks because young people, those who live alone, those who work, and those who live in large cities are less available to interviewers and therefore less likely to be found at home (T. Smith, 1983). City dwellers are disproportionately likely to refuse to be interviewed. Those who are not able to be interviewed tend to be elderly or people with limited mental capacities.[12] Research on who responds to mail surveys shows a consistent negative correlation between educational level and likelihood of returning a questionnaire (Kanuk and Berenson, 1975). To the extent that the underrepresented subgroups hold different WTP values than the overrepresented subgroups, a potential for sample nonresponse bias exists. Fortunately, there are procedures which may be used to compensate, at least in part, for this source of bias.

When nonresponse bias is a result of unit nonresponse, no information is available for the individual missing cases. The primary correction technique is to weight the cases in the realized sample so that the weighted sample statistics for key demographic variables correspond to known population parameters. Typically, information from the Census Bureau's Current Population Survey is used as a source of the population parameters in

[12] See T. Smith (1983) for a review of the research on who refuses in-person surveys. This is a difficult research area, and one where the findings are contradictory and inconclusive.

developing the weights for each case. Thus the cases in the realized sample from an underrepresented category, such as young black males, would receive relative weights sufficiently greater than 1 to enable them to represent their missing counterparts, while middle-aged white women, who are usually overobserved relative to the population, would receive relative weights of less than 1.[13] Weighting techniques are based on the assumption that the nonrespondents have similar WTP values to those with similar characteristics who were interviewed.

As with unit nonrespondents, item nonrespondents are not randomly distributed among the different subgroups. Among item nonrespondents, the underrepresented categories include those having low incomes, less education, and greater age, and there is reason to assume that people in these categories hold different WTP values than others. A broader range of techniques is available to correct for item nonresponse because information is available from the respondents' answers to the other questions. In addition to information on the item nonrespondents' demographic characteristics, these data usually include answers to attitude questions which are frequently related to the good in question.

Although weighting can be used to correct for item nonresponse, the preferred techniques use the available information from other questions answered by the respondents to help impute the missing values.[14] The most common approach defines ad hoc imputation classes, such as young white women, and assigns values to the missing cases in each class that are based on the WTP amounts given by those observations in the same class for whom valid values are available. The values used for this purpose are sometimes the mean or median value for the class, or a randomly assigned value from the pool of valid observations for the class. The mean or median approach tends to reduce the variance, whereas use of random values maintains the variance properties of the distribution. The Census Bureau frequently uses a third method to determine values for the missing cases in a given imputation class. Known as the "hot deck" technique (Bailar and Bailar, 1978), this approach takes advantage of the spatial autocorrelation of answers to questions that typically occurs in areal probability surveys that use stratification and clustering.

[13] See Anderson, Basilevsky, and Hum (1983), and especially Kalton (1983, chap. 3), for reviews of these procedures. It must be emphasized that the use of weighting procedures is often a crude corrective, and runs the risk of substantially inflating the variances of the survey estimates unless a sufficient number of respondents (25 or more) are available in each weighting class. The valid use of weighting for CV purposes requires response rates of the level achieved by professional researchers using current state-of-the-art techniques.

[14] For a useful description of imputation procedures, see Kalton (1983). David and his coauthors (1986) also compare a number of imputation class and regression procedures for estimating missing values.

Another method for imputing missing values is to use a model-based regression or maximum-likelihood procedure (for example, the EM algorithm) (Orchard and Woodbury, 1972; Dempster, Laird, and Rubin, 1977) which predicts the expected values of the missing observations, conditional on the structural model used to estimate the equation and the values of the variables used to predict the missing WTP responses. This estimated structural model is essentially a valuation function. To see how a valuation function can be used to correct for sample nonresponse bias, consider the estimates of a very simple relationship of the form

$$\text{WTP} = a + b^{*}\text{INCOME} + c^{*}\text{USE} + \epsilon, \qquad (12\text{-}2)$$

using the realized sample. The estimated mean WTP amount for the population can be obtained by

$$\overline{\widehat{\text{WTP}}}_{(\text{POP})} = \hat{a} + \hat{b}^{*}\,\overline{\text{INCOME}}_{(\text{POP})} + \hat{c}^{*}\overline{\text{USE}}_{(\text{POP})} \qquad (12\text{-}3)$$

where $\overline{\text{INCOME}}_{(\text{POP})}$ and $\overline{\text{USE}}_{(\text{POP})}$ are the population mean values for the right-hand side variables in the valuation function.[15] Using this approach to correct for sample nonresponse bias is dependent on two key assumptions. First, it must be assumed that the researcher knows the correct specification of the mathematical form of the valuation function and the correct specification for the distribution of error terms, except for a small number of unknown coefficients which can be estimated from the data; and second, that the data do not suffer from sample selection bias.[16]

Still another imputation method is CART (Breiman and coauthors, 1984), a tree-structured classification and regression procedure. CART combines features of both the model-based and ad hoc imputation class approaches in that it first estimates an optimal model (in the sense of predicting the WTP amounts) in terms of imputation class definitions, and then uses any of the imputation methods (such as the hot deck) to assign values to the missing cases.[17]

[15] DuMouchel and Duncan (1983) argue that weights that make the realized sample representative of the population are not needed for estimating the regression (that is, the valuation) function if the regression model assumes constant coefficients for all observations and there is no sample selection bias.

[16] It would also be possible to use this approach, with modifications, to account for the sample selection bias problem. However, the two problems together imply a large loss of information, particularly if the nonresponse problem is of the unit variety.

[17] CART is similar in many respects to the Michigan Automatic Interaction Detection (AID) program (Sonquist, Baker, and Morgan, 1974), which has been shown to work well in imputing missing values (Kalton, 1983). CART has solved several of the AID program's problems, however, by using tenfold cross validation to prune a tree grown from the bottom up. This computer-intensive procedure results in both an honest estimate of the predictive power of the tree and an optimally sized tree.

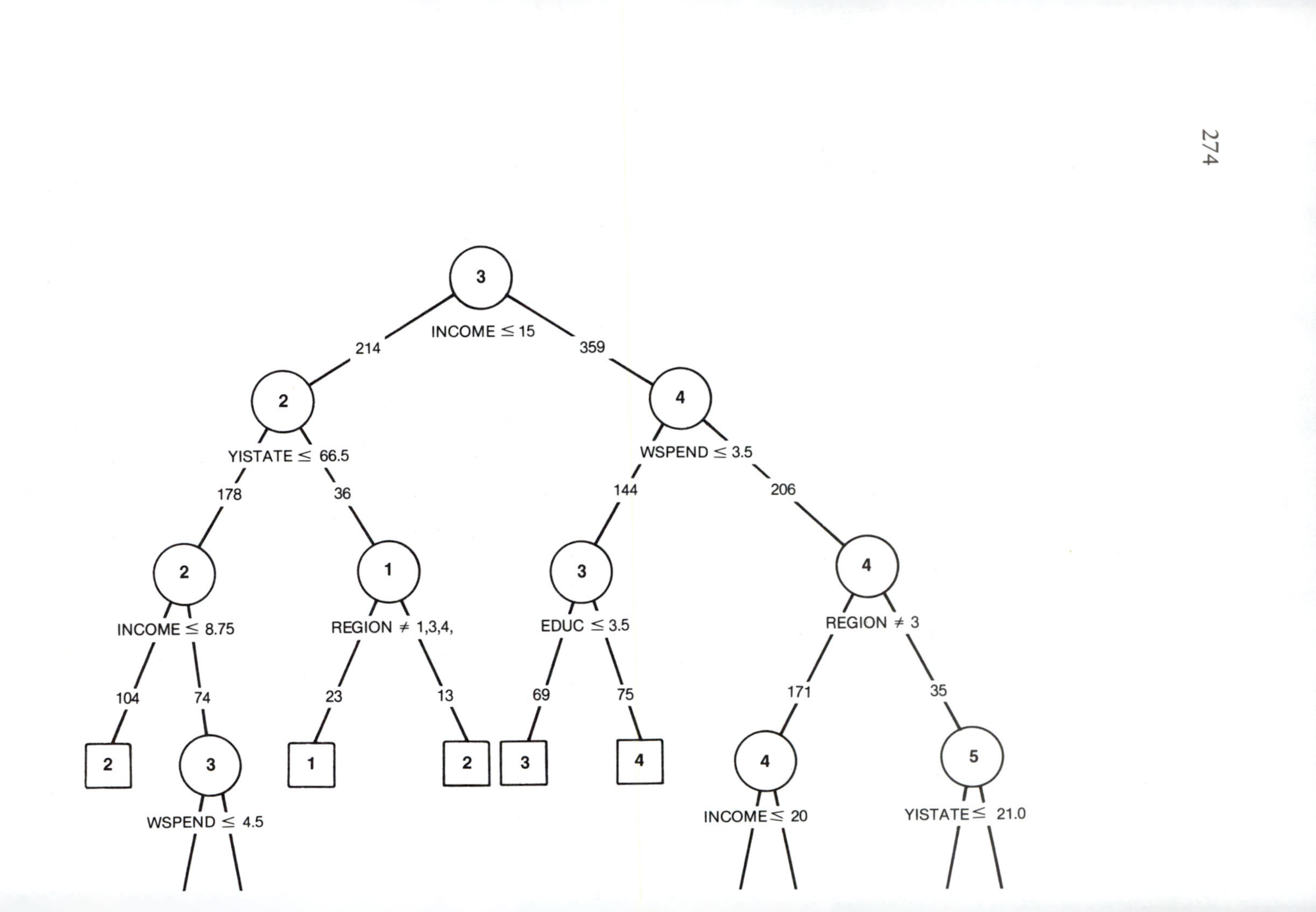
3
INCOME ≤ 15
214
359
2
4
YISTATE ≤ 66.5
WSPEND ≤ 3.5
178
36
144
206
2
1
3
4
INCOME ≤ 8.75
REGION ≠ 1,3,4,
EDUC ≤ 3.5
REGION ≠ 3
104
74
23
13
69
75
171
35
2
3
1
2
3
4
4
5
WSPEND ≤ 4.5
INCOME ≤ 20
YISTATE ≤ 21.0

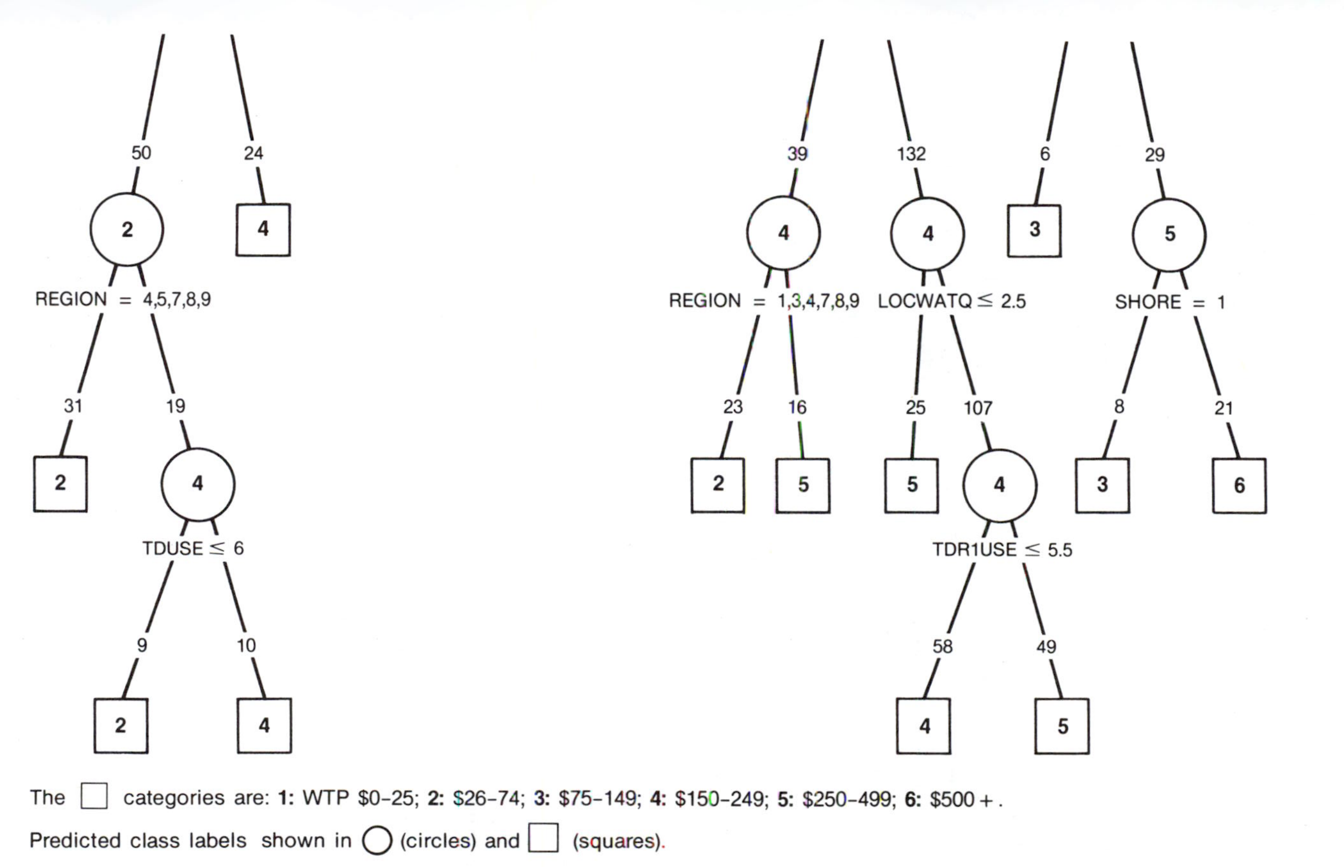

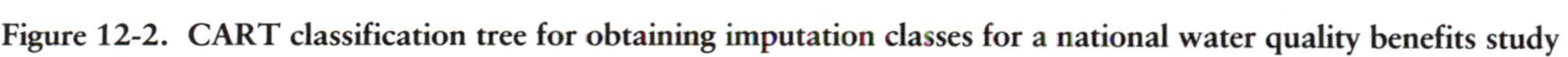

Figure 12-2. CART classification tree for obtaining imputation classes for a national water quality benefits study

Exploration of the properties of different imputation methods is an active area of current research (Panel on Incomplete Data, 1983). Carson (1984, 1986b) has compared several methods of imputing missing WTP responses for a CV study of national freshwater benefits (Carson and Mitchell, 1986). His study highlights the importance of imputing values, as he estimates that the failure to do so with this data set would result in approximately a 20 percent overestimate of the benefits. The methods he examined included the use of imputation class means and medians, the hot deck, the EM procedure, CART, and no imputation. CART was found to be the most satisfactory imputation method.

Figure 12-2 provides an example of the CART model used to impute missing values in the water benefits study. Based on those characteristics of the realized sample which best predicted their WTP amounts (income, education, region of the country, use of fresh water for recreation, and attitudes toward water quality improvements and environmental quality), predictions were made for those who did not give WTP amounts. CART's surrogate split procedure is a particularly useful feature for contingent valuation surveys. If a value for a predictor split (such as income or education) is missing for a given respondent, the program substitutes a value for the available variable, which produces a split most similar to that of the best predictor.

While we recommend the use of imputation procedures to correct for item nonresponse, every effort should be made to keep the nonresponses to a minimum, as it is far preferable to obtain valid values than to fabricate them, as one does when one imputes values. When imputation procedures are used, they should be applied cautiously, with full recognition that their utility rests on the quality of the assumptions made about the imputation classes. It is also important to recognize that for the purpose of calculating the sample size, imputed values should not be treated as if they were real values. As Kalton (1983:66) points out, "imputation serves to *reduce,* not to increase, the effective sample size" (emphasis in the original). Imputation procedures are bias-reduction, not variance-reduction, techniques.

Sample Selection Bias

The nonresponse biases discussed so far refer to the imperfect representation of identifiable categories of people among those in the chosen sample for whom valid data are obtained. The correction procedures for those biases rely on random nonresponse within the categories. Another type of nonresponse bias, which may occur independently or simultaneously with a sample nonresponse bias, is sample selection bias. Here nonresponse is nonrandom—the individuals who do not respond to the survey or the elicitation question hold different expected values for the amenity than comparable individuals who do respond. This type of bias is most likely to occur when the CV survey format gives people the opportunity to select for themselves whether they will participate in the survey or not. For example, those who have a stronger interest in an amenity (and usually a high value for it) may be more inclined to fill out and return a mail questionnaire asking them to value the amenity than those with similar demographic characteristics who are less interested and whose WTP amounts, if they were obtained, would tend to be lower. Sample selection bias cannot generally be compensated for by weighting and imputation procedures, which assume that those in a given population subgroup who did not respond have the same expected value for the amenity as those in the same subgroup who did respond (Anderson, Basilevsky, and Hum, 1983).

It is useful to define sample selection bias more precisely in mathematical terms. To do this we follow Heckman (1979) and introduce the notion of a population ($i=1, \ldots, N$) regression for a variable(s) of interest, Y, in terms of an $N \times k$ matrix of exogenous regressors, X. The relationship between Y and X, assuming linearity, is given by

$$Y_i = X_i\beta + U_i \quad (12\text{-}4)$$

where $E(U)$ is assumed to be zero for all i. The U_i may represent random behavior or the influence of unobserved true regressor variables. The population regression function for this relationship is given by

$$E(Y_i | X_i) = X_i\beta \; /\forall_i \quad (12\text{-}5)$$

The regression function for the available subsample of data is

$$E(Y_i | X_i, \text{SAMPLE SELECTION RULE}) = X_i\beta + E(U_i | \text{SAMPLE SELECTION RULE}) \quad (12\text{-}6)$$

If the conditional expectation of U_i in equation (12-3) is zero, then a sample selection bias is said not to exist.

Because sample selection bias affects the distribution of error terms (equation [12-6]) rather than the observed variables, correcting for it requires the use of statistical techniques for dealing with censored or truncated error distributions. Unfortunately, these methods may be of limited use in contingent valuation studies when little or no information is available on factors affecting the probability of responding to the survey. Correction of sample selection bias usually requires a second equation to predict the probability of having an observed response, and very strong assumptions about the relevant error distributions in the two equations.

We know of no CV study that has attempted to use these techniques to correct for sample selection bias.[18]

In the absence of workable procedures based on formal assumptions, we offer a somewhat ad hoc approach to the sample selection problem which will provide useful information for a policymaker who wishes to assess the effect of this type of bias on benefit estimates for a given policy.[19] For ease of exposition, we assume that the data suffer only from item nonresponse.[20] The first step is to estimate a valuation function and impute the missing values. Assume that there are r observations with actual WTP amounts and m observations with only imputed WTP amounts. Calculate

$$\overline{\text{WTP}}_{(1)} = \frac{1}{r+m}\left[\sum_{r} \text{WTP}_i + \sum_{m}(1)\text{WTP}_i\right] \qquad (12\text{-}7)$$

$\overline{\text{WTP}}_{(1)}$ is the appropriate WTP estimate if there is no *sample* selection bias.[21] Now calculate $\overline{\text{WTP}}_{(.75)}$ in the same fashion, replacing (1) with (.75). This is the appropriate WTP estimate if each nonrespondent would have given, on average, 75 percent of what a similar individual who did respond would have given. Use multipliers of .5, .25, and 0 (and any others) and calculate $\overline{\text{WTP}}_{(.5)}$, $\overline{\text{WTP}}_{(.25)}$, and $\overline{\text{WTP}}_{(0)}$. $\overline{\text{WTP}}_{(0)}$ is the solution for sample selection bias recently used by Bishop and Boyle (1985) in a mail survey of a recreational site amenity in Illinois. It represents the most extreme case, as it assumes that those who do not respond are not willing to pay anything for the amenity. If a policy passes a benefit-cost test based on $\overline{\text{WTP}}_{(0)}$, sampling selection bias is irrelevant. If it does not pass, the cost of the policy may be compared with a table of estimates based on the various $\overline{\text{WTP}}_{(\cdot)}$ assumptions. This procedure would ensure that the policymaker is clearly aware of the assumptions about the nonrespondents' values implied by any particular benefit-cost comparison.

[18] Ziemer and coauthors (1982) used these statistical techniques in a travel cost study. Most of the relevant work on this topic has been in labor economics, where it has been long realized that the characteristics which differentiate those in the labor force from those who are not currently employed extend beyond the standard observed demographic variables. For theoretical work and examples, see Heckman (1976, 1979), Amemiya (1985), and Little (1985).

[19] This procedure can be shown to be approximately equivalent to trying out various values for the unknown covariance term which could be estimated if the appropriate information for estimating a formal system of (sample selection) equations were available.

[20] It would be possible to use an analogous, but cruder, procedure to correct for sample nonresponse bias.

[21] There is also no nonresponse bias, $\sum_{r} \text{WTP}_i / r = \sum_{m} \text{WTP}_i / m$.

Effects of Nonresponse and Sample Selection Biases on Mail, Telephone, and In-Person Surveys

In chapter 5 we reviewed the advantages and disadvantages of in-person, telephone, and mail surveys from a nonsampling point of view. In this section we examine their vulnerability to nonresponse and sample selection biases. In comparison to the other two methods, mail surveys are especially prone to errors from these sources, and particularly from sample selection bias. This raises important questions about the growing use of mail surveys in contingent valuation studies.

No matter which survey method is used, unit nonresponse rates can be reduced by applying follow-up methods, such as multiple callbacks to locate those not found at home by in-person or telephone interviewers and follow-up mailings to mail survey nonrespondents.[22] The higher the number of callbacks or mailings, the higher the quality of the data, but at the cost of a higher expense for each completed interview. The most important way to reduce item nonresponse rates is by careful pretesting to enhance the instrument's clarity and plausibility. The special nonresponse problems of mail surveys stem from (1) their relatively high nonresponse rates, and (2) their vulnerability to sample selection bias. The latter makes it very difficult, if not impossible, to use weighting and imputation methods to correct for bias from nonresponses.

Although techniques are now available that can increase response rates to mail surveys to a level of 60 percent (or higher for certain types of populations),[23] these levels are still below those achieved in professionally conducted in-person and phone surveys of the same populations (Schuman and Kalton, 1985). High mail survey response rates require an up-to-date list of names and addresses for the sampling frame,[24] a well-designed and carefully pretested questionnaire, and a sponsor who has a noncommercial identity, such as a university or a government body (Dillman, 1978; Heberlein and Baumgartner, 1978). Bishop and Heberlein's CV study of deer hunters showed that by meeting these conditions (as well as by making small monetary payments as incentives) a mail survey can increase response rates to the exceptionally high level of 90 percent or more when the respondents perceive the CV survey's purpose to be *directly connected* with their interests (Bishop, Heberlein, and Kealy, 1983). Where this

[22] The level of unit nonresponse in mail surveys is affected by the design of the mail package (which includes the envelope, cover letter, and questionnaire).

[23] Since survey methodologists usually speak of response rates rather than nonresponse rates, we will adopt their usage in the following discussion. A 60 percent response rate is equivalent to a 40 percent nonresponse rate.

[24] The use of telephone directories for this purpose obviously poses the same types of sample bias problems for mail as for telephone surveys.

connection is attenuated even moderately, however, lower response rates occur. In a CV study of the value of an Illinois nature preserve, Bishop and Boyle (1985) used $2 payments and a carefully designed series of follow-up mailings (including a final mailing containing a replacement questionnaire, sent to nonrespondents by certified mail) to samples of Illinois residents. Despite these efforts, only 66 percent of the nearby residents and 61 percent of the non-neighbor residents who received their questionnaires returned them. In many cases, CV mail surveys have experienced considerably lower response rates than these.

The correct way to calculate the response rate for a mail survey is to divide the number of completed questionnaires returned by the number in the original sample. This method, which includes in the nonreturns the sample members who no longer live at the addresses listed for them, as well as those who received the questionnaire but failed to return it, shows how well one has done in reaching all potential respondents (Dillman, 1978:49).[25]

Response rates calculated by this method for a number of mail CV studies are presented in table 12-3. Most of these surveys sent at least two follow-up mailings to those with valid addresses who did not respond to the first mailing. Some, such as Welle's and Bishop and Boyle's, used certified letters for their final follow-up mailing. Overall, the unit coverage for many of these studies is much lower than the 60 percent benchmark noted earlier for studies that use state-of-the-art mail survey techniques. In a number of instances, half or more of the chosen sample is not represented with valid responses.

Not only are unit response rates for mail surveys lower than those for telephone or in-person surveys; the potential for sample selection bias in mail surveys is higher. This problem stems from the self-administered character of mail surveys and the concomitant lack of control the researcher has over the process of getting the respondent's cooperation and eliciting his answers.

With telephone and in-person surveys, it is normally possible to assume that the nonresponses are not related to the subject matter of the survey. There are three considerations here. First, failure to interview people who are not found at home or who are incompetent to be interviewed has

[25] The common practice of calculating the response rate for a mail survey from the number of questionnaires mailed less those returned by the post office because the addressee was no longer at a given address is only appropriate as an indicator of the instrument's success in inducing respondents to return the questionnaire (Dillman, 1978). Such a practice masks possibly large sample selection bias effects due to the loss of the more mobile population elements. According to Berdie and Anderson (1976), this practice originated with marketing research firms as a way to suggest to clients that the researcher had reached a larger percentage of the promised population than was actually the case.

Table 12-3. Mail Survey Response Rates for Various CV Studies

Study	Total mailed	Returned[a]	Response rate (percentage)
Hammack and Brown (1974)	4,900	2,455	50
Cicchetti and Smith (1976a)[b]	600	240	40
Bishop and Heberlein (1980)[c]	257	221	93
Bishop and Heberlein (1980)[c]	237	190	80
Loehman and De (1982)	1,800	404	22
Schulze and coauthors (1983)	2,103	166	8
Brookshire, Eubanks, and Randall (1983)[b]	3,000	900	30
Stoll and Johnson (1985)[b]	1,800	650	36
Bergstrom, Dillman, and Stoll (1985)	600	250	42
Bishop and Boyle (1985)	400	232	58
Randall and coauthors (1985)	2,206	718	33
Sellar, Stoll, and Chavas (1985)[b]	2,000	1,260	63
Sutherland and Walsh (1985)	280	171	61
Walsh, Sanders, and Loomis (1985)	600	214	36
Welle (1985)	1,000	689	69
Tolley and Babcock (1986)	103	42	41

[a]Includes questionnaires that were mostly completed, but which were missing responses to the CV elicitation questions. The number of questionnaires usable for directly estimating the WTP amount was therefore often lower than the number given here.

[b]Approximate numbers.

[c]This study contains two independent samples.

nothing to do with their personal reaction to the survey's topic. Second, those who refuse to be interviewed for these surveys usually reject participation before the specific topic of the survey is made known to them.[26] And third, studies of people who refuse personal or telephone interviews (Stinchcombe, Jones, and Sheatsley, 1981; Smith, 1983) suggest that refusals occur because of general, rather than survey-specific, reasons.

These assumptions cannot be made for those who receive a mail survey and fail to return it. Unless the recipient throws out the package without opening it, his decision on whether to respond or not (which includes the decision to put the package aside for a while) is likely to be influenced by his examination of the cover letter and the questionnaire. Research has shown that the less salient a mail questionnaire is to a potential respondent, the less likelihood there is that the respondent will take the time to fill it out and send it back (Heberlein and Baumgartner, 1978; Tull and

[26] We presume here that the interview topic is described in general terms at the point when the respondent's cooperation is first requested, as it is in many surveys in order to avoid this type of bias. The interviewer would say, for example, that he was "conducting a study of people's views about certain kinds of environmental issues," rather than saying that the study was specifically about "how much people are willing to pay to reduce the risk of cancer from trihalomethane contamination in their drinking water."

Hawkins, 1984).[27] Since in the case of public goods interest in the subject matter is likely to be correlated with the value the good has for the respondent, it is likely that nonrespondents to mail surveys will hold lower or even zero value for the good compared with those of equivalent demographic categories who do answer.

In short, mail surveys have a strong potential for sample selection bias, which means that information from those who happen to give valid WTP answers cannot be used to infer or to impute WTP values for the nonrespondents. This is one of the reasons why market research texts do not recommend their use for general populations (Tull and Hawkins, 1984, for example).[28] The lack of information about why mail questionnaires are not returned also means that the additional information necessary to use the censored and truncated distribution techniques to correct for sample selection bias is absent. The most defensible way to handle the sample selection problem in mail surveys (if a lower bound on WTP is desired) appears to be the procedure used by Bishop and Boyle (1985) in their study of the economic value of an Illinois nature preserve, in which they made the conservative assumption that the nonrespondents (excluding those for whom addresses were incorrect) to their mail survey (more than 30 percent of the sample) valued the preserve at $0. If they had not made this adjustment in their population estimates, it is likely that they would have substantially overestimated the value of the amenity.

Valuation Functions and Sampling Design

One potential alternative to the use of probability samples in mail, telephone, and in-person interviews is to estimate a valuation function based on a nonprobability sample. The debate between the model-based and

[27] Undoubtedly some of those who neglect to respond to mail surveys do so for reasons unrelated to the topic. The nature of mail surveys is such, however, that no interviewer is present to record that a potential respondent is sick or has traveled abroad for a month, so nonresponses on the part of such persons cannot be distinguished from nonresponses from those who refuse to answer surveys.

[28] Some CV researchers have argued, on the basis of a study conducted by Wellman and his coauthors (1980), that nonresponse bias is not likely to be statistically significant. The Wellman study compared early and late respondents to a mailed non-CV, outdoor recreation survey which achieved a 70 percent response rate. The authors argued, on the basis of apparent similarities among these groups on a number of characteristics, that the time and effort spent on intensive follow-ups to increase survey response rates might be better spent on other phases of the research process. This finding provides an insufficient basis for assuming random nonresponse, as Wellman and his colleagues did not study the 30 percent of their sample who failed to respond to their survey. There are no a priori grounds for believing that late respondents to mail surveys such as theirs are a valid surrogate for the nonrespondents, and there exists empirical evidence to the contrary (Anderson, Basilevsky, and Hum, 1983:479–480).

probability based sampling approaches has a long history in the statistics literature (Hansen, Madow, and Tepping, 1983). Although there may be a large gain in the efficiency of the estimated coefficients of the valuation function from using a model-based nonprobability sampling scheme, that gain is conditional on knowing the correct model. If the model is seriously misspecified, the data obtained through a nonprobability sampling design may be close to useless, as the researcher may even be unable to estimate benefits at the specific quantity points for which the WTP amounts were solicited. The potential for such misspecification is large in a CV survey because the exact mathematical form of a valuation function is unlikely to be known. This makes the use of a nonprobability sample a risky strategy indeed.

In the only study we know that used CV data gathered in a nonprobability sample to estimate a valuation function, the researchers (Tolley and Randall, 1985) hedged their bets by using a two-stage approach. In the first stage they used a probability sampling design to measure the population's willingness to pay for air visibility improvements in five different eastern cities, thus guaranteeing that they could obtain WTP estimates for those specific cities. They then combined the data from the five cities to estimate a valuation function for the entire eastern United States.[29]

Inference Biases

If one wants to use the results of a CV study to infer something about a policy change other than that explicitly valued in the original study, the possibilities of additional biases must be taken into account. The first of these is *temporal selection bias,* which will occur if the public's valuation of the policy change shifts significantly over time. The second, *sequence aggregation bias,* may occur in trying to add the value of independently obtained "small" policy changes to obtain the value of a "large" policy change. Such aggregation may involve either a bigger geographic area or more public goods. Under most conditions the components of this larger policy will be somewhat complementary, and the value of the larger policy will be smaller than the sum of its component small policies. The issue of how to transfer a benefit estimate from one CV study to a policy situation which is similar in some ways but significantly different in others is taken up in the next chapter.

[29] A troublesome aspect of Tolley and Randall's valuation function for the eastern United States was the presence of significant coefficients on dummy variables for the particular cities. These coefficients should have been zero if, as assumed, the respondents valued air visibility uniformly across the eastern United States. It is hard to say whether this assumption was false or whether the valuation function was in some way misspecified.

Temporal Selection Bias

Benefit estimates based on individual preferences, whether the preferences were expressed in actual behavior or in verbal expressions of intended behavior, are necessarily dependent upon the distribution of preferences at the time when the study was conducted. *Temporal selection bias* occurs when benefit estimates are based on outdated preferences held by the economic agents of interest. If preferences are relatively immutable, as some economists assume (Becker, 1976, for example), users of CV findings would not have to worry about their currency. Once obtained, a CV-based benefit estimate for a particular amenity would have a very long shelf life. Our view is that preferences can and do change (Etzioni, 1985), and that the possibility of temporal selection bias should not be ignored. For example, public preferences for environmental amenities, particularly those that result from pollution control programs, appear to have changed substantially between 1965 and 1975. One longtime observer of the polls called the speed and urgency with which environmental issues burst into the American consciousness in the late 1960s a "miracle of public opinion" (Erskine, 1971). Similarly, public willingness to tolerate the risk of continuing to build nuclear reactors is distinctly more negative in the mid 1980s than it was ten years before the accident at Three Mile Island, near Harrisburg, Pennsylvania (Freudenburg and Baxter, 1985).

How vulnerable are contingent valuation studies to temporal selection bias? For most goods for which the evidence is available, it appears that survey-based preferences are sufficiently stable to yield benefit estimates which are usable over a reasonable period of time. The first source of evidence for this assertion is public opinion data, particularly data based on survey questions that use the spend more-same-less format in valuing a particular policy or program. The second source of evidence is in CV studies themselves. We will examine each of these sources of evidence, and then discuss newly emergent public concerns which, in contrast, may be vulnerable to temporal selection bias.

If environmental preferences have undergone a significant change, this appears to have been a sea change rather than a temporary shift of interest. Since 1970 numerous polls have documented consistent high public support for environmental protection (Mitchell, 1984; Gillroy and Shapiro, 1986). This level of support has persisted despite the oil embargo, the energy crisis, various serious problems with the nation's economy, the expenditure of large amounts of money on environmental regulation, and the efforts of the Reagan administration to redefine the scope and direction of the federal environmental regulatory programs. Data from the National Opinion Research Center's General Social Survey (National Opinion Research Center, 1983), a study which annually replicates a large number of key policy-relevant questions for a national sample, show that public preferences for other types of public goods, ranging from crime protection to the space program, are similarly stable over the time period for which these data are available (1972 to 1983).[30]

The evidence for temporal stability that is in the contingent valuation literature shows a similar pattern. We are aware of six comparisons of contingent valuation WTP estimates for similar goods measured at different points in time. Most of these comparisons are crude because the scenarios used differed to a greater or lesser degree. Rowe, d'Arge, and Brookshire (1980) compared the findings of three iterative bidding game studies of visibility reductions in the southwest United States. Despite the fact that these studies differed considerably in details of method (but not in basic design) and in the populations sampled, they found "very comparable" results across the three studies. Walsh, Sanders, and Loomis (1985) report that researchers at Colorado State University have found roughly equivalent dollar values for water quality in Colorado rivers in studies done over approximately a ten-year period. Mitchell and Carson (1985) report a similar finding for two somewhat equivalent CV surveys of national water quality benefits taken four years apart (1980 and 1984). Tolley and Randall (1985) have replicated (with some changes) an air visibility survey they conducted in 1984 in different cities, obtaining almost identical values. Sorg (1982) and Sorg and Nelson (1986) found quite similar values for elk hunting in studies conducted several years apart in Wyoming and Idaho. Bishop and Heberlein (1986) found similar values for deer-hunting permits in Wisconsin in their 1983 and 1984 CV willingness-to-pay surveys, which used different elicitation methods, and again in a partial replication a year later that used some of the same respondents (Heberlein, 1986). Carson, Hanemann, and Mitchell (1986), in two Field California Polls taken a month apart, found similar aggregate voting intentions on their referenda simulation on a water quality bond issue. Finally, Jones-Lee, Hammerton, and Philips (1985) successfully replicated a CV risk-reduction valuation question after a month's interval.

Most of the above examples involved amenities with which the respondents were long familiar. Big game hunters have long sought elk, and the problems associated with water pollution are hardly novel. But when problems and amenities are newly recognized, or when older problems

[30] The format used to measure these preferences is sensitive to major spending initiatives, as it asks for peoples' preferences for increasing or decreasing the amount the country is currently spending on various programs. The one exception to the stability generalization relates to national defense preferences. From 1973 to 1980 support for spending more on defense increased, only to drop sharply as the 1980s wore on. Rather than capturing a decrease in peoples' preferences for a strong national defense, however, the drop may simply reflect the fact that the Reagan administration initiated substantial increases in defense spending.

and amenities suddenly receive widespread publicity (as did the risks of nuclear power following the accident at the Soviet Union's Chernobyl reactor in 1986), stability of preferences is more problematic. Public concern about newly recognized problems, such as health risks from hazardous waste dumps or exposure to radon gas inside homes, tends to go through a cycle from heightened concern at the beginning to routinized concern as the media turn to other topics and more becomes known about the problem. For example, over a period of three and a half years, from October 1980 to February 1984, the percentage of New Jersey residents who said they were personally "very concerned" about the problem of chemical wastes in the state declined from 78 percent to 59 percent (Eagleton Institute of Politics, 1984). In a parallel fashion, those who felt that cleaning up "chemical dumps and hazardous wastes . . . must be done immediately regardless of the cost" dropped from 58 percent to 42 percent. While concern about this problem was still high in 1984 relative to other problems, as was support for spending money to clean up the wastes, it is likely that a formal CV study measuring the benefits of reducing the risks from hazardous waste sites would have obtained lower values in 1984 than in 1980. Barring some dramatic event associated with toxic waste dumps in the state, it is also likely that benefit estimates obtained in a 1984 CV survey would have a longer shelf life than estimates obtained at the time when hazardous wastes were considered by the public to be a new and largely unknown problem.

Sequence Aggregation Bias

Sequence aggregation bias was introduced in chapter 2 when we discussed the problem of adding together WTP amounts for quantity changes in different public goods. In that context we reviewed the findings of Randall and his colleagues (Hoehn and Randall, 1982; Tolley and coauthors, 1983, 1985), who used the economic concepts of complements and substitutes to show why, under many conditions, component benefits measured in separate CV studies cannot be combined without overcounting. They also demonstrated that when the components are sequentially valued in the same study, the order of appearance in the sequence affects the results. The combined entities might be the benefits for a given amenity in different geographical areas, such as air quality in the Grand Canyon, the Rocky Mountain region, and the eastern United States, in which case *geographical sequence aggregation bias* might occur; or they might be the benefits for different components of an overall program, such as the air and water pollution control components of the national environmental pollution control program, which might result in *multiple public good sequence aggregation bias.*

Respondents in an area having several lakes whose water quality is polluted will value cleaning up the first lake more than they value cleaning up a second lake, for several reasons. First, the cleaned-up first lake is available as a substitute for a second cleaned-up lake.[31] Second, the individual's allocation of money for the first lake reduces the money she has available for cleaning up the second lake. If separate CV studies value the lakes individually, however, respondents will treat the lake they are asked to value as if it were the first lake. An overvaluation of the benefits of a combined cleanup will occur if the separate values are added up. If the lakes are valued in sequence in a single study by a sample of respondents, the benefit estimates for the individual lakes (but not for the entire set of lakes) will be biased unless the valuation sequence replicates the actual sequence in which the cleanup would occur.[32] It should be clear that any good being valued has a place in a sequence relative to some other goods; the other goods are provided either before, at the same time, or later than the good being valued.

Randall and his colleagues (1985) have considered the problem of how to disaggregate WTP estimates for all national environmental programs into specific program components, such as air and water pollution control. They show that unless the sequence in which the components will be implemented is known, or unless some very strong separability conditions on the utility function are met, there is no unique disaggregation of the component values. The essential problem is that the substitution and income elasticities are unknown. There are some techniques which may be used to obtain an upper or lower bound on the WTP amounts independent of sequence if the researcher is prepared to make some fairly weak assumptions about the curvature of the marginal benefit curves. These techniques are discussed in Randall and coauthors (1985).

The Fallacy of Motivational Precision: Measuring Benefit Categories and Subcategories

In the preceding two sections we discussed the problems of aggregating benefits across geographic areas and different goods. However, in the contingent valuation literature much more attention has been paid to the issue of how to disaggregate a single CV estimate into different benefit components. The framework presented in chapter 3, in the context of considering the multidimensional character of benefits, decomposed the total value of a freshwater quality improvement into two classes, use and

[31] Where the income effect is small, the second good in the sequence may be valued more than if it had been the first, if the two goods are complements.

[32] In their CV study of the benefits of preserving up to fifteen wild and scenic rivers in Colorado, Walsh, Sanders, and Loomis (1985) clearly demonstrate the decreasing benefits of preserving additional rivers.

existence, which were further subdivided into a total of thirteen subcategories, ten of which, we hypothesized, potentially influence citizens' values for the amenity. The purpose of that analytic exercise was to identify the full range of benefits so that researchers could be aware of what they were likely to measure and not to measure in a given scenario. Since decision makers tend to inadequately take into account relevant benefits (Abelson and Levi, 1985:285), researchers have to design each scenario in such a way that respondents are reminded of the full range of benefits they are intended to consider in making their WTP judgments. We argued that although the utility that people receive from an amenity may stem from some or all of these benefit dimensions, the WTP judgment is based on a holistic assessment rather than on a conscious summing of the several components to reach a total value.

This view of the valuation decision leads us to be skeptical of attempts to ask respondents to separately value the several benefit categories or dimensions for a given amenity. It is difficult enough for respondents to evaluate the desirability of an amenity. To ask them to say how much they are willing to pay just to know that others, but not themselves, can use the amenity is even more difficult. Consider a consumer analogy. Most consumers, if asked to value each of the following aspects of a newly purchased car in an open-ended fashion, would have trouble saying with any degree of precision how much they were willing to pay for the styling, the horsepower, the comfort of the seats, the size of the trunk, and the prestige associated with ownership. William James (cited in Fischer, 1970:209) labeled the "psychologists' fallacy" the error of assuming that a person who has a given psychic experience knows it when he has it, in the same way that an observing psychologist would know the experience occurred. Similarly, the error of assuming that respondents are aware, to the degree of precision desired by the researcher, of what motivates their value judgments may be called the "fallacy of motivational precision."

Four Measurement Strategies

In view of the difficulty of separately valuing benefit components, how can these be meaningfully measured? There are four measurement strategies which have been or might be used to obtain meaningful estimates of the various types of benefits that the respondent receives from a given amenity. The first three are based on subjective judgments made by the respondent, whereas the third uses self-reported information about the respondent's use of a given amenity to infer a lower bound on aggregate existence value. Strategies I and II are particularly prone to the fallacy of motivational precision.

In strategy I, each benefit category is individually described to the respondents, who are asked directly how much each dimension of the amenity is worth to them. If separate values for direct use, indirect use, and existence are obtained in this way, the total WTP amount for the good could theoretically be obtained by adding the values. Although this strategy has the advantage of simplicity, and studies using this approach have succeeded in getting respondents to give dollar answers to the questions, the potential for invalid or meaningless answers is high because of the fallacy of motivational precision. It is also a dangerous strategy because it stands a chance of grossly overestimating total WTP.

Strategy II, a decomposition strategy, involves asking respondents to separate a previously obtained total WTP amount into values for one or more benefit components. Each respondent might be asked, "Of this [total] amount, how much of it would you pay for ____ [description of category or subcategory]?" This strategy is preferable to strategy I because it first obtains a presumably valid total WTP amount before attempting to break that amount into potentially invalid component values. Obtaining the total WTP amount first also helps respondents to grasp the idea that the component values are a subset of the overall value. Strategy II is consistent with one of the definitions of existence value given in chapter 3. To empirically measure that value, it is first necessary to measure a respondent's total value for the amenity. This WTP amount will include both the use and existence values the respondent has for the good. The existence component, at least in theory, is equivalent to the amount which would be obtained by asking the respondent to say how much he would be willing to pay for the good if, for some reason, his household could not use or have access to it. To the extent that the new contingency was plausible to the respondent, this approach might avoid the fallacy.

The experience of CV researchers who have attempted to use these strategies underscores the methodological difficulties associated with their use. Employing strategy I, one set of researchers (Greenley, Walsh, and Young, 1982) asked respondents to take their chances of future recreational use into account and say how much they would pay to postpone deterioration in the quality of a river basin in Colorado. The respondents were then asked how much they would pay just to know that clean water existed at a given level, "if it were certain you would not use the South Platte River Basin for water-based recreation" (stewardship). Respondents were asked additional separate questions to measure bequest values for future generations and quasi-option values. Although Greenley, Walsh, and Young believed their respondents were able to sufficiently comprehend their questions to give meaningful answers, the amounts they obtained for existence values were sufficiently similar to those obtained for the bequest and quasi-option values to suggest that the respondents may not have been able to adequately differentiate these dimensions in their minds (Mitchell and Carson, 1985).[33] Walsh, Sanders, and Loomis (1985) used strategy II

[33] As given by the water fee vehicle used, the mean value in dollars per month for all resident households in the South Platte River Basin was $1.16, $1.23, and $.90 for the three types of benefits measured (Greenley, Walsh, and Young, 1982).

in their study of the benefits of protecting wild and scenic rivers in Colorado. After obtaining each respondent's total value, they asked the respondent what proportion she would assign to the following types of benefits: use, "an insurance premium . . . to guarantee your choice of recreation use of these rivers in the future," bequest, and "inherent." The mean WTP amounts were diverse,[34] suggesting that repetition of presumably meaningless values was avoided. Whether the respondents could grasp the three types of nonuse benefits well enough to reliably assign them shares of the total value is open to question.

Strategy III involves posing two or more scenarios to separate subsamples, or possibly to the same respondents.[35] The scenarios differ only with respect to the specific benefit component the researcher desires to measure. The difference between the total WTP amounts for the two scenarios yields an estimate of the desired quality (such as a natural area being available or unavailable for recreational purposes). This strategy avoids the fallacy of motivational precision because it asks respondents to give a total value only for a given scenario. To the extent that the different scenarios are plausible, this strategy offers a potentially useful, though expensive, approach. A variant of strategy III would be to create a single scenario in which the only possible motive for valuing the amenity lies in a specific benefit class or category. WTP for the second scenario (the absence of the benefit class) in this variant would be implicitly assumed to be zero.

Strategy IV avoids the fallacy of motivational precision because it does not rely at all on respondents' subjective assessments of the benefit components. Instead, it uses reported use and/or anticipated future use of the good to indirectly estimate existence value. On the basis of self-reports, the respondents are divided into those who use the amenity and those who do not. The WTP amounts are measured in the conventional manner. The WTP amounts for the amenity given by the nonusers are treated as a relatively pure expression of nonuse or existence value, whereas the users' WTP amounts include some combination of use and implicit values. Since no defensible external criteria are available to determine which portion of the users' WTP amounts should be assigned to the existence category, estimating existence values by this approach will result in a lower bound for existence value (Fisher and Raucher, 1984). Despite its methodological limitations, to be described shortly, this strategy has provided suggestive evidence about the magnitude of existence values.

[34] On a per household basis, Walsh and his colleagues obtained $19, $16, $28, and $36 for the four types of benefits for the "fifteen most valuable rivers" (Walsh, Sanders, and Loomis, 1985:72).

[35] Use of the same respondents would be much less expensive than two surveys, but raises questions about contamination across scenarios. Can respondents value a second scenario without being influenced by the content of the first?

Existence Value Estimates

The magnitude of existence values for environmental amenities is of substantial interest to those who wish to estimate benefits. Despite his evident discomfort with the contingent valuation method, Freeman (1982), for example, made an estimate of national water quality existence benefits based on the sketchy evidence available as of 1980. Fisher and Raucher (1984) undertook a comprehensive review of the later evidence for water benefits, most of it based on CV studies of varying quality, and concluded that "intrinsic [existence] values are positive and nontrivial." We agree with this general conclusion. Nonetheless, there are problems associated with the use of strategy IV to make more precise estimates of the magnitude of water quality existence values. The problems are such that great caution should be exercised in using the available estimates of existence value. We will use our own research to describe some of these problems.[36]

In our 1980 pilot study of national water quality benefits (Mitchell and Carson, 1981), we attempted to estimate, in a preliminary way, the existence value for different levels of national water quality by treating the values given by those respondents who had not engaged in in-stream recreation in the past two years as representing almost pure nonuse benefits. In doing so we assumed that past use is predictive of future use, an assumption that can only approximate the truth at best. Water recreators were assumed to have average existence values of the same magnitude as those of non-water recreators. This allowed use values for water recreators to be obtained by subtracting existence values from total willingness to pay.

This effort, which suggested that existence benefits constituted a large percentage of the total WTP amount given by the respondents (about 55 percent), was flawed in several respects. In the first place, our procedure for assigning existence benefits to users was arbitrary, and may have resulted in under- or overestimating existence benefits. Second, the questions asked to identify the users did not inquire about recreational water use by other household members, nor did they measure indirect-use behavior. When we obtained household use information in our subsequent water quality study (Carson and Mitchell, 1986), the new estimates of existence value ranged from 40 percent to as low as 19 percent of the total WTP amount.[37] It is also possible, however, that some nonusers had a

[36] We will not attempt to critique here all the applications of strategy IV, since they are well covered in Fisher and Raucher's review. For a good example of the use of strategy IV to value non-water benefits, see Brookshire, Eubanks, and Randall's (1983) study of the benefits of preserving grizzly bear and bighorn sheep.

[37] The highest percentage was found when we counted only the respondent's direct use, and the lowest when we counted any household member's direct or indirect use as constituting use.

use value for higher-than-prevailing levels of water quality. If so, our assignment of all their WTP amounts to existence values may have overstated their true existence value for water quality at these levels.[38]

The shortcomings of Mitchell and Carson (1981) suggest several guidelines for the use of strategy IV to measure implicit benefits. First, in studies that measure household benefits, the identification of users should be done in such a way that household use is measured. Second, as many categories of use as possible (indirect as well as direct, for example) should be measured. Where some categories are unmeasured, they should be noted in the report on the study, and their possible effects estimated. Third, where relevant, potential use of the amenity under improved conditions should be identified. Fourth, it should be kept in mind that the only really valid measure of existence value is the WTP amounts of nonusers, since there is no nonarbitrary way to specify that portion of the users' WTP which is attributable to existence values.

Contingent valuation's ability to include existence values for the amenities it is used to value is one of its most important advantages over the alternate methods, such as travel cost, which are restricted to use benefits. We agree with Fisher and Raucher (1984) that the available evidence indicates that exclusion of existence values from total water quality benefits would substantially understate their magnitude. It is too early, however, to use the handful of CV studies they review as a firm basis for estimating existence benefits in benefit-cost analysis. The quality of the present existence estimates, ours included, are not yet adequate to the task.[39]

Summary and Conclusions

The aggregation of CV findings to obtain benefit estimates for different levels of provision of public goods is a necessary step in applying these

[38] Given the widespread availability to most consumers of substitute sites at present levels of water quality, any overstatement arising from this factor was likely to be small in this particular case. This is not likely to be true when "new" amenities are to be provided, for which no current substitutes are available.

[39] A case in point is the Environmental Defense Fund's (1984) benefit-cost analysis of the proposed development of the Toulumne River, which calculated existence benefits for its preservation at 60 percent of the measured use value. Why 60 percent? This figure represents a weighted average of the nine studies for which Fisher and Raucher could obtain an existence value estimate of any kind, a procedure which treats each study as equivalent to the others despite the fact that the quality of these studies—in terms of sampling procedures, sample size, and tests for bias—varies greatly. Given the poor data base on which this ratio is based, the true existence value for this resource could easily range from 20 to 120 percent of the user benefits. More work is needed before the use of a reasonably narrow percentage range for this purpose can be extrapolated from the CV literature.

findings to public policy decisions. The magnitude of the aggregation error resulting from sample and nonresponse biases is potentially very large. In this chapter we have summarized some of the considerations involved in selecting a probability based sample, examined the sampling error trade-offs involved in using various survey methods to gather data, and discussed the techniques available to compensate for sample nonresponse bias and sample selection bias. We have seen that both the failure to obtain interviews from some of the people selected for the sample (sample nonresponse) and the failure to obtain valid WTP amounts from some of those who are interviewed (item nonresponse) may produce bias because of the propensity of certain people (for example, older, less educated people) to be disproportionately underrepresented in the final data set. We recommended the use of statistical techniques to weight the data set and, in the case of item nonresponse, to impute missing values as a way to compensate in part for the bias which would otherwise result. Unfortunately, these techniques are not applicable to the problem of sample selection bias, where, with respect to WTP amounts, nonresponses are nonrandom within identifiable groups of respondents. Mail surveys are particularly vulnerable to sample selection bias because of the potential for nonrespondents to be consciously self-selected.

We argued that contingent valuation surveys which measure the benefits of well-established amenities are not likely to be vulnerable to temporal selection bias. Estimates for newly identified goods, however, were found to be more time-sensitive. The adding up of values for separately measured programs or amenities was seen to pose the problem of sequence aggregation bias. This bias is difficult to overcome unless the programs or amenities are measured in a single study in the sequence in which they would be provided. We believe that this bias is underrecognized, and is responsible for some of the grossly large and implausible CV estimates sometimes seen.

The last section of the chapter was devoted to the question of whether classes or categories benefits, such as use or existence values, can be meaningfully measured in CV surveys. Four strategies for measuring these values were described. A key problem of the first three strategies was seen to be the fallacy of motivational precision, whereby the error is made of assuming that respondents are as precisely aware of the motives behind their value judgments as the researcher desires them to be. The fourth strategy, which avoids reliance on respondents to disentangle their own motives, offers a more promising basis for measuring benefits categories. This last strategy estimates existence values on the basis of the values given for an amenity by nonusers. Its empirical application poses a number of largely unresolvable problems that make it difficult to offer a definitive assessment of the relative magnitude of existence values at this time.

13

Conclusion

The Promise of Contingent Valuation

Contingent valuation shows promise as a powerful and versatile tool for measuring the economic benefits of the provision of nonmarketed goods. It is potentially capable of directly measuring a broad range of economic benefits for a wide range of goods, including those not yet supplied, in a manner consistent with economic theory. Other available methods, in contrast, are capable of measuring only some of those benefits and are limited to valuing existing goods and existing quantity and quality levels, and researchers employing them must make a number of unverifiable assumptions in the course of deriving benefit estimates from the available data.

But can CV surveys actually measure values that are sufficiently reliable and valid for use in benefit estimation? Our conclusion is basically affirmative. Certainly the prophecy that respondents will usually act strategically and will deliberately bias their values is not supported by the evidence reviewed in chapters 6 and 7. Our assessment in this regard is similar to that reached by Arrow (1986:183), who states: "Neither the empirical evidence nor the theoretical arguments convinced me that strategic bias is liable to be significant (in CV studies)." Nor is the hypothetical character of CV surveys necessarily an impediment to their usefulness. Unlike ordinary surveys, which often measure unconstrained attitudes toward vaguely defined goods, CV surveys elicit specific behavioral expectations—for example, "I would be willing to pay an additional $X a year in higher taxes out of my current income in exchange for the specified improvement in local air quality." On the basis of our reading of the literature on attitudes and behavior, we believe that the properties which have been found to maximize behavioral predictions are largely compatible with the fundamental structure of most contingent valuation scenarios. Moreover, comparisons of the outcomes of identical simulated and hypothetical markets, in the controlled experiments reviewed in chapter 9, have shown that hypothetical markets for quasi-private goods are able to predict market outcomes when real payments are involved. These data lie behind the emerging consensus of skeptics (Kahneman, 1986, and Freeman, 1986, for example) and practitioners that CV studies are able to measure meaningful values for "familiar" goods such as local recreational amenities.

What about the use of contingent valuation to measure the benefits of less familiar goods, such as air quality improvements or risk reductions of various kinds? This is more difficult terrain because here we cannot test the accuracy of CV surveys against a criterion. Air visibility cannot be sold in simulated markets in the way that Bishop and Heberlein sold deer-hunting permits for Wisconsin's Sandhill preserve. Those who harbor the most qualms about the contingent valuation method's ability to obtain meaningful values for pure public goods hold the view that CV surveys should replicate a consumer market in familiar goods. According to this notion, the method can only work when respondents either have well-defined preference orderings for the goods being valued at the point that they are asked to take part in a study (see Freeman, 1986; Department of the Interior, 1986:27721), or when the valuation procedure provides an extended learning process by which firm preference orderings can be acquired; otherwise, in these views, valid WTP amounts cannot be obtained. Those who hold the latter view doubt that even a 45-minute personal interview could provide such a learning experience.

While sympathetic to the concerns raised by these skeptics, we believe their view that meaningful valuation requires previously "well-exercised" preferences is based on an inappropriate market model. In our view, the appropriate model for CV surveys of pure public goods—goods that citizens are least likely to have direct experience in valuing—is the referendum, by which citizens make binding decisions about the provision of public goods. From this perspective, instead of falling short of the relevant market model, well-conducted CV surveys offer significant improvements over actual referenda as a means of measuring consumer preferences.

First, CV surveys can elicit a more informed decision than referenda. Studies of voter decisions show that people typically face an informational vacuum when they vote on noncontroversial propositions, which leads them to rely on endorsements by political leaders or to make snap judgments (Magleby, 1984). In sharp contrast, relatively detailed and focused information is presented to respondents in the course of a CV interview. Second, CV surveys are based on a more representative set of responses than most referenda. In actual referenda those who vote are often not particularly representative of the entire voting population owing to low turnouts and the tendency of some categories of voters, such as those with low levels of education, to be underrepresented among those who do vote.

The use of probability sampling, callbacks, and imputation procedures makes it possible for CV surveys to reach a more representative audience.[1]

Provided that respondents can be motivated to carefully follow the contingent market described in the scenario, and find it sufficiently plausible, CV surveys offer the possibility of obtaining meaningful information about consumer preferences for nonmarketed amenities. During the course of the interview, respondents make a decision about how much they are willing to pay for the amenity based on the material presented in the scenario, any prior information they might have, and their preferences regarding what they would like the government to do with their tax dollars.

Relevance and Quality

Although the contingent valuation method is a promising technique, the fact remains that the methodological challenge in conducting a CV study is considerable because it is often difficult to convey to respondents what a policymaker wants them to take into account in a way that is both theoretically and technically correct *and also* understandable and plausible. This problem has often been underestimated by CV practitioners and ignored by benefit analysts, who tend to treat CV studies as if they were all of equal quality. But how can a potential user of a CV study know when to place credence in its findings? The answer is, only by assessing the study's relevance to the policy change the user wishes to value as well as the study's quality.

Relevance

As the number of contingent valuation studies increases, it is likely that studies whose scenarios were designed for one situation will be used to infer something about other situations. Evaluating a study's relevance or transferability for a particular purpose requires a careful mapping of the changes the study valued against the changes implied by the policy. One issue here is the correspondence between the provision of the amenity described in the scenario and the amenity changes implied by the policy changes. Another is the context in which the amenity was valued. In a CV study, the improvements that respondents value are specific to the scenario presented to the respondents. It is to be expected that the WTP amounts will be sensitive to the method of provision, the payment vehicle, and the other features of the hypothetical market used to elicit the values. The

[1] The real issue when using the political market model is whether the researcher desires to predict the relatively uninformed, unrepresentative vote which would occur if a referendum about the amenity valued in the CV survey was actually held, or whether the researcher prefers to predict what would happen if a relatively informed and representative vote occurred.

correspondence between the sample interviewed for the study and the population affected by the policy must also be assessed.

Consider the use of findings from a CV study of drinking water risk reductions in a southern Illinois community (Mitchell and Carson, 1986c) to infer the value of a program to reduce low-level risks from pesticide contamination of a California farming community's drinking water. The Illinois study valued risk reductions from trihalomethanes (THMs), a probable carcinogen caused by the interaction of the chlorine used to purify drinking water with humic matter in the raw water; reductions were from various levels slightly above the risk implied by the EPA-set maximum contaminant level (MCL) to the risk of 1 in 100,000 implied by the MCL. Let us presume that the mortality risk reductions to be valued in California are reductions of the same size as those valued in Illinois, and that the procedures to reduce the risk in both places involve changes in the community drinking water treatment plants. Among the questions which would have to be answered in making the comparison are: (1) Would pesticide risks have the same meaning to the Californians as THM risks had to the Illinois respondents? Specifically, would the Californians regard pesticide risks as a regrettable but unavoidable side effect of necessary practices, or would they be perceived differently, with possible consequences for the benefit estimates? (2) Is the level to which the California risks are reduced an officially promulgated standard comparable to the maximum contaminant level for THMs? If this is not the case, it would be important for Californians to believe they have a comparable level of security. (3) Would the California program employ a water bill payment vehicle, as was used in the Illinois study? Again, a difference in payment vehicles might affect the WTP amount. (4) In those demographic characteristics shown to influence the respondents' WTP amounts in the Illinois study, is the California population comparable? If not, have the differences been taken into account by obtaining a California estimate through, say, the use of a valuation function?

Quality

In addition to assessing a given CV study's relevance, anyone wishing to use its estimates must also judge its quality. In the absence of information about the benefits of a policy, the findings of even a flawed CV study may provide some crude insights into the possible range of values for the policy, but *only if the shortcomings of the CV study are recognized and taken into account.* The findings of a good-quality CV study, on the other hand, may dramatically narrow the range of uncertainty.

How can the quality of a contingent valuation study be assessed? One potentially attractive approach would be to consider valuable only those CV studies that meet certain preestablished criteria, such as a sample size of at least 200 respondents and use of the bidding game elicitation method, to name two criteria recommended for CV studies by the Water Resources Council's much-cited *Principles and Standards for Planning* (Water Resources Council, 1979; Department of the Interior, 1986:27722). In our view, the adoption of a prescriptive approach of this kind would be counterproductive. First, such criteria are too inflexible. The design features that are appropriate for a given CV study depend a great deal on the particular amenity the study values. What may be desirable in one context may be totally inappropriate in another. Similarly, the appropriate sample size for a given survey depends on the natural variance of the public's WTP amounts, the contribution made by the survey instrument to the observed variance, and the degree of precision desired by the policymaker. Second, it would be premature, given the present state of our knowledge, to make definitive judgments about many important issues. For example, since the Water Resources Council recommended the use of bidding games in 1979, evidence has accumulated that this elicitation technique is seriously prone to starting point bias. Such is the weight of the council's recommendations, however, that government contract officers have been known to continue to insist on use of the bidding game despite the protests of CV researchers. Third, minimum standards set by government guidance documents tend to become ends in themselves. On the basis of the council's recommendations, researchers and reviewers may feel satisfied with a study whose sample size is 200, whereas for many purposes this sample would be much too small. Finally, there are so many possible sources of error that may beset a CV survey that prescriptive criteria cannot adequately address them all. The presence of a few criteria may generate overconfidence in those studies that meet them.

A Best Practice Standard

An alternative approach, which we prefer, is to evaluate studies (or research proposals) on a study-by-study basis against a "best practice" standard. This requires the exercise of judgment by those capable of making such assessments, an admittedly subjective practice, but one which is, in the form of peer review, a familiar and workable way to evaluate research. This book is an attempt to contribute to an understanding of what constitutes best practice in contingent valuation surveys. The framework we present for identifying the potential sources of error in CV surveys and minimizing their effects is based on our synthesis of available knowledge about how respondents behave in CV surveys. Some of our judgments are based on a general consensus among CV researchers; others are more speculative.

Our views on the measurement of reliable and valid WTP amounts (presented primarily in chapters 11 and 12 and in appendix C) are best understood as working hypotheses, to be tested and refined. It is inevitable, given the complexity of the issues and the partial state of our knowledge, that the researcher will face the situation where no one alternative clearly represents best practice. If a particular factor, such as the payment vehicle, is expected to be especially problematic in a given study, its effect can be assessed by conducting experiments using split samples, either in a large pretest or in the final survey. If the subsamples are randomly selected and sufficiently large to provide an appropriately sensitive test, the results obtained by using payment vehicle A can be compared with those using payment vehicle B. If no difference is found, the factor may be disregarded as a source of bias; if a difference is found, the interpretation of the findings will have to address that issue.

Given the many extraneous factors that may influence a CV study's estimates, and the incomplete state of knowledge about these effects, we recommend that researchers employ an explicit design criterion—a rule for making choices when design decisions have to be made and there is no clearly unbiased alternative. In many cases it will be desirable to use a conservative rule under which decisions are consistently made in such a way that if one reasonable choice about a scenario element would potentially increase the WTP amount and the other reasonable choice would potentially reduce it, the latter is chosen. For example, where there is a choice between asking for the WTP amount in the form of a monthly or an annual payment, the annual payment would be preferred (according to this criterion) on the grounds that it would maximize the respondent's awareness of the financial implications of his responses. If such a rule is followed with reasonable consistency, and the study is otherwise well conducted, the use of a conservative design rule offers some assurance that the resulting WTP estimate will be a credible lower bound of the amenity's true value. There are situations, however, where a liberal design rule would be advisable. A researcher might be commissioned to conduct a study for use in a court case whose aim is to demonstrate a credible upper-bound estimate of an amenity's value.

In order to facilitate the review of their studies, CV researchers should include in their reports basic information about the sampling plan and its execution; the pretesting procedures used to develop the instrument and the assumptions behind the various design decisions; the values of any other variables used to estimate a valuation function; and which if any observations were omitted from the analysis, and the rationale for their deletion. It is essential that the report include a copy of the survey instrument,[2] including all materials shown to the respondents, and a rank-

[2] We recommend that journal editors ask those who submit articles based on CV studies for copies of the complete questionnaire, and that these be provided to the reviewers. If a CV-based article is accepted, it is desirable that key portions of the scenario be included in the text or in an appendix.

ordered list of the individual WTP amounts on which the estimates were based.[3]

Evaluating the Findings of a CV Study

Answers to the list of questions that follow will provide an analyst or a reviewer with the basic knowledge needed to evaluate a given CV study. The answers are based on the discussion presented in this book; where appropriate, the chapter that considers the topic is indicated.

1. *Background*

Who was the sponsor of the study, and what interests, if any, does the sponsor have in the provision of the amenity? (chapter 11)

When were the data gathered? Have there subsequently been any major changes in public opinion which are likely to affect the benefit estimates? (chapter 12)

2. *Sampling and Aggregation Procedures* (chapter 12)

What population did the study wish to represent in the sample?

What sampling plan was used to draw the sample from the population of interest? Was it probability based? How well was it executed?

What were the original sample size, the sampling response rates, and the usable number of respondents whose WTP amounts were employed to estimate the benefits?

What were the nonresponse rates to the valuation questions?

What effect did the nonresponses have on the benefits estimates?

3. *Scenario*

In evaluating the scenario, three dimensions should be considered: (A) whether the hypothetical market makes sense from the standpoint of economic theory, (B) whether the scenario is relevant to the policy being valued, and (C) whether respondents are likely to understand the scenario. Some key questions are:

How was the amenity described? Could an average person understand the description?

What property right was assumed? Were the measures used (for example, willingness to pay or willingness to accept) appropriate for the property right and meaningful to the respondents? (chapter 2)

Was the amenity being valued distinguished from related amenities with which respondents might confuse it? (chapter 11)

What types of benefits (use or existence, for example) were likely to be included in the respondents' WTP amounts for the amenity? (chapters 3 and 11)

[3] Greenley, Walsh, and Young (1982) and Desvousges, Smith, and McGivney (1983) are models of how to report the results of a CV survey.

Was the researcher aware of possible sequencing effects? For instance, if a deliberate sequencing effect was not desired and more than one good or level of provision was being valued, were the respondents informed of what they would be asked *before* they valued the first improvement? Were respondents given a chance at a later point in the interview to revise their amounts if they wished to do so?

Were key scenario elements such as the payment vehicle and probability of provision appropriate to the policy being valued? Were respondents provided with sufficient information to enable them to make an informed decision?

Was the description of the amenity accurate? To what extent are the descriptions of the amenity and the changes in the magnitude of its provision relevant for policy use? (chapter 11)

What provisions were made in the wording of the scenario to ensure that the potential sources of bias from instrument effects were minimized? (chapters 11 and 12)

4. *Survey Procedures* (chapters 3, 4, and 12)

What method was used to gather the data? (If a telephone or a mail survey was used, have the special problems posed by these methods been addressed?)

What procedures were used to develop and pretest the instrument?

How was the survey administered? This information will vary somewhat according to the survey method. Of particular importance are such questions as: How was the survey explained to the respondents? Who was described as the sponsor? Who executed the interviews or conducted the mail survey? What procedures were used to ensure that prevailing standards of survey practice were followed?

5. *Data Analysis*

What procedures were used to identify and handle outliers and protest responses? Is sufficient information provided about the cases dropped to permit a judgment about the validity of this procedure? (chapter 10)

What methods were used to compensate for missing data? (chapter 12)

If a valuation function is estimated, have alternative specifications been considered? Is the valuation function robust to violations of the assumptions made in estimating it? (chapters 2 and 12)

Are the data available for independent analysis?

6. *Evidence of Reliability and Validity* (chapters 9 and 10)

In order to adequately evaluate a CV study, it is necessary to examine the complete questionnaire that was used to gather the data. The key questions are:

Was the questionnaire (including introductory material and all materials shown to the respondent) clearly worded throughout? Was the descriptive material presented in a way likely to maintain the respondent's interest?

Did the questionnaire contain any material that might lead respondents to place a greater or lesser value on the amenity than would be the case in a genuinely neutral instrument? Did the wording overemphasize the hypothetical nature of the study or the impact it would have on public policy in a way that might lead respondents to give strategic responses? Did the information provided about the amenity include all the characteristics necessary for the valuations to be meaningful?

Was any consistent design rule used to make decisions about the sample and scenario design? If yes, what is the implication for the findings?

What evidence is there that the respondents understood the questions as intended by the researcher? Does the researcher discuss those response patterns of various groups of respondents which are consistent (or inconsistent) with the respondents' understanding of the scenario? Are the results of a meeting held to debrief the interviewers at the conclusion of the study reported?

What evidence is there of the effects of potential biases—especially those anticipated to be most troublesome—from the sources identified in this book or elsewhere in the literature? Of particular importance, are the results of an experiment built into the design of the survey using, for example, split samples? (chapter 5)

What are the results of a regression analysis of the WTP amounts on a set of theoretically relevant predictor variables? (This would provide evidence of reliability and validity.)

Were sensitivity analyses conducted, and if so, what were their findings? (These will aid in assessment of the findings' stability.)

What are the statistical confidence intervals for the WTP estimates, based on sampling variability? Has the role of nonsampling errors been satisfactorily addressed, and have appropriate warnings been provided?

New Applications

Because we still have much to learn about contingent valuation's limitations and possibilities, future applications of that method should routinely include methodological experiments as part of their study design. We will not attempt to outline a research agenda for such experiments here, but refer readers to the many problematic methodological issues identified in the preceding chapters. We will, however, suggest another type of agenda consisting of extensions of CV research beyond the field of environmental benefit assessment where it has seen the most use.

One research area in which contingent valuation could be put to greater use is the valuing of "local public goods" (Stiglitz, 1977), particularly those in an urban setting. These include such amenities as city parks, museums, libraries, major changes in elementary or secondary education programs, increases in police and fire protection, and programs to help the homeless.[4] The referendum market model seems particularly suited for valuing goods of this kind.

Heretofore, contingent valuation has been used primarily to value a single good. It may be useful to employ it to systematically value a much larger basket of public goods, and to estimate the elasticities of substitution between the goods in that basket.[5] Randall and his coauthors (1985) recently initiated CV work in this area by asking people to value simultaneous changes in a number of environmental amenities. Related research, using non-CV survey findings, has been conducted in four areas. The first is an effort by some researchers (Citrin, 1979, for example) to examine the implications of public opinion surveys which appear to show that the public wants more of everything.[6] Closely related is the systematic effort by Shanks and his coauthors (1981) to look at citizen reasoning about policy tradeoffs, particularly as they relate to the transfer of resources from one program to another. The third is the work on allocation games with tax refunds (Hockley and Harbour, 1983), briefly noted in chapter 4, while the fourth is a recent effort to estimate a system of demand equations from a set of spend more-same-less survey questions for different public goods (Ferris, 1983).

Another intriguing research area would be the determination of how the public's willingness to pay for a given public good differs under various payment schemes. This focus would turn the concept of payment vehicle bias on its head by accepting the fact that people have strong preferences for the ways in which they make their payments. The questions that might be answered by CV studies include: Are people willing to pay a constant amount or percentage more (or less) if the collection vehicle is an income tax rather than sales tax? Are people willing to pay more if taxation is progressive rather than proportional or regressive? To what extent are people willing to pay more if the collection scheme is regarded as "fair"? Are there potential payment mechanisms that are capable of capturing most (if not all) of the public's willingness to pay for a particular good?

[4] A number of such studies already exist; see, for instance, Devine and Marion (1979), Garbacz and Thayer (1983), and Throsby (1984).

[5] Knowledge of these elasticities is often necessary for making a correct aggregation of independently derived benefit estimates for different public goods.

[6] Not surprisingly, such an apparent finding makes many economists suspicious of public opinion surveys on policy issues. However, the typical public opinion poll questions on which this observation is based ask respondents about their preferences for specified programs without invoking the additional costs involved to the taxpayers, much less specifying what these costs might be. This finding is probably best interpreted as showing that people want more of most public goods if their provision does not cost very much.

The issue of what discount rate is appropriate for use in estimating the benefits of public projects can also be profitably examined through contingent valuation techniques. For projects which generate benefits (or costs) over a long time horizon, the choice of discount rate may be the most important factor in determining whether the benefits are sufficiently large to justify undertaking the project (Mishan, 1976). It has long been argued that the discount rate used by the government (often called the social discount rate) should be lower than the one used in private markets (Lind, 1982). Beyond this controversy, the standard theoretical framework for discounting has come under attack on several fronts. Machina (1984) has shown that it suffers from inconsistencies that may be accounted for by nonexpected utility theory, which allows for a broader range of "rational" behavior than is currently accepted by economists. Hausman (1979) has presented empirical evidence that individuals do not use the "accepted" method of discounting in the purchase of energy durables, and Thaler (1981) has obtained similar results in a more controlled experimental setting. By using an experimental design and exposing respondents to alternative scenarios that differ only on the time stream of the public good's provision, contingent valuation can be used to explore these issues (Carson, Machina, and Horowitz, 1987). Determining how the public in different time periods discounts lives saved would seem to be a particularly interesting research topic.

Still another new area of application is the valuation of risk reductions. A beginning has been made in this difficult research area,[7] and further research is suggested by Slovic, Fischhoff, and Lichtenstein's (1980) finding that people perceive risk in a multidimensional fashion. The important question for benefit-cost analysis is the degree to which risk characteristics (other than the size of the risk reduction and the initial risk level) influence the value people place on risk reductions. According to standard economic theory, such characteristics as the type of quick death should not affect the value for a given risk reduction; yet some research (for example, Torrance, Boyle, and Horwood, 1982), showing that people have strong preferences in regard to these risk characteristics, would suggest otherwise. Systematic investigation of this topic might help to explain some of the wide variation in the value of a statistical life found by CV and non-CV researchers (Blomquist, 1982).

There are undoubtedly many more potential new uses for contingent valuation, but these should suffice to show how the method offers those studying public goods an extremely powerful and versatile tool. It was

[7] For recent examples, see Smith, Desvousges, and Freeman's (1985) study of the benefits of reduced risks from hazardous waste sites, and Mitchell and Carson's (1986c) study of the benefits of risk reductions from THM contamination in drinking water.

only a little more than twenty years ago that Davis asked people on vacation in the Maine woods to place a dollar value on their recreational site. Since then more than 120 contingent valuation studies, to our knowledge (see appendix A), have been conducted. The method's promise is apparent. Nevertheless, much work remains to be done before we can regard its use as routine. In the meantime, the findings of individual contingent valuation surveys should receive careful, critical scrutiny before they become the basis for policy decisions affecting the provision of public goods.

Appendix A

Summary of the Content, Form, and Methodologies of Selected Contingent Valuation Surveys

Key to the numbered items in the surveys listed in the following pages, with key to abbreviations included:

1. Good being valued
2. Year survey was conducted
3. Research procedure(s) used

Mail	ML
Telephone	TLP
Personal interview	PI
Focus group	FG

4. Sample size (usable sample in parentheses); multiple numbers indicate that the study used multiple samples.
5. Willingness to pay (WTP) and/or willingness to accept (WTA)?
6. Were separate benefit categories estimated? Yes or No
7. Type of geographic area from which sample was taken

Local	L
State	S
Nationwide	N

8. Regression equation/valuation function? Yes or No
9. Were aggregate policy estimates made? Yes or No
10. Was a comparison with other techniques made? Yes or No
11. Elicitation method(s) used

Iterative bidding or bidding game	BG
Direct or open-ended question	DQ
Payment card or checklist	PC

Take-it-or-leave-it TILI
Contingent ranking CR

12. Were tests for biases made? Yes or No

13. How much of the questionnaire is provided in the source?
    None N
    Part P
    All A

Example:

Hanemann (1978). 1. Water quality / 2. 1973 / 3. PI / 4. 467 / 5. WTP / 6. No / 7. L / 8. Yes / 9. Yes / 10. Yes / 11. PC / 12. No / 13. A

Selected CV Surveys

Not all numbered items appear in all studies. Absence of a numbered item in a given study indicates that the item was not ascertainable from the source document.

Acton (1973). 1. Heart attack programs / 2. 1970 / 3. PI, ML / 4. 181 (60) / 5. WTP / 6. Yes / 7. L / 8. Yes / 9. Yes / 10. No / 11. DQ / 12. No / 13. A

Anderson and Devereaux (1986). 1. Artificial fishing reef / 2. 1985 / 3. PI, TLP, ML / 4. 201 (55) / 5. WTP / 6. No / 7. L / 8. Yes / 9. No / 10. No / 11. PC, DQ / 12. Yes / 13. P

Banford, Knetsch, and Mauser (1977). 1. Freshwater fishing sites / 2. 1975 / 3. ML, TLP / 4. 785 / 5. WTP, WTA / 6. No / 7. S / 8. Yes / 9. No / 10. No / 11. DQ / 12. No / 13. N

Beardsley (1971). 1. Recreation benefits / 2. 1966 / 3. PI / 5. WTP / 6. No / 7. L / 8. No / 9. Yes / 10. Yes / 11. BG / 12. No / 13. P

Bell and Leeworthy (1985). 1. Saltwater beaches / 2. 1984 / 3. PI / 4. 1051, 870 / 5. WTP, WTA / 6. No / 7. S / 8. Yes / 10. No

Bergstrom, Dillman, and Stoll (1985). 1. Environmental amenities / 2. 1981–82 / 3. ML / 4. 600 (250) / 5. WTP / 6. No / 7. L / 8. Yes / 9. Yes / 10. No / 11. PG / 12. Yes / 13. N

Berry (1974). 1. Urban parks / 5. WTP / 7. S / 10. No / 11. DQ

Bishop and Boyle (1985). 1. Illinois Beach State Natural Preserve / 2. 1985 / 3. ML / 4. 600 (359) / 5. WTP / 6. Yes / 7. S / 8. Yes / 9. Yes / 10. No / 11. TILI / 12. No / 13. A

Bishop and Heberlein (1980). 1. Hunting permits / 2. 1978 / 3. ML / 4. 353 (332) / 5. WTP, WTA / 6. No / 7. N / 8. Yes / 9. No / 10. Yes / 11. TILI / 12. No / 13. N

Blomquist (1984). 1. View-related amenities / 2. 1981 / 3. PI / 4. 208 / 5. WTP, WTA / 6. No / 7. L / 8. Yes / 9. No / 10. Yes / 11. BG / 12. Yes / 13. N

Bohm (1984). 1. Government statistics / 2. 1982 / 3. M / 4. 279 (274) / 5. WTP / 6. No / 7. N / 8. No / 9. Yes / 10. No / 11. DQ / 12. Yes / 13. No

Boyle and Bishop (1984b). 1. Scenic beauty / 2. 1982 / 3. PI / 4. 188 / 5. WTP / 6. No / 8. Yes / 10. No / 11. BG / 12. Yes

Brookshire, Eubanks, and Randall (1983). 1. Grizzly bear, bighorn sheep / 3. ML / 4. 3,000 / 5. WTP / 6. Yes / 7. S / 8. Yes / 9. No / 10. No / 11. DQ / 12. No / 13. N

Brookshire, Ives, and Schulze (1976). 1. Air visibility / 2. 1975 / 3. PI / 4. 104 (82) / 5. WTP / 6. No / 7. L / 8. Yes / 9. Yes / 10. No / 11. BG / 12. Yes / 13. A

Brookshire, Randall, and Stoll (1980). 1. Elk hunting / 2. 1977, 1978 / 3. PI / 4. 108 / 5. WTP, WTA / 6. No / 7. L / 8. Yes / 9. No / 10. No / 11. BG / 12. Yes / 13. N

Brown, Charbonneau, and Hay (1978). 1. Fish and wildlife recreational values / 2. 1975 / 3. TLP, ML / 4. 106,000, 20,000 / 5. WTP / 6. No / 7. N / 8. Yes / 9. No / 10. No / 11. DQ / 12. No / 13. N

Burness and coauthors (1983). / 1. Disposal of toxic wastes / 2. 1982 / 3. PI / 4. 74, 84 / 5. WTP / 6. N / 7. L / 8. Yes / 9. Yes / 10. No / 11. PC, BG / 12. Yes / 13. N

Cameron and James (1987). 1. Recreational fishing / 2. 1984 / 3. PI / 4. 4,161 / 5. WTP, WTA / 6. Yes / 7. S / 8. Yes / 9. Yes / 10. No / 11. TILI / 12. No / 13. A

Carson, Hanemann, and Mitchell (1986). 1. Water quality bond issue / 2. 1984 / 3. TLP / 4. 1,022 / 5. WTP / 6. No / 7. S / 8. Yes / 9. No / 10. No / 11. TILI / 12. No / 13. P

Cicchetti and Smith (1973). 1. Congestion in wilderness recreation / 3. ML / 4. 600 (195) / 5. WTP / 6. No / 7. L / 8. Yes / 9. No / 10. No / 11. DQ / 12. No / 13. N

Cocheba and Langford (1978). 1. Waterfowl hunting / 2. 1976 / 3. ML / 4. (169) / 5. WTP / 6. No / 7. L / 8. Yes / 9. Yes / 10. No / 11. PC / 12. No / 13. P

Conrad and LeBlanc (1979). 1. Development rights / 3. PI / 4. 22 / 5. WTA / 6. No / 7. L / 8. Yes / 9. No / 10. No / 11. DQ / 12. No / 13. N

Cummings, Schulze, Gerking, and Brookshire (1986). 1. Municipal infrastructure / 2. 1980 / 3. PI / 4. 486 / 5. WTP / 6. No / 7. L / 8. No / 9. No / 10. Yes / 11. BG / 12. No / 13. A

d'Arge (1985). 1. Water quality / 3. PI / 4. 20 / 5. WTP, WTA / 6. No / 7. L / 8. Yes / 9. No / 10. Yes / 11. DQ / 12. No / 13. A

Darling (1973). 1. Urban water parks / 3. PI / 5. WTP / 6. No / 7. L / 8. Yes / 9. Yes / 10. Yes / 11. BG, DQ / 12. No / 13. N

Daubert and Young (1981). 1. Instream flows / 2. 1978 / 3. PI / 4. 134 / 5. WTP / 6. No / 7. L / 8. Yes / 9. Yes / 10. No / 11. BG / 12. Yes / 13. P

Davis (1963b). 1. Recreation site / 2. 1961 / 3. PI / 5. WTP / 6. No / 7. S / 8. Yes / 9. No / 10. No / 11. BG

Davis (1980). 1. Water quality / 2. 1973 / 3. PI / 4. 2,000 (1,600) / 5. WTP / 6. No / 7. L / 8. Yes / 9. Yes / 10. No / 11. DQ / 12. Yes / 13. A

Deaton, Morgan, and Anschel (1982). 1. Rural-urban migration / 2. 1971 / 3. PI / 4. 396 / 5. WTA / 6. No / 7. S / 8. Yes / 9. No / 10. No / 11. DQ / 12. No / 13. P

Dennis and Hodgson (1984) 1. Boating facilities / 3. PI / 4. 120, 150 / 5. WTP / 6. No / 7. L / 8. Yes / 9. No / 10. No / 11. BG / 12. Yes / 13. A

Devine and Marion (1979). 1. Price comparison information for supermarkets / 2. 1974 / 3. TLP, ML / 4. 1,800, 1,500 (507, 363) / 5. WTP / 6. No / 7. S / 8. Yes / 9. Yes / 10. No / 11. DQ / 12. No / 13. P

Dickie, Fisher, and Gelkin (1987). 1. Strawberries / 2. 1984 / 3. PI / 4. 72 / 5. WTP / 6. No / 7. L / 8. Yes / 9. No / 10. Yes / 11. DQ / 12. Yes / 13. A

Donnelly and coauthors (1985). 1. Steelhead fishing / 2. 1983 / 3. TLP / 4. 427 / 5. WTP / 6. No / 7. S / 8. Yes / 9. Yes / 10. Yes / 11. BG / 12. No / 13. A

ECO Northwest (1984). 1. Fish populations / 2. 1984 / 3. PI / 4. 920 / 5. WTP, WTA / 6. No / 7. L / 8. Yes / 9. Yes / 10. Yes / 11. DQ / 12. Yes / 13. A

Foster, Halstead, and Stevens (1982). 1. Agricultural land / 2. 1981 / 3. PI / 4. (85) / 5. WTP / 6. No / 7. L / 8. Yes / 9. Yes / 10. Yes / 11. BG / 12. No / 13. A

Frankel (1979). 1. Value of life (airline accident) / 3. PI / 4. 169 / 5. WTP / 6. No / 10. No / 11. DQ

Garbacz and Thayer (1983). 1. Senior Companion program / 3. PI / 4. 60 (29) / 5. WTP / 6. No / 7. L / 8. No / 9. Yes / 10. No / 11. DQ / 12. No / 13. N

Gramlich (1977). 1. Water quality / 2. 1973 / 3. TLP, PI / 4. 165 / 5. WTP / 6. No / 7. L / 8. Yes / 9. Yes / 10. No / 11. TILI, DQ / 12. No / 13. P

Greenley, Walsh, and Young (1981). 1. Water quality / 2. 1976 / 3. PI / 4. 202 / 5. WTP / 6. Yes / 7. L / 8. Yes / 9. Yes / 10. No / 11. BG / 12. Yes / 13. P

Gregory Research (1982). 1. Museum exhibits / 3. PI / 4. 141 / 5. WTP, WTA / 6. No / 7. L / 10. No / 11. DQ / 12. Yes

Hageman (1985). 1. Marine mammals / 2. 1984 / 3. ML / 4. 1,000 (180, 175, 174, 174) / 5. WTP / 6. Yes / 7. S / 8. Yes / 9. Yes / 10. No / 11. PC / 12. Yes / 13. A

Halstead (1984). 1. Nonmarket values of agricultural land / 3. PI / 4. 85 / 5. WTP / 6. No / 7. S / 8. Yes / 9. Yes / 10. No / 11. BG / 12. Yes / 13. N

Hammack and Brown (1974). 1. Migratory waterfowl / 2. 1969 / 3. ML / 4. 4,900 (2,455) / 5. WTP, WTA / 6. No / 7. N / 8. Yes / 9. No / 10. No / 11. PC / 12. No / 13. A

Hammerton, Jones-Lee, and Abbott (1982). 1. Statistical life / 3. PI / 4. 120 / 5. WTP / 6. No / 7. L / 8. No / 9. No / 10. No / 11. DQ / 12. Yes / 13. P

Hammitt (1986). 1. Foodborne risks / 2. 1985 / 3. FG / 4. 45 (43) / 5. WTP / 6. No / 7. L / 8. No / 9. Yes / 10. Yes / 11. DQ / 12. No / 13. A

Hanemann (1978). 1. Water quality / 2. 1973 / 3. PI / 4. 467 / 5. WTP / 6. No / 7. L / 8. Yes / 9. Yes / 10. Yes / 11. PC / 12. No / 13. A

Harris (1984). 1. Water pollution control program / 3. PI / 5. WTP / 6. No / 7. L / 8. No / 9. Yes / 10. No / 11. BG / 12. No / 13. N

Hoehn and Randall (1985b). 1. Urban viewing site / 2. 1981 / 3. PI / 4. 319, 87, 147 / 5. WTP / 6. No / 7. L / 8. Yes / 9. No / 10. Yes / 11. BG / 12. No / 13. N

Horvarth (1974). 1. Recreation benefits / 3. ML / 4. 12,068 (9,322) / 5. WTA / 6. No / 7. N / 8. Yes / 9. Yes / 10. No / 11. DQ / 12. No

Jackson (1983). 1. Environmental quality / 2. 1969 / 3. PI / 4. 1,248 / 5. WTP / 6. No / 7. N / 8. Yes / 9. No / 10. No / 11. BG / 12. No / 13. A

Johnson, Shelby, and Bregenzer (1986). 1. White water recreation / 2. 1984 / 3. ML / 4. 300, 300 (193, 200) / 5. WTP, WTA / 6. No / 7. S / 8. Yes / 9. No / 10. No / 11. DQ, TILI / 12. No / 13. A

Jones-Lee (1976). 1. Value of life / 2. 1975 / 3. ML / 4. 30 / 5. WTA, WTP / 6. No / 8. Yes / 9. No / 10. No / 11. DQ / 12. No / 13. A

Jones-Lee, Hammerton, and Philips (1985). 1. Safety / 2. 1982 / 3. PI / 4. 1,103 (1,057) / 5. WTP, WTA / 6. No / 7. N / 8. Yes / 9. No / 10. No / 11. DQ, BG / 12. Yes / 13. P

Kealy, Dovidio, and Rockel (1986). 1. Private market good / 3. FG / 4. 240 (145) / 5. WTP / 6. No / 8. Yes / 9. No / 10. No / 11. BG, DQ, TILI / 12. Yes / 13. A

Lah (1985). 1. Work of Calif. Conservation Corps / 2. 1984 / 3. PI / 4. 212, 151 (186, 115) / 5. WTP / 6. No / 7. S / 8. Yes / 9. Yes / 10. No / 11. TILI / 12. No / 13. N

Lareau and Rae (1985). 1. Diesel odors / 2. 1984 / 3. PI / 4. 290 / 5. WTP / 6. No / 7. S / 8. Yes / 9. Yes / 10. No / 11. DQ, CR / 12. No / 13. A

Loehman (1984). 1. Air quality / 3. PI / 4. 412 / 5. WTP / 6. Yes / 7. L / 8. Yes / 10. Yes / 11. PC

Loehman and De (1982). 1. Air pollution control / 2. 1977 / 3. ML / 4. 1,800 (404) / 5. WTP / 6. No / 7. L / 8. Yes / 9. Yes / 10. No / 11. PC / 12. No / 13. P

Loomis (1987a). 1. Water quality / 2. 1986 / 3. ML / 4. 603 / 5. WTP / 6. Yes / 7. S / 8. No / 9. Yes / 10. No / 11. TILI, DQ / 12. No / 13. N

Majid, Sinden, and Randall (1983). 1. Public parks / 2. 1982 / 3. PI / 4. 140 / 5. WTP / 6. No / 7. L / 8. Yes / 9. No / 10. No / 11. BG / 12. No / 13. A

Mathews and Brown (1970). 1. Salmon fisheries / 2. 1968 / 3. ML / 4. 5,000 (2,146) / 5. WTP, WTA / 6. No / 7. N / 8. Yes / 9. No / 10. No / 11. CL / 12. No / 13. N

McConnell (1977). 1. Day at the beach / 2. 1974 / 3. PI / 4. 229 / 5. WTP / 6. No / 7. L / 8. Yes / 9. No / 10. No / 11. BG / 12. No / 13. N

Meyer (1974a). 1. Salmon-oriented recreation / 2. 1971–72 / 3. PI / 4. 3,617 / 5. WTP / 6. Yes / 7. N / 8. Yes / 9. Yes / 10. No / 11. DQ / 12. No / 13. A

Meyer (1980). 1. Fish and wildlife / 3. PI / 4. (504, 514, 510) / 5. WTA / 6. No / 7. L / 8. Yes / 9. Yes / 10. No / 11. DQ / 12. No / 13. A

Michalson and Smathers (1985). 1. Public campground use / 2. 1982 / 3. PI / 4. 350 / 5. WTP / 6. No / 7. L / 8. Yes / 9. Yes / 10. Yes / 11. BG / 12. Yes / 13. N

Milon (1986). 1. Marine artificial reef / 2. 1985 / 3. ML / 4. 3,600 / 5. WTP / 6. No / 7. L / 8. Yes / 9. Yes / 10. Yes / 11. BG / 12. Yes / 13. N

Mitchell and Carson (1981). 1. Freshwater quality / 2. 1980 / 3. PI / 4. 1,516 / 5. WTP / 6. Yes / 7. N / 8. Yes / 9. No / 10. No / 11. PC / 12. Yes / 13. A

Mitchell and Carson (1984). 1. Freshwater quality / 2. 1983 / 3. PI / 4. 813 (564) / 5. WTP / 6. Yes / 7. N / 8. Yes / 9. Yes / 10. Yes / 11. PC / 12. Yes / 13. A

Mitchell and Carson (1986c). 1. Drinking water / 2. 1985 / 3. PI / 4. 286 (238) / 5. WTP / 6. No / 7. L / 8. Yes / 9. No / 10. No / 11. DQ / 12. Yes / 13. A

Mulligan (1978). 1. Nuclear plant accidents / 3. PI / 5. WTP / 6. No / 10. No / 11. BG

O'Neill (1985). 1. Two river sites / 2. 1984 / 3. PI / 4. 517, 392 / 5. WTP, WTA / 6. No / 7. S / 8. Yes / 9. No / 10. Yes / 11. DQ, BG / 12. Yes / 13. A

Oster (1977). 1. Freshwater pollution / 2. 1973, 1974 / 3. TLP / 4. 200 / 5. WTP / 6. No / 7. L / 8. No / 9. No / 10. No / 11. DQ / 12. No / 13. A

Randall, Ives, and Eastman (1974). 1. Environmental damage / 2. 1972 / 3. PI / 4. 747 / 5. WTP / 6. No / 7. L / 8. Yes / 9. Yes / 10. No / 11. BG / 12. No / 13. N

Randall, Blomquist, and coauthors (1985). 1. National air and freshwater pollution control / 2. 1983 / 3. ML, TLP, PI, FG / 4. 991 / 5. WTP / 6. No / 7. N / 8. Yes / 9. Yes / 10. Yes / 11. TILI / 12. Yes / 13. A

Randall, Grunewald, and coauthors (1978). 1. Surface coal mine reclamation / 2. 1977 / 3. PI / 4. 220 / 5. WTP / 6. No / 7. L / 8. No / 9. Yes / 10. No / 11. BG / 12. No / 13. N

Reizenstein, Hills, and Philpot (1974). 1. Air quality / 3. PI / 4. 400 (376) / 5. WTP / 6. No / 7. L / 8. Yes / 9. No / 10. No / 11. DQ / 12. No / 13. N

Roberts, Thompson, and Pawlyk (1985). 1. Offshore diving platforms / 2. 1982 / 3. ML, PI, TLP / 4. 144 / 5. WTP / 6. No / 7. S / 8. Yes / 9. Yes / 10. No / 11. BG / 12. Yes / 13. P

Rowe, d'Arge, and Brookshire (1980). 1. Atmospheric visibility / 3. PI / 5. WTP, WTA / 6. No / 7. L / 8. Yes / 9. No / 10. No / 11. BG / 12. Yes / 13. N

Rowe and Chestnut (1984). 1. Reduction in asthma days / 2. 1983 / 3. PI / 4. 82 (65) / 5. WTP / 7. L / 8. Yes / 11. PC

Samples, Dixon, and Gower (1986). 1. Humpback whales / 3. FG / 4. 240 / 5. WTP / 6. No / 8. Yes / 9. No / 10. No / 11. DQ / 12. Yes / 13. P

Schulze, Cumming, and coauthors (1983). 1. Visibility / 2. 1980 / 3. PI / 4. 600 / 5. WTP / 6. No / 7. N / 8. Yes / 9. Yes / 10. No / 11. BG / 12. No / 13. N

Scott and Company (1980). 1. Air pollution / 3. PI / 4. 752 / 5. WTP / 6. No / 7. S / 8. No / 9. No / 10. No / 11. DQ, PC / 12. No / 13. A

Sellar, Stoll, and Chavas (1985). 1. Recreational boating / 2. 1981 / 3. ML / 4. 2,000 (275, 211) / 5. WTP / 6. No / 7. S / 8. Yes / 9. No / 10. Yes / 11. DQ, TILI / 12. No / 13. P

N. Smith (1980). 1. Recreation site / 2. 1979 / 3. PI / 4. 200 / 5. WTP / 6. No / 8. Yes / 10. Yes / 11. BG / 12. No / 13. A

V. K. Smith and Desvousges (1986a). 1. Demand for distance from disposal sites / 2. 1984 / 3. PI / 4. 609 / 5. WTP / 6. No / 7. L / 8. Yes / 9. No / 10. No / 11. DQ / 12. No / 13. N

V. K. Smith and Desvousges (1986b). 1. Freshwater quality benefits / 2. 1981 / 3. PI / 4. 303 / 5. WTP / 6. Yes / 7. L / 8. Yes / 9. Yes / 10. Yes / 11. BG, DQ, PC, CR / 12. Yes / 13. A

Sorg and coauthors (1985). 1. Fishing / 2. 1983 / 3. TLP / 4. 1,758, 194 / 5. WTP / 6. No / 7. S / 8. Yes / 9. Yes / 10. Yes / 11. BG / 12. No / 13. A

Sorg and Nelson (1986). 1. Elk hunting / 2. 1983–84 / 3. TLP / 4. 1,629 / 5. WTP / 6. No / 7. S / 8. Yes / 9. Yes / 10. Yes / 11. DQ, BG / 12. No / 13. A

Stoll and Johnson (1985). 1. Whooping crane / 2. 1982–83 / 3. ML, PI / 4. 1,800, 800 / 5. WTP / 6. Yes / 7. N / 8. Yes / 9. Yes / 10. No / 11. BG / 12. No / 13. A

Sutherland (1983). 1. Recreation facility and travel time / 2. 1980 / 3. TLP / 4. 3,000 (364, 1,327) / 5. WTP / 6. No / 7. N / 8. Yes / 10. No / 11. DQ / 12. No / 13. A

Sutherland and Walsh (1985). 1. Freshwater quality / 2. 1981 / 3. ML / 4. 280 (171) / 5. WTP / 6. Yes / 7. S / 8. Yes / 9. Yes / 10. No / 11. DQ / 12. No / 13. P

Thayer (1981). 1. Environmental damage / 2. 1976, 1977 / 3. PI / 4. 112 (106) / 5. WTP / 6. No / 7. L / 8. Yes / 9. No / 10. No / 11. BG / 12. Yes / 13. N

Throsby (1984). 1. Public good output of the arts / 2. 1982 / 3. PI / 4. 827 / 5. WTP / 6. No / 7. S / 8. Yes / 9. No / 10. No / 11. DQ / 12. Yes / 13. P

Tolley and Babcock (1986). 1. Health risks / 2. 1985 / 3. ML, PI / 4. 199 (176) / 5. WTP / 6. No / 7. L / 8. Yes / 9. No / 10. No / 11. BG / 12. No / 13. P

Tolley, Randall, and coauthors (1985). 1. Air visibility / 2. 1982 / 3. PI / 4. 792 (538) / 5. WTP / 6. No / 7. N / 8. Yes / 9. Yes / 10. Yes / 11. BG, DQ, PC / 12. Yes / 13. P

Tyrrell (1982). 1. Beaches / 2. 1981 / 3. PI / 4. 251 / 5. WTP / 6. No / 7. N / 8. Yes

Walsh and Gilliam (1982). 1. Congestion in wilderness area / 2. 1979 / 3. PI / 4. 126 / 5. WTP / 6. No / 7. L / 8. Yes / 9. Yes / 10. No / 11. BG, DQ / 12. Yes / 13. N

Walsh, Loomis, and Gillman (1984). 1. Wilderness protection / 2. 1980 / 3. ML / 4. 218 (195) / 5. WTP / 6. Yes / 7. S / 8. Yes / 9. Yes / 10. No / 11. DQ / 12. Yes / 13. N

Walsh, Miller, and Gilliam (1983). 1. Congestion in ski area / 2. 1980 / 3. PI / 4. 236 / 5. WTP / 6. No / 7. S / 8. Yes / 9. Yes / 10. No / 11. BG / 12. Yes / 13. N

Walsh, Sanders, and Loomis (1985). 1. Wild and scenic rivers / 2. 1983 / 3. ML / 4. 214 / 5. WTP / 6. Yes / 7. S / 8. Yes / 9. Yes / 10. Yes / 11. DQ / 12. No / 13. A

Wegge, Hanemann, and Strand (1985). 1. Marine recreational fishing / 2. 1984 / 3. ML / 4. 2,915 (1,383) / 5. WTP / 7. N / 8. Yes / 9. Yes / 10. No / 11. TILI / 12. No / 13. A

Welle (1985). 1. Environmental degradation from acid rain / 2. 1985 / 3. ML / 4. 910 (669) / 5. WTP, WTA / 6. Yes / 7. S / 8. Yes / 9. No / 10. No / 11. TILI / 12. No / 13. A

Whittington and coauthors (1986). 1. Rural water services in Haiti / 2. 1986 / 3. PI, FG / 4. 345 / 5. WTP / 6. No / 7. L / 8. Yes / 9. No / 10. No / 11. DQ, BG / 12. Yes / 13. A

Willis and Foster (1983). 1. Water quality / 3. PI / 5. WTP / 11. DQ / 13. N

Appendix B

Survey Instrument for a National Freshwater Benefits Survey

The following is the complete text of the instrument used by Mitchell and Carson in their study of the benefits of national freshwater quality improvements (Mitchell and Carson, 1984; Carson and Mitchell, 1986), described in chapter 1. The instrument was administered by Opinion Research Corporation interviewers to a national stratified sample of 813 respondents in personal interviews.

INTERVIEWER: Hand Respondent Filled out "Thank You" pamphlet

Study # 65450

Line # ______

Location # ______

Supervisor's Name: ______

Respondent's
Name: Mr. Mrs. Miss ______
(Circle)

Address: ______

City: ______ State: ______ Zip Code ☐☐☐☐☐

Telephone # ☐☐☐ – ☐☐☐ – ☐☐☐☐
Area Code

Date of Interview: ______ Time: ______ AM PM (Circle)

Length of Interview: ______ Minutes

INTERVIEWER: IF RESPONDENT REFUSES TO GIVE YOU HIS/HER TELEPHONE NUMBER, SAY:

"I need your telephone number in order for my supervisor to confirm that this interview was conducted properly and that I performed my job in a courteous and businesslike fashion. No one else will ever have access to your number."

INDICATE: 1 TELEPHONE NUMBER OBTAINED
2 REFUSED

I hereby certify that this is an honest interview taken in accordance with my instructions.

____________________ Interviewer's Signature

____________________ Date

FOR OFFICE USE ONLY

DATE	TIME	RESULT	COMMENTS	VERIFIED BY

LOCATION #: __________ 65450
LINE #: __________ 110383

WATER BENEFITS SURVEY

INTERVIEWER: __________ TIME ENDED: __________
INTERVIEWER ID. #: __________ TIME STARTED: __________
DATE: __________ INTERVIEW LENGTH: ______(MINUTES)

Hello, I'm __________________ of Opinion Research Corporation in Princeton, New Jersey. We are talking to a cross-section of people in the United States about how much public programs are worth to them. Your views will be used to help policy makers make informed decisions.

First, let me begin by saying that most of the questions have to do with <u>your</u> attitudes and opinions, and there are no right or wrong answers.

This interview is completely confidential; your name will never be associated with your answers.

1. First, I'm going to read a list of several issues which, over the years, have been of concern to the taxpayers. For each, please tell me whether you feel the amount of money we are spending as a nation is too much, just about the right amount, or too little.

	Too Much	About the Right Amount	Too Little	DON'T KNOW	REFUSED
a. Reducing air pollution	1	2	3	4	5
b. Fighting crime	1	2	3	4	5
c. Reducing water pollution in freshwater lakes, streams, and rivers	1 ASK Q.2	2 ASK Q.4	3 ASK Q.3	4 ASK Q.4	5 ASK Q.4

IF Q.1c IS "TOO MUCH", ASK:

2. You said that we are spending "Too much money" on reducing water pollution in freshwater lakes, streams, and rivers. In your opinion, do you think we should be spending a great deal less or only a little less on reducing water pollution?

   1 Great deal less
   2 A little less
   3 DON'T KNOW
   4 REFUSED

   → SKIP TO Q.4

IF Q.1c IS "TOO LITTLE", ASK:

3. You said that we are spending "Too little money" on reducing water pollution in freshwater lakes, streams and rivers. In your opinion, do you think we should be spending a great deal more or only a little more on reducing water pollution?

   1 Great deal more
   2 A little more
   3 DON'T KNOW
   4 REFUSED

ASK EVERYONE
(HAND RESPONDENT BOOKLET)

4. I'd like you to look at this booklet that contains several cards. Please look at Card 1. It contains three statements regarding pollution control and costs of pollution control. Please follow along as I read these statements to you, and then tell me which statement you agree with most. (READ EACH STATEMENT TO RESPONDENT.)

   1 Protecting the environment is so important that pollution control requirements and standards cannot be too strict and continuing improvement must be made regardless of cost, or
   2 We have made enough progress on cleaning up the environment that we should now concentrate on holding down costs rather than requiring stricter controls, or

3 Pollution control requirements and standards have gone too far and they already cost more than they are worth.
4 BETWEEN 1 AND 2 (VOLUNTEERED)
5 DON'T KNOW
6 REFUSED

5. Some national goals are more important to people than others. How important to you personally is a national goal of protecting nature and controlling pollution? Is it very important, somewhat important, or not very important to you?

1 VERY IMPORTANT
2 SOMEWHAT IMPORTANT
3 NOT VERY IMPORTANT
4 DON'T KNOW

IF "1" ON Q.5, ASK:

6. You said a national goal of protecting nature and controlling pollution is "very important" to you. Would you say it is one of your very top priorities or is it of somewhat less importance to you?

1 VERY TOP PRIORITY
2 SOMEWHAT LESSER IMPORTANCE
3 DON'T KNOW

7. Please turn to Card 2. It contains a list of six different sources of water pollution in freshwater lakes, rivers and streams. Tell me which one or two sources you feel probably cause the most water pollution in the nation. Just read me the numbers.

1 Runoff from agriculture
2 Sewage from cities and towns
3 Drainage from mines
4 Runoff from roads and highways
5 Seepage from garbage dumps
6 Dumping of factory waste into waterbodies
7 NONE
8 DON'T KNOW
9 REFUSED

SECTION B: HOUSEHOLD ACTIVITIES GRID

INTRODUCTION: The next few questions concern participation in outdoor recreational activities by members of this household.

8. First, how many people—both adults and children—live in this household, including yourself?

01 Respondent only ———→ SKIP TO Q.10
___ Number in household including Respondent
98 DON'T KNOW
99 REFUSED

9. How many of these people are under 18 years of age?
___ Number under 18 yrs. old
98 DON'T KNOW
99 REFUSED

10. Now about you. Please tell me your age at your last birthday. RECORD IN HOUSEHOLD GRID IN "AGE" COLUMN. CIRCLE APPROPRIATE SEX.

IF MORE THAN ONE HOUSEHOLD MEMBER, ASK Q.11, OTHERWISE SKIP TO Q.12.

11. Starting with the oldest member of this household, please tell me the sex and age of the other household members, and their relationship to you. RECORD IN HOUSEHOLD GRID.

INTERVIEWER CHECK: MAKE CERTAIN THAT THE NUMBER OF RESPONDENTS LISTED IN THE GRID IS THE SAME AS THE NUMBER OF HOUSEHOLD MEMBERS IN Q.8.

ASK EVERYONE

12. During the past 12 months, that is, since November, 1982, did you (or any member of this household over five years old) boat, fish, swim, wade or waterski in a freshwater river, lake, pond or stream anywhere in the U.S. for recreational purposes? Please keep in mind that this does not include swimming in swimming pools or boating, fishing or swimming in the ocean.

1 Yes ———→ GO TO INSTRUCTIONS FOR ACTIVITY GRID
2 No
3 DON'T KNOW ———→ SKIP TO Q.19
4 REFUSED

INSTRUCTIONS FOR ACTIVITY GRID

ASK Q.13–15 IN A SERIES FOR EACH HOUSEHOLD MEMBER OVER FIVE YEARS OLD, STARTING WITH THE RESPONDENT. THEN ASK Q.13–15 FOR EACH REMAINING MEMBER OVER 5 YEARS OLD.

13. During the past 12 months, did (you/HOUSEHOLD MEMBER) use freshwater lakes, rivers or streams in this state or any other state for recreational boating? By boating, I mean canoeing, kayacking, rafting, motorboating, sailing, windsurfing, and waterskiing.

14. During the past 12 months, did (you/HOUSEHOLD MEMBER) use freshwater lakes, rivers or streams in this state or any other state for recreational fishing?

15. During the past 12 months, did (you/HOUSEHOLD MEMBER) use freshwater lakes, rivers or streams in this state or any other state for recreational swimming?

FOR EACH "YES" IN Q.13–15, ASK Q.16 AND Q.17 IN A SERIES STARTING WITH THE RESPONDENT. THEN ASK Q.16 AND Q.17 FOR EACH REMAINING HOUSEHOLD MEMBER OVER 5 YEARS OLD. RECORD NUMBER OF DAYS ON GRID. RECORD "998" FOR "DON'T KNOW", "999" FOR "REFUSED" AND "000" FOR "NONE". PROBE NUMBER OF DAYS WITH: Your best estimate will do.

16. About how many days did (you/HOUSEHOLD MEMBER) go freshwater (boating/fishing/swimming) in this state?

17. About how many days did (you/HOUSEHOLD MEMBER) go freshwater (boating/fishing/swimming) out-of-state?

IF ANY HOUSEHOLD MEMBER FISHED, ASK Q.18; OTHERWISE SKIP TO Q.19
(ASK Q.18 ABOUT HOUSEHOLD MEMBER WHO FISHED THE MOST DAYS BOTH IN-STATE AND OUT-OF-STATE. IF MORE THAN ONE QUALIFIES, ASK ABOUT OLDEST MEMBER OF HOUSEHOLD.)

18. How important to (you/HOUSEHOLD MEMBER) is freshwater fishing as a recreational activity? Would you say it is . . . ?

 1 Very important
 2 Somewhat important,
 3 or Not at all important?
 4 DON'T KNOW → DON'T READ
 5 REFUSED → DON'T READ

ASK EVERYONE
19. Did you (or any member of your household) swim in a swimming pool or in the ocean in this state during the past 12 months?

 1 Yes
 2 No
 3 DON'T KNOW
 4 REFUSED

SECTION B: HOUSEHOLD ACTIVITIES GRID

	Q.11				BOATING			FISHING			SWIMMING		
	RELATIONSHIP TO RESPONDENT	SEX M	F	AGE	Q.13	Q.16 # DAYS IN-STATE	Q.17 # DAYS OUT-OF-STATE	Q.14	Q.16 # DAYS IN-STATE	Q.17 # DAYS OUT-OF-STATE	Q.15	Q.16 # DAYS IN-STATE	Q.17 # DAYS OUT-OF-STATE
1	RESPONDENT	1	2		1 YES 2 NO 3 DK 4 REF			1 YES 2 NO 3 DK 4 REF			1 YES 2 NO 3 DK 4 REF		
2		1	2		1 YES 2 NO 3 DK 4 REF			1 YES 2 NO 3 DK 4 REF			1 YES 2 NO 3 DK 4 REF		
3		1	2		1 YES 2 NO 3 DK 4 REF			1 YES 2 NO 3 DK 4 REF			1 YES 2 NO 3 DK 4 REF		
4		1	2		1 YES 2 NO 3 DK 4 REF			1 YES 2 NO 3 DK 4 REF			1 YES 2 NO 3 DK 4 REF		
5		1	2		1 YES 2 NO 3 DK 4 REF			1 YES 2 NO 3 DK 4 REF			1 YES 2 NO 3 DK 4 REF		
6		1	2		1 YES 2 NO 3 DK 4 REF			1 YES 2 NO 3 DK 4 REF			1 YES 2 NO 3 DK 4 REF		
7		1	2		1 YES 2 NO 3 DK 4 REF			1 YES 2 NO 3 DK 4 REF			1 YES 2 NO 3 DK 4 REF		
8		1	2		1 YES 2 NO 3 DK 4 REF			1 YES 2 NO 3 DK 4 REF			1 YES 2 NO 3 DK 4 REF		

20. During the past 12 months, did you (or any member of this household) take part in recreational activities on the shore of or near any freshwater lakes, river, or streams anywhere in the U.S.? These could be activities like picnicking, camping, bird watching, duck hunting, or living in a vacation cottage?

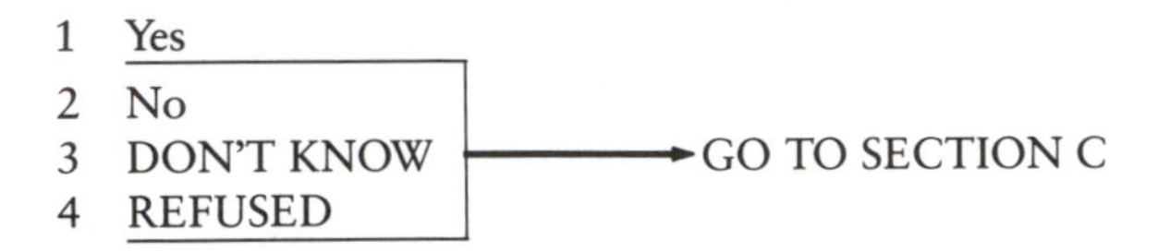

IF "YES" TO Q. 20, ASK:

21. Were these activities done in-state, out-of-state, or both?

    1 In-state
    2 Out-of-state
    3 Both
    4 DON'T KNOW
    5 REFUSED

SECTION C: WATER QUALITY LEVELS

This next series of questions is about different levels of water quality in the nation's lakes, rivers, and streams and about how much different levels of water quality in those freshwater bodies is worth to you (and all other members of this household).

In these questions, I will not be talking about saltwater, or water that is underground, or about drinking water. For the remainder of the interview, I will always be referring to the freshwater in lakes, rivers and streams across the country.

Because of growing water pollution problems nationwide, Congress passed strict water pollution control laws in 1972 and 1977 and provided money to pay most of the costs for building new sewage plants for communities. These laws also required many industries to install and pay for expensive water pollution control equipment.

The laws Congress passed are intended to improve the quality of water. One way of thinking about different levels of water quality is to use a ladder like the one shown on Card 3 of the booklet.

The top of the water quality ladder stands for the best possible quality of water, and the bottom of the ladder stands for the worst. On the ladder you can see the different levels of water quality. For example:

Level "D" (POINT) is so polluted that it has oil, raw sewage and other things like trash in it; it has no plant or animal life, smells bad, and contact with it is dangerous to human health.

Water at level "C" (POINT) is boatable. Water of this quality would not harm you if you happened to fall into it for a short time while boating or sailing.

In the United States today, because of water pollution control programs, this is now the minimum national quality level. In other words, the present quality of more than 99 percent of all the nation's freshwater lakes, rivers and streams is at least at this level. Those water bodies which can only be used for boating at the present time are mostly located in areas with a lot of industry and also where large numbers of people live. If we stopped spending money for water pollution control, the quality of these and many other water bodies would fall below the boatable level.

Level "B" (POINT) is fishable. Although some kinds of fish can live in boatable water, it is only when water gets this clean that game fish like bass can live in it. Today many of the nation's freshwater bodies are as clean as this.

Level "A" (POINT) is swimmable. Today perhaps 70–80% of the nation's freshwater is as clean as this.

22. Perhaps as I have talked, you have thought about the quality of water in this area. Think about the nearest freshwater lake, river, stream, pond or creek that is large enough so that game fish might live in it. It does not matter if it is manmade or not, how would you rate its quality of water? Choose a letter on the water quality ladder which you think best describes the water quality of this lake or pond. (PROBE: Your best estimate will do.)

LETTER ON LADDER	CORRESPONDING NUMBER ON LADDER
1 D	(0–less than 2)
2 C	(2–less than 3)
3 B	(3–less than 6)
4 A	(6–less than 8)
5 More than A	(8–10)
6 DON'T KNOW	
7 REFUSED	

23. Now I'd like you to think about how much having clean water in the United States, including this state, is worth to you and (all members of your household). Some people believe that controlling water pollution is of great value, while other people do not feel that control of water pollution is very important to them. Card 4 in your booklet shows various reasons why some people might value water quality. Please read it over.

    Which two of these reasons, if any, for reducing water pollution are most important to you personally? Just read me the numbers.

    1 Your (Your household's) use of freshwater for fishing, boating or swimming
    2 Your (Your household's) use of areas surrounding freshwater for picnicking, bird watching, or staying in a vacation cottage
    3 You (Your household) get satisfaction from knowing other people may use and enjoy freshwater
    4 You (Your household) get satisfaction from knowing that the nation's water is cleaner
    5 NONE/I DO NOT VALUE WATER QUALITY
    6 DON'T KNOW
    7 REFUSED

SECTION D: WATER QUALITY EVALUATION

In this next section of the questionnaire, I am going to ask you how much it is worth to you in real dollars and cents to reach three different national water quality goals. Since this is not something we usually think about, it may be helpful for you to know what the average household like yours pays in taxes and higher prices for some other types of public programs. In order to do this, would you please look at the next card, Card 5, in the booklet and give me the letter next to the category which includes your (household's) total, yearly gross income from all sources, that is, before taxes in 1982. Once again, I'd like to remind you that this interview is completely confidential and your name will never be associated with your answers. (CIRCLE LETTER OF PAYMENT CARD CHOSEN.)

	COLOR OF PAYMENT CARD
1 A Under $10,000	WHITE
2 B $10,000–$19,999	YELLOW
3 C $20,000–$29,999	BLUE
4 D $30,000–$49,999	GREEN
5 E $50,000 or more	PINK
6 F REFUSED ——→	GIVE RESPONDENT BLUE PAYMENT CARD, AND SAY: If you would look at this payment card which reflects the middle range of incomes in the United States.

GIVE RESPONDENT APPROPRIATE PAYMENT CARD FOR HIS/HER INCOME RANGE.

The payment card I have given you lists many different amounts. It also gives an estimate of how much households in your income range paid in 1982 in taxes and product prices for programs like the space program, police and fire protection, roads and highways, public education, and the defense program.

As you may also know, programs to control air and water pollution are also something we all pay for. We pay for water pollution control in two ways, as shown on the next card, Card 6.

First, part of the money we pay in federal and state taxes goes to construct sewage treatment plants, conduct research on water pollution and to enforce the water pollution laws. Any local taxes and sewer fees which are often part of your water bill help to pay the cost of running these plants.

The second way involves the price of things we buy. A small amount of the money you pay for many products goes for the water pollution control equipment the government requires industries to install. In order to pay for this equipment, companies increase somewhat the cost of the products they sell to consumers.

GIVE RESPONDENT WORKSHEET AND PENCIL. RESPONDENT SHOULD ALSO HAVE COLORED PAYMENT CARD. REFER TO WORKSHEET AS YOU READ.

Here are (POINTING TO THE LEVELS ON THE WORKSHEET) three national water pollution goals. The lowest one is goal C which is where we are today with 99 percent or more of all freshwater bodies at least at the boatable quality level, although many are higher in quality.

Goal B would be to raise the minimum level to where 99 percent or more of the freshwater bodies would at least be at the fishable level so game fish like bass could live in them.

Goal A would further raise the minimum level to where 99 percent or more of the freshwater bodies would be swimmable.

I'm going to ask you to say how much (you are/your household is) willing to pay each year, if anything, to reach each of these three goals. In doing this, I want you to keep in mind:

- First, imagine that if the amount you are willing to pay is more than you are currently paying in taxes and higher prices for this purpose, your taxes would be raised to cover the cost. Of course, if the amount you are willing to pay is lower, you would receive a refund. In this way, every household in the country, including yours, has the opportunity to say how much they are willing to pay for water pollution control.

- Second, no matter what amount you give for water pollution control, you will also continue to pay for the nation's other environmental programs such as air pollution, and that air quality will remain at its present level or improve slightly.

Do you have any questions?

(IF RESPONDENT ASKS HOW MUCH HE OR SHE IS CURRENTLY PAYING): I can't give you that information at this point in the interview, because we need to know how much water pollution control is really worth to you without any reference to what you are currently paying for it. However, in order to help you understand how much you are already paying for things the government provides, the payment card gives information about how much you are paying for other types of government programs. At the end of the interview, I will be glad to give you information about your actual payments for water pollution control.

24. First, Goal C. What amount on the payment card, or any amount in between, is the most you (your household) would be willing to pay in taxes and higher prices each year to continue to keep the nation's freshwater bodies from falling below the boatable level where they are now? In other words, what is the highest amount you (your household) would be willing to pay for Goal C each year before you would feel you are spending more than it's really worth to you (all members of your household)?

    ___ ENTER DOLLAR AMOUNT HERE, ON FLAP AND ON WORKSHEET
    000 ZERO OR "NOTHING"
    998 DON'T KNOW
    999 REFUSED

25. Would it be worth anything (more) to you (your household) to achieve goal B, where 99 percent or more of the freshwater bodies are clean enough so game fish like bass can live in them?

    1 Yes → SKIP TO Q.26
    2 No
    3 DON'T KNOW
    4 REFUSED
    → SEE Q. 24; IF DOLLAR AMOUNT GIVEN ON Q.24 THEN SKIP TO Q.27. IF "ZERO", "NOTHING" GIVEN ON Q.24 AND "NO" ON Q.25 THEN SKIP TO Y1; ALL OTHERS SKIP TO Y3.

IF "ZERO", "NOTHING" TO Q.24 AND "NO" TO Q.25, ASK Q.Y1

Y1. People have different reasons for saying zero dollars or nothing. For some people that is all water pollution control is worth to them. They don't want to continue to pay anything for it as they are now in taxes and prices. Other people give different reasons for saying this. Did you say zero dollars because that is what water quality is worth to you (your household) or because of other reasons?

1 That is what it is worth to me (my household) → SKIP TO Q.37
2 Did not realize I am currently paying for it, I thought that the money I gave would be in addition to what I am paying now
3 Some other reason (Specify): ______ → SKIP TO Q.Y3a

4 DON'T KNOW
5 REFUSED
→ SKIP TO Q.37

IF "2" ON Q.Y1, ASK:

Y2. You are already paying some amount for water pollution control in your taxes and prices. It is very important to us to learn what value you place on achieving the water quality goals when you are given the chance to make the choice yourself. Would you be willing to answer these questions if I later tell you how much you are currently paying in taxes and prices and give you the chance to make any changes in your answers you would like to make?

1 Yes → GO BACK TO Q.24

2 No
3 DON'T KNOW
4 REFUSED
→ SKIP TO Q.37

IF "DON'T KNOW" OR "REFUSED" TO Q.24, AND "DON'T KNOW", OR "REFUSED" TO Q.25, ASK Q.Y3

Y3. People have different reasons for saying they don't know or can't answer these questions. I'm going to read you some reasons. Please tell me whether or not they represent your feelings about this question.

Y3a. Did you give this answer because you are (your household is) paying too much in taxes already and don't want to spend more?

1 Yes → SKIP TO Q.Y4

2 No
3 DON'T KNOW
4 REFUSED
→ SKIP TO Q.Y5

IF "YES" ON Q.Y3a, ASK:

Y4. I'd like to remind you that you are (your household is), already paying some amount for water pollution control in your taxes and prices. It is very important to us to learn what value you place on achieving the water quality goals when you are given the chance to make the choice yourself. Would you be willing to answer these questions if I later tell you how much you are (your household is) currently paying in taxes and prices and give you the chance to make any changes in your answers you would like to make?

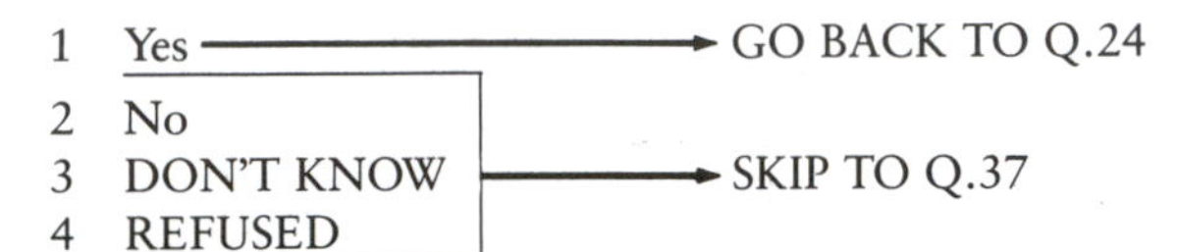

IF "NO", "DON'T KNOW" OR "REFUSED" ON Q.Y3a, ASK:

Y5. Did you give this answer because you think the government should be able to meet this goal with the money they have or because you think the government wastes too much money? (CIRCLE ALL THAT APPLY.)

1 Yes, government should be able to meet goal with the money they have
2 Yes, government wastes too much money
3 No
4 DON'T KNOW → SKIP TO Q.Y7
5 REFUSED

IF "YES," 1 OR 2 ON Q.Y5, ASK:

Y6. It is very important to us to learn what value you (your household) place on achieving the water quality goals when you are given the chance to make the choice yourself. This value is the highest amount you are (your household is) willing to pay for an efficient and worthwhile program to reach each of the water quality goals. Would you be willing to answer these questions if I noted here that the amounts you give are based on the assumption that the water pollution programs would be efficient and well run?

1 Yes → GO BACK TO Q.24
2 No
3 DON'T KNOW → SKIP TO Q.37
4 REFUSED

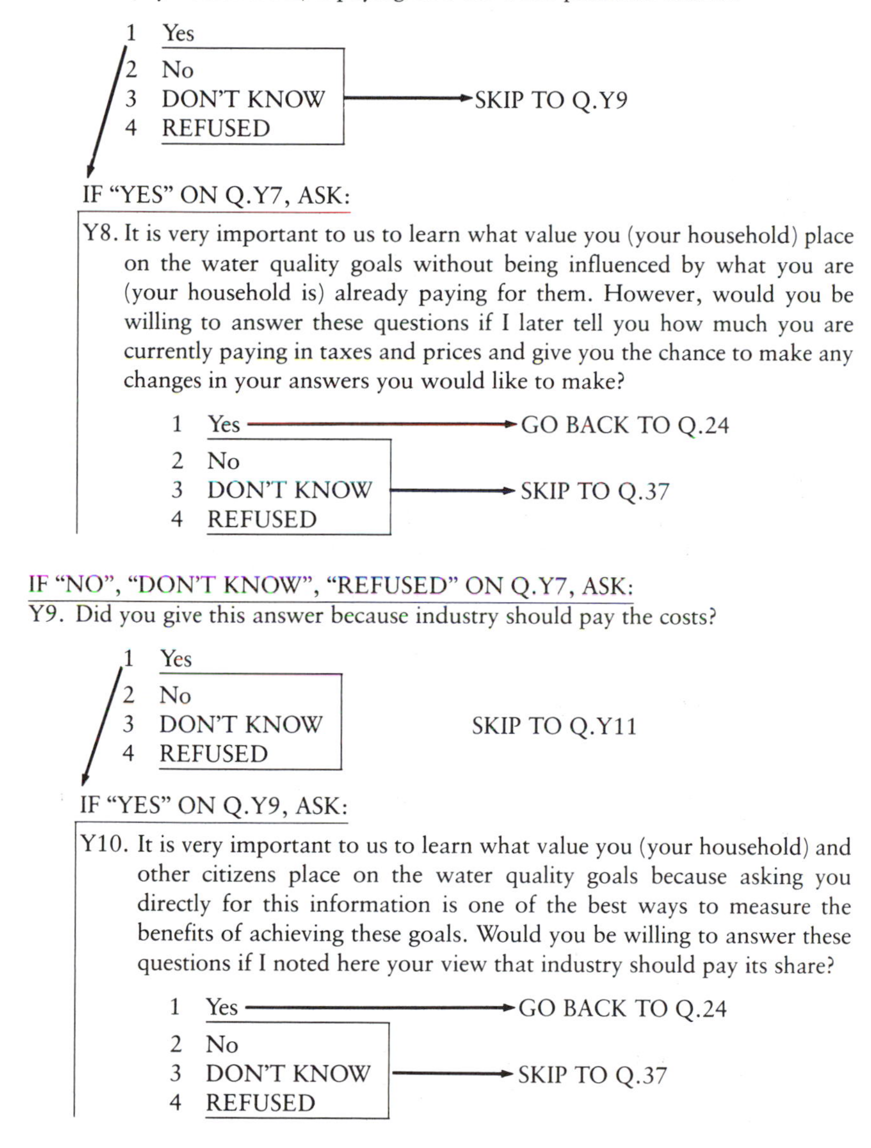

IF "NO", "DON'T KNOW", "REFUSED" ON Q.Y5, ASK:

Y7. Did you give this answer because it is too hard to say without knowing what I am (my household is) is paying now for water pollution control?

1 Yes
2 No
3 DON'T KNOW → SKIP TO Q.Y9
4 REFUSED

IF "YES" ON Q.Y7, ASK:

Y8. It is very important to us to learn what value you (your household) place on the water quality goals without being influenced by what you are (your household is) already paying for them. However, would you be willing to answer these questions if I later tell you how much you are currently paying in taxes and prices and give you the chance to make any changes in your answers you would like to make?

1 Yes → GO BACK TO Q.24
2 No
3 DON'T KNOW → SKIP TO Q.37
4 REFUSED

IF "NO", "DON'T KNOW", "REFUSED" ON Q.Y7, ASK:

Y9. Did you give this answer because industry should pay the costs?

1 Yes
2 No
3 DON'T KNOW SKIP TO Q.Y11
4 REFUSED

IF "YES" ON Q.Y9, ASK:

Y10. It is very important to us to learn what value you (your household) and other citizens place on the water quality goals because asking you directly for this information is one of the best ways to measure the benefits of achieving these goals. Would you be willing to answer these questions if I noted here your view that industry should pay its share?

1 Yes → GO BACK TO Q.24
2 No
3 DON'T KNOW → SKIP TO Q.37
4 REFUSED

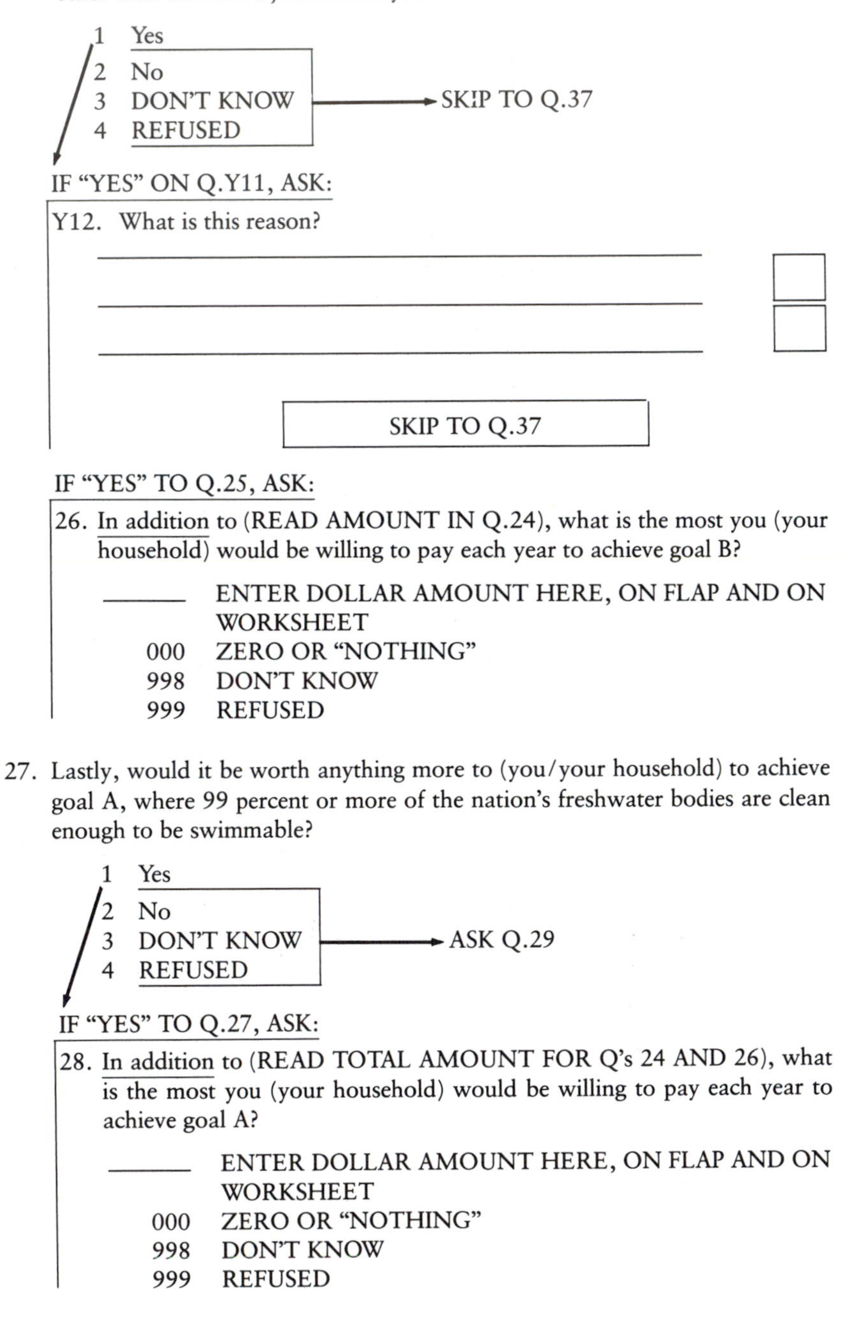
IF "NO", "DON'T KNOW", "REFUSED", ON Q.Y9, ASK:

Y11. Is there a reason why you gave this answer (ANSWER TO Q.24 AND Q.25) other than the ones I just read to you.

1 Yes
2 No
3 DON'T KNOW
4 REFUSED

2–4 → SKIP TO Q.37

IF "YES" ON Q.Y11, ASK:

Y12. What is this reason?

______

______

______

SKIP TO Q.37

IF "YES" TO Q.25, ASK:

26. In addition to (READ AMOUNT IN Q.24), what is the most you (your household) would be willing to pay each year to achieve goal B?

______ ENTER DOLLAR AMOUNT HERE, ON FLAP AND ON WORKSHEET
000 ZERO OR "NOTHING"
998 DON'T KNOW
999 REFUSED

27. Lastly, would it be worth anything more to (you/your household) to achieve goal A, where 99 percent or more of the nation's freshwater bodies are clean enough to be swimmable?

1 Yes
2 No
3 DON'T KNOW
4 REFUSED

2–4 → ASK Q.29

IF "YES" TO Q.27, ASK:

28. In addition to (READ TOTAL AMOUNT FOR Q's 24 AND 26), what is the most you (your household) would be willing to pay each year to achieve goal A?

______ ENTER DOLLAR AMOUNT HERE, ON FLAP AND ON WORKSHEET
000 ZERO OR "NOTHING"
998 DON'T KNOW
999 REFUSED

INTERVIEWER: IF RESPONDENT VOLUNTEERS AT ANY POINT UP TO NOW THAT HE/SHE WANTS TO CHANGE AN ANSWER PLEASE GO BACK AND DO SO. JUST MAKE SURE THE ANSWERS ARE CHANGED ON THE QUESTIONNAIRE, THE FLAP AND THE WORKSHEET.

29. ADD UP THE AMOUNTS THE RESPONDENT GAVE FOR Q.24, 26 AND 28 AND ENTER THE AMOUNT ON FLAP AND ON WORKSHEET.

At this point in the interview, I want to review what you have just said and give you the chance to make adjustments and changes. We often find when we ask questions like these that people don't realize that we are going to ask them about three different goals until after we have asked all the questions. Looking at the worksheet, you said you were willing to pay $_______ for goal C, $_______ more for goal B and $_______ more for goal A. This gives $_______ total dollars as the <u>maximum</u> annual amount (you/your household) would be willing to pay to reach the nation's water quality goals. If you would like to make <u>any</u> changes, please don't hesitate to do so. We want to get your best judgment about how much each of these goals is worth to your household. There are no right or wrong answers. Would you like to shift any amounts around or raise or lower the total amount?

1 Yes, make changes → HELP RESPONDENT CHANGE AMOUNTS ON QUESTIONNAIRE AND ON WORKSHEET INCLUDING TOTAL. RECORD NEW AMOUNTS ON FLAP UNDER COLUMN HEADED Q.29.
2 No
3 DON'T KNOW
4 REFUSED

<u>VERSION A</u>

INTERVIEWER NOTE: THE DOLLAR VALUES TO BE INSERTED IN QUESTIONS 30, 31, AND 32 ARE THE FINAL DOLLAR VALUES GIVEN BY THE RESPONDENT UP TO THIS POINT. THEREFORE, IF RESPONDENT CHANGED DOLLAR AMOUNTS ON QUESTION 29, USE THOSE FIGURES WHEN ASKING QUESTIONS 30, 31, AND 32.

30. You said you were willing to pay (READ TOTAL AMOUNT ON WORKSHEET OF Q.24 AND Q.26) to achieve the goal of a fishable level of water quality and (READ AMOUNT ON WORKSHEET AT Q.28) for a further improvement to swimmable.

Would you still be willing to pay (READ AMOUNT AT Q.28) if the best we could do was to raise the minimum only halfway from fishable to swimmable? (POINT TO MIDWAY BETWEEN LEVELS B AND A ON WORKSHEET.) At halfway, more water bodies would be improved over the fishable level, and some additional, but not all, water bodies would even be improved to the swimmable level.

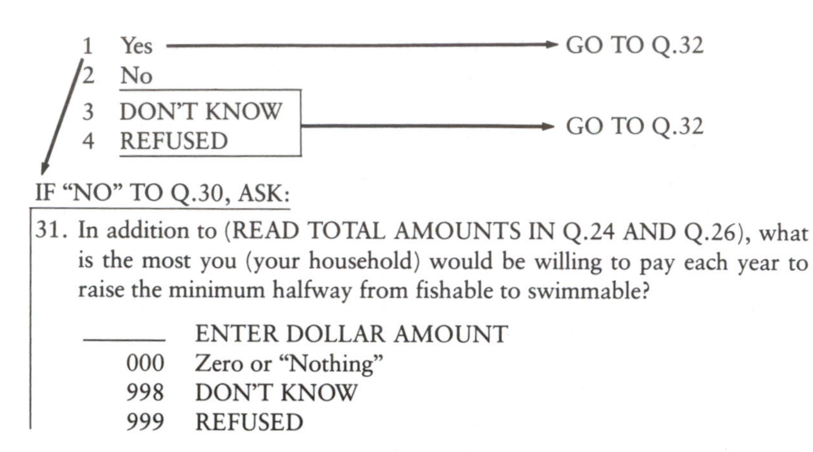

1 Yes → GO TO Q.32
2 No
3 DON'T KNOW
4 REFUSED → GO TO Q.32

IF "NO" TO Q.30, ASK:

31. In addition to (READ TOTAL AMOUNTS IN Q.24 AND Q.26), what is the most you (your household) would be willing to pay each year to raise the minimum halfway from fishable to swimmable?

______	ENTER DOLLAR AMOUNT
000	Zero or "Nothing"
998	DON'T KNOW
999	REFUSED

IF ANY DOLLAR AMOUNTS IN Q.24, 26, OR 28, ASK:

32. You said that you (your household) would be willing to pay a total of (TOTAL AMOUNT FOR Q.24, 26, 28) to reach the nation's water quality goals. Presuming that people in other states would also divide their money honestly, how many dollars or what percent of this amount would you give to (THIS STATE) and how many dollars or what percent to the rest of the nation for water improvement?

			DON'T KNOW	REFUSED
THIS STATE	$_____	_____%	9998	9999
REST OF NATION	$_____	_____%	9998	9999

VERSION B

INTERVIEWER NOTE: THE DOLLAR VALUES TO BE INSERTED IN QUESTIONS 30, 31, AND 32 ARE THE FINAL DOLLAR VALUES GIVEN BY THE RESPONDENT UP TO THIS POINT. THEREFORE, IF RESPONDENT CHANGED DOLLAR AMOUNTS ON QUESTION 29, USE THOSE FIGURES WHEN ASKING QUESTIONS 30, 31, AND 32.

30. Now I would like to ask you about a slightly different situation. Please turn to Card 6a. You said you were willing to pay (READ TOTAL AMOUNT OF Q.24 AND Q.26 ON WORKSHEET) $_______ to achieve the goal of having 99% or virtually all of the nation's water be at least at the fishable level. If that were not possible, would you still be willing to pay (READ AMOUNT AT Q.28) to have five percent of the nation's water bodies remain at the boatable level while the other 95% improve to a fishable quality? (POINT TO PLACE ON LADDER WHERE THE 99% IS MARKED OUT AND 95% SUBSTITUTED.) The lakes, rivers and streams comprising this five percent would all be located in heavily industrial and/or urban locations where a lot of people live.

1 Yes, worth the same amount → GO TO Q.32
2 No, worth less
3 DON'T KNOW
4 REFUSED → GO TO Q.32

IF "NO" TO Q.30, ASK:

31. How much less would it be worth each year to (you/your household)?

______	ENTER DOLLAR AMOUNT
998	DON'T KNOW
999	REFUSED

IF ANY DOLLAR AMOUNTS IN Q.24, 26, OR 28, ASK:

32. You said that you (your household) would be willing to pay a total of (TOTAL AMOUNT FOR Q.24, 26, 28) to reach the nation's water quality goals. Presuming that people in other states would also divide their money honestly, how many dollars or what percent of this amount would you give to (THIS STATE) and how many dollars or what percent to the rest of the nation for water improvement?

			DON'T KNOW	REFUSED
THIS STATE	$_____	_____%	9998	9999
REST OF NATION	$_____	_____%	9998	9999

VERSION A

Please look at the water quality ladder again (Card 3). A major purpose of this survey is to learn the value people place on reaching the three national water pollution goals. Because so many people find it hard to say just how much these goals are worth to them in dollars, they sometimes ask us to tell them how much they are currently paying for water pollution control. We don't provide this information early in the interview because we want people to think about how much the goals are really worth to them without being influenced by information such as this.

Now that you have had a chance to think about this, we would like to tell you the dollar range paid for water pollution control by households in your income bracket and offer you the chance to revise your dollar amounts for water pollution, if you should wish to do so for any reason.

Before doing this you need to know two things. First, the actual amount people pay varies according to the size of their household and other factors.

Second, it is uncertain whether paying this amount of money each year will provide enough money to reach any of the goals higher than boatable.

GIVE RESPONDENT APPROPRIATE CARD A9 FOR HIS/HER INCOME. Last year, households like yours paid between (READ RANGE FROM BELOW FOR RESPONDENT'S INCOME GROUP) for the nation's water pollution control programs.

INCOME GROUP	COLOR CARD	WATER POLLUTION AMOUNT
UNDER $10,000	WHITE	$10 to $100
$10,000–$19,999	YELLOW	$70 to $150
$20,000–$29,999	BLUE	$175 to $300
$30,000–$49,999	GREEN	$400 to $600
$50,000 OR MORE	PINK	$1,200 to $1,500

POINT TO WORKSHEET.

33. Here are the amounts you said you would be willing to pay for the three goals. Please feel free to change any of these amounts, up or down. Remember, what we want is your realistic estimate of the highest amount of money each of these goals is worth to you whether or not you are currently paying that amount. Would you like to make any changes? (PAUSE; IF RESPONDENT APPEARS HESITANT, ENCOURAGE RESPONDENT BY REPEATING RELEVANT PARTS OF THE QUESTION.)

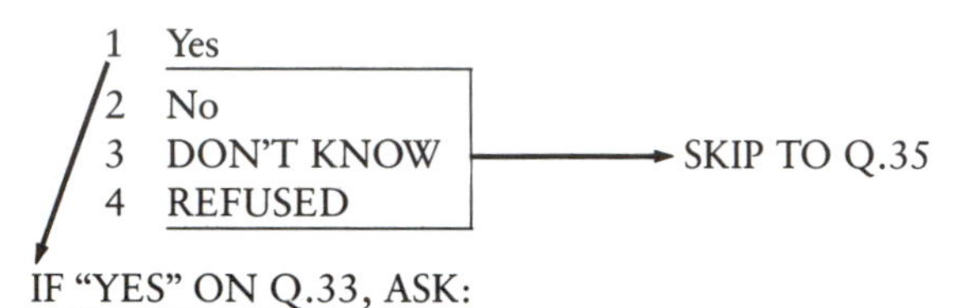

IF "YES" ON Q.33, ASK:

34. What are the new amounts? (HELP RESPONDENT CHANGE THE AMOUNTS ON THE WORKSHEET INCLUDING TOTAL. RECORD THE NEW AMOUNTS ON FLAP.)

VERSION B

Please look at the water quality ladder again (Card 3). A major purpose of this survey is to learn the value people place on reaching the three national water pollution goals. Because many people find it hard to say just how much these goals are worth to them in dollars, they sometimes ask us to tell them how much they are currently paying for water pollution control. We don't provide this information early in the interview because we want people to think about how much the goals are really worth to them without being influenced by information such as this.

Now that you have had a chance to think about this, we would like to tell you the dollar range paid for both water and air pollution control by households in your income bracket and offer you the chance to revise your dollar amounts for water pollution, if you should wish to do so for any reason.

Before doing this you need to know two things. First, the actual amount people pay varies according to the size of their household and other factors.

Second, it is uncertain whether paying this amount of money each year will provide enough money to reach any of the goals higher than boatable.

GIVE RESPONDENT APPROPRIATE CARD B9 FOR HIS/HER INCOME. Last year, households like yours paid between (READ RANGE FROM BELOW FOR RESPONDENT'S INCOME GROUP) for the nation's water pollution control programs. In addition, last year you also paid between (READ RANGE FROM BELOW FOR RESPONDENT'S INCOME GROUP) in higher prices and taxes for air pollution control programs for the entire country, including this state. This amount of money will be enough to maintain present air quality in the country or perhaps slightly improve it.

INCOME GROUP	COLOR CARD	WATER POLLUTION		AIR POLLUTION
UNDER $10,000	WHITE	$10 to $100	+	$15 to $150
$10,000–$19,999	YELLOW	$70 to $150	+	$100 to $195
$20,000–$29,999	BLUE	$175 to $300	+	$265 to $420
$30,000–$49,999	GREEN	$400 to $600	+	$650 to $850
$50,000 OR MORE	PINK	$1200 to $1500	+	$1775 to $2200

POINT TO WORKSHEET.

33. Here are the amounts you said you would be willing to pay for the three goals. Please feel free to change any of the amounts you gave for the three water quality goals, up or down. Remember, what we want is your realistic estimate of the highest amount of money each of these water quality goals is worth to you whether or not you are currently paying that amount. Would you like to make any changes? (PAUSE; IF RESPONDENT APPEARS HESITANT, ENCOURAGE RESPONDENT BY REPEATING RELEVANT PARTS OF THE QUESTION.)

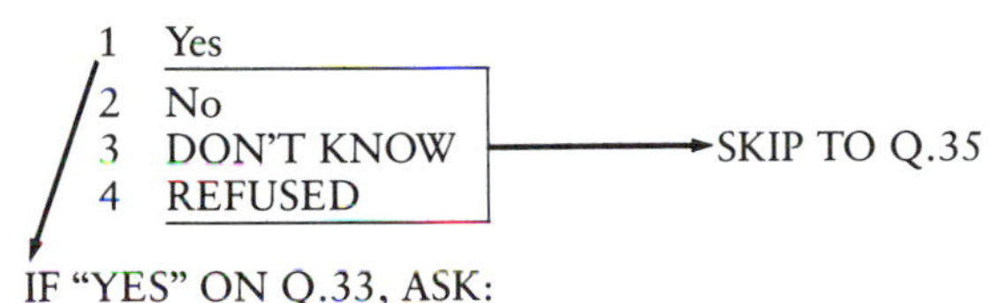

IF "YES" ON Q.33, ASK:

34. What are the new amounts? (HELP RESPONDENT CHANGE THE AMOUNTS ON THE WORKSHEET INCLUDING TOTAL. RECORD THE NEW AMOUNTS ON FLAP.)

ASK EVERYONE:

35. One last question about the amounts you gave on the worksheet. What if the amounts you gave here were not enough to reach any of these three goals, including goal C, the boatable level where we are now. Would you (your household) be willing to pay anything more to try to reach any or all of these goals or are these amounts the most you (your household) would realistically give to reach each of them? (PAUSE, IF RESPONDENT APPEARS HESITANT ENCOURAGE RESPONDENT BY REPEATING RELEVANT PARTS OF THE QUESTION.)

1 Yes, willing to pay more
2 No, not willing to pay more
3 DON'T KNOW
4 REFUSED
→ SKIP TO Q.37 (for 2–4)

IF "YES" ON Q.35, ASK:

36. What is the most you (your household) would pay each year to reach each of goals C, B, and A before you feel you are spending more than it's really worth to you (all members of your household)? (HELP RESPONDENT CHANGE THE AMOUNTS ON THE WORKSHEET INCLUDING TOTAL. RECORD THE NEW AMOUNTS ON FLAP.)

SECTION E: BACKGROUND INFORMATION

This last section asks a few questions about you.

37. What was the last grade of regular school that you completed? Do not include specialized schools like secretarial, art, or trade schools.

    1 Grade school or less (0–8)
    2 Some high school (9–11)
    3 High school graduate (12)
    4 Some college or junior college
    5 College graduate (4 or 5 year degree)
    6 Post graduate work or degree
    7 DON'T KNOW
    8 REFUSED

38. How many years have you lived in THIS STATE? (PROBE: Your best estimate will do. IF LESS THAN 1, ENTER 1.)

    ____ Number of Years
    98 DON'T KNOW
    99 REFUSED

39. ASK ONLY IF NOT OBVIOUS: How would you describe your racial or ethnic background? READ CHOICES.

    1 White
    2 Black
    3 Hispanic
    4 Asian or Pacific Islander
    5 Or some other race (SPECIFY)
    6 DON'T KNOW
    7 REFUSED

    INTERVIEWER NOTE:
    White & Black = Black
    White & Hispanic = Hispanic
    Black & Hispanic = Hispanic

40. Please turn to the last card in the book—Card 7. For classification purposes only, please tell me which category best describes the total income that you (and all other members of this household) earned during 1982 before taxes. Please be sure to include each member's wages and salaries, as well as net income from any business, pensions, dividends, interest, tips, or other income. Just tell me the number that best describes your household's income.

A	1	UNDER $5,000
B	2	$5,000 to less than $10,000
C	3	$10,000 to less than $15,000
D	4	$15,000 to less than $20,000
E	5	$20,000 to less than $25,000
F	6	$25,000 to less than $30,000
G	7	$30,000 to less than $35,000
H	8	$35,000 to less than $40,000
I	9	$40,000 to less than $45,000
J	10	$45,000 to less than $50,000
K	11	$50,000 to less than $100,000
L	12	$100,000 and over
	13	DON'T KNOW
	14	REFUSED

IF THIS IS A RESPONDENT-ONLY HOUSEHOLD, SKIP TO Q.42

41. How much of this total household income is income that you personally make? Is your share 75% or less of the total household income or is your share more than 75% of the total household income?

    1 75% (3/4) or less
    2 More than 75%
    3 DON'T KNOW
    4 REFUSED

ASK EVERYONE:

42. I would like you to think back to the questions I asked you about how much your household is willing to pay to reach each of the three water quality goals, C, B, and A. We find that some people are more sure than others about the amounts they gave for Goals C, B, and A. How about yourself? Would you say you are very sure, somewhat sure, somewhat unsure or very unsure about the amount you gave for these goals?

    1 Very sure
    2 Somewhat sure
    3 Somewhat unsure
    4 Very unsure
    5 DON'T KNOW
    6 REFUSED

CLOSING: Thank you for your time and cooperation.

SECTION F: INTERVIEWER'S EVALUATION

INTERVIEWER: COMPLETE THESE QUESTIONS AS SOON AS POSSIBLE AFTER THE INTERVIEW.

These two questions are only concerned with how the respondent answered Questions 24–29, which asked the respondent to value the three levels of water quality.

43. Irrespective of whether or not the respondent answered Q.24–29, in your judgment, how well did the respondent understand what he or she was asked to do in these questions?

    1 Understood completely
    2 Understood a great deal
    3 Understood somewhat
    4 Understood a little
    5 Did not understand very much
    6 Did not understand at all
    7 Other (SPECIFY):

    ____
    ____
    ____

44. Which of the following descriptions best describe the degree of effort the respondent made to arrive at a value for the three levels of water quality?

    1 Gave the questions prolonged consideration in an effort to arrive at the best possible value
    2 Gave the questions careful consideration, but the effort was not prolonged
    3 Gave the questions some consideration
    4 Gave the questions very little consideration
    5 Other (SPECIFY):

    ____
    ____
    ____

WATER BENEFITS SURVEY
EXHIBIT BOOKLET

CARD 1

STATEMENTS REGARDING POLLUTION CONTROL

1—Protecting the environment is so important that pollution control requirements and standards cannot be too strict, and continuing improvement must be made regardless of cost,

OR

2—We have made enough progress on cleaning up the environment that we should now concentrate on holding down costs rather than requiring stricter controls,

OR

3—Pollution control requirements and standards have gone too far, and they already cost more than they are worth.

CARD 2

SOURCES OF WATER POLLUTION

1 RUNOFF FROM AGRICULTURE

2 SEWAGE FROM CITIES AND TOWNS

3 DRAINAGE FROM MINES

4 RUNOFF FROM ROADS AND HIGHWAYS

5 SEEPAGE FROM GARBAGE DUMPS

6 DUMPING OF FACTORY WASTE INTO WATERBODIES

CARD 3

WATER QUALITY LADDER

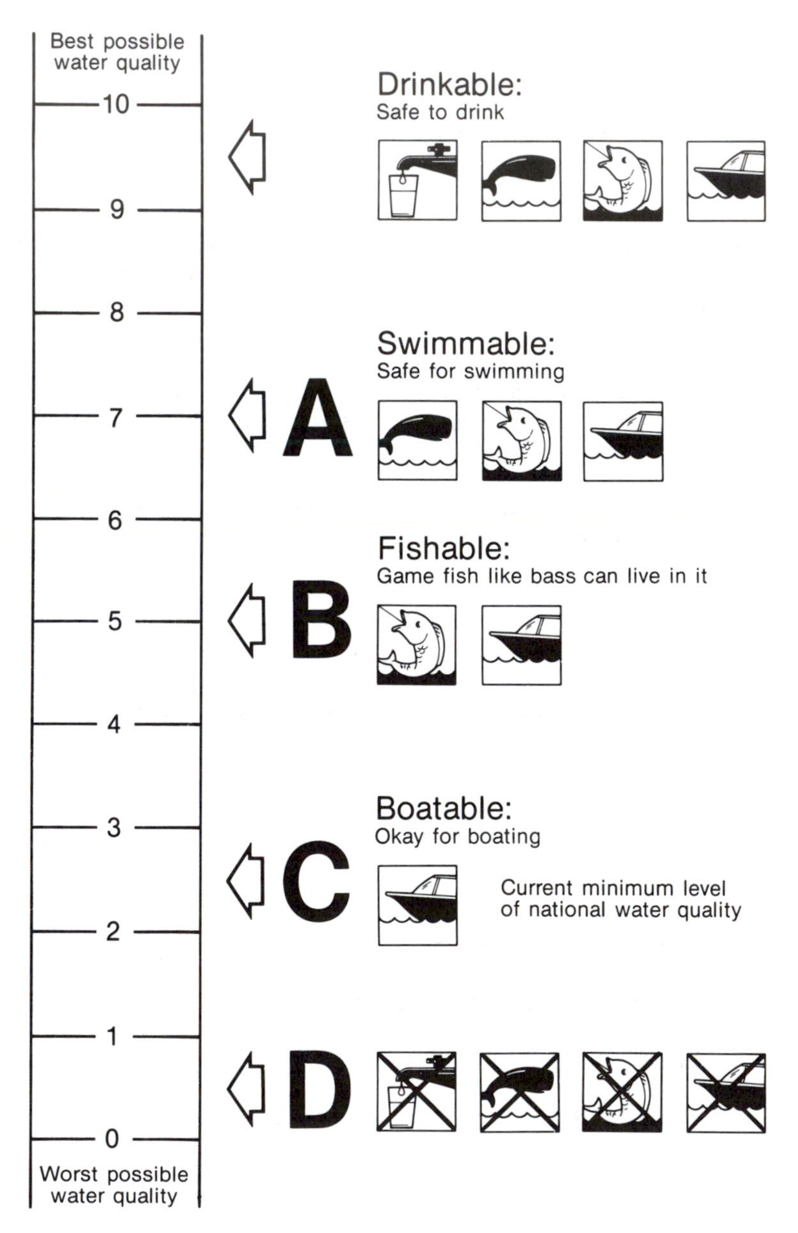

CARD 4

WHY MY HOUSEHOLD MIGHT VALUE NATIONAL FRESHWATER QUALITY

1. I (MY HOUSEHOLD) USE FRESHWATER FOR:

   FISHING
   BOATING, OR
   SWIMMING

2. I (MY HOUSEHOLD) USE AREAS SURROUNDING FRESHWATER FOR:

   PICNICKING
   BIRDWATCHING, OR
   STAYING IN A VACATION COTTAGE

3. I (MY HOUSEHOLD) GET SATISFACTION FROM KNOWING OTHER PEOPLE MAY USE AND ENJOY FRESHWATER

4. I (MY HOUSEHOLD) GET SATISFACTION FROM KNOWING THAT THE NATION'S WATER IS CLEANER

CARD 5

1982 HOUSEHOLD INCOME BEFORE TAXES

A UNDER $10,000
B $10,000–$19,999
C $20,000–$29,999
D $30,000–$49,999
E $50,000 AND OVER

CARD 6

EVERYONE PAYS FOR WATER POLLUTION CONTROL THROUGH:

1 A PORTION OF YOUR TAXES

LOCAL SEWER/WATER TAXES STATE FEDERAL	To build, maintain, and run community sewage plants, conduct research, enforce water pollution laws, etc.

2 A PORTION OF THE PRICES YOU PAY

ON PRODUCTS SOLD TO CONSUMERS BY COMPANIES	To build, maintain, and run waste disposal plants the government requires industries to install in order to meet water pollution standards.

CARD 6A

WATER QUALITY LADDER

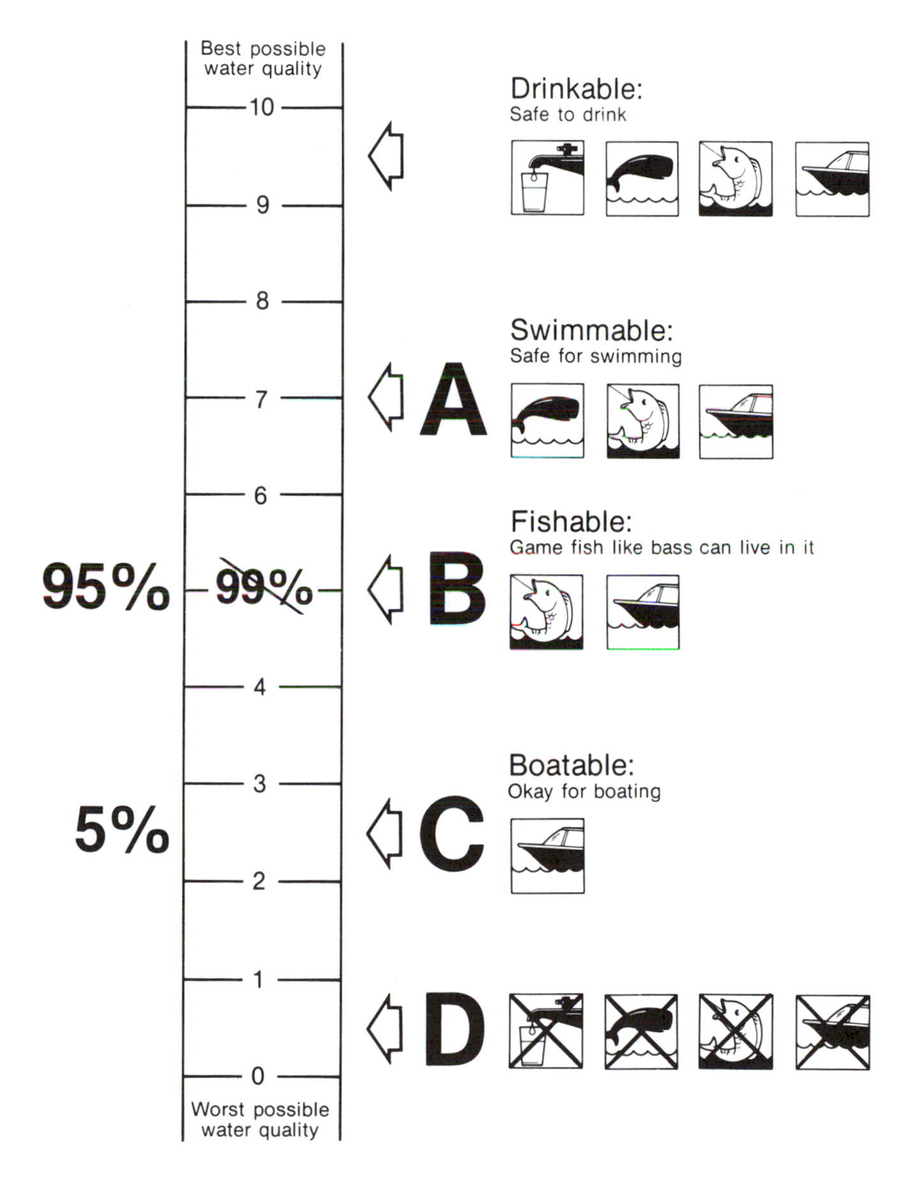

PAYMENT CARD

ANNUAL HOUSEHOLD INCOME BEFORE TAXES
UNDER $10,000

(AVERAGE ANNUAL AMOUNT IN 1982 TAXES AND PRICES
PAID FOR SOME PUBLIC PROGRAMS)

$ 0		$ 45	—POLICE	$120		$270	
1		50	AND FIRE	130		280	
2		55	PROTECTION	140		290	
3		60		150		300	
4		65		160		320	
5		70		170		340	
10	—SPACE	75		180		360	
15	PROGRAM	80		190		380	
20		85		200		400	—DEFENSE
25		90		220		420	PROGRAM
30		95		240	—PUBLIC	440	
35		100	—ROADS AND	250	EDUCATION	460	
40		110	HIGHWAYS	260		480	

PAYMENT CARD

ANNUAL HOUSEHOLD INCOME BEFORE TAXES
$10,000–$19,999

(AVERAGE ANNUAL AMOUNT IN 1982 TAXES AND PRICES
PAID FOR SOME PUBLIC PROGRAMS)

$ 0		$ 90	—POLICE AND FIRE PROTECTION	$295		$550	
5		100		310		565	
10		110		325		580	
15		120		340		595	
20		130		355		615	
25	—SPACE PROGRAM	140		370		635	
30		150		385		655	
35		160		400		675	
40		170		415		695	
45		180	—ROADS AND HIGHWAYS	430		715	
50		190		445		735	
55		205		460		755	
60		220		475		775	
65		235		490	—PUBLIC EDUCATION	795	
70		250		505		815	—DEFENSE PROGRAM
75		265		520		835	
80		280		535		855	

PAYMENT CARD

ANNUAL HOUSEHOLD INCOME BEFORE TAXES
$20,000–$29,999

(AVERAGE ANNUAL AMOUNT IN 1982 TAXES AND PRICES
PAID FOR SOME PUBLIC PROGRAMS)

$ 0		$190		$ 620		$1140	
10		210	—POLICE AND FIRE PROTECTION	650		1180	
20		230		680		1220	
30		250		710		1260	
40		270		740		1300	
50		290		770		1340	
60	—SPACE PROGRAM	310		800		1380	
70		330		830		1420	
80		350		860		1460	
90		380	—ROADS AND HIGHWAYS	890		1500	
100		410		920	—PUBLIC EDUCATION	1540	
110		440		950		1580	
120		470		980		1620	
130		500		1010		1660	
140		530		1040		1700	
150		560		1070		1740	—DEFENSE PROGRAM
170		590		1100		1780	

PAYMENT CARD

ANNUAL HOUSEHOLD INCOME BEFORE TAXES
$30,000–$49,999

(AVERAGE ANNUAL AMOUNT IN 1982 TAXES AND PRICES PAID FOR SOME PUBLIC PROGRAMS)

$ 0		$ 450		$1445		$2720	
15		480	—POLICE	1520		2805	
30		510	AND FIRE	1595		2890	
45		540	PROTECTION	1670		2975	
60		570		1745		3060	
90		600		1820		3145	
120	—SPACE PROGRAM	630		1895		3230	
150		695		1970		3315	
180		770	—ROADS AND HIGHWAYS	2045		3400	
210		845		2120		3485	
240		920		2195		3570	
270		995		2270		3655	
300		1070		2345		3740	
330		1145		2420		3825	
360		1220		2495	—PUBLIC EDUCATION	3910	
390		1295		2570		3995	
420		1370		2645		4080	—DEFENSE PROGRAM

PAYMENT CARD

ANNUAL HOUSEHOLD INCOME BEFORE TAXES
$50,000 AND OVER

(AVERAGE ANNUAL AMOUNT IN 1982 TAXES AND PRICES
PAID FOR SOME PUBLIC PROGRAMS)

$ 0		$1150		$3860		$ 7410	
25		1250	—POLICE AND FIRE PROTECTION	4060		7660	
50		1350		4260		7910	
75		1450		4460		8160	
100		1550		4660		8410	
150		1660		4860		8660	
200		1760	—ROADS AND HIGHWAYS	5060		8910	
250		1860		5260		9160	
300	—SPACE PROGRAM	2060		5460		9410	
450		2260		5660		9660	
450		2460		5860		9910	
550		2660		6060		10160	
650		2860		6260		10410	
750		3060		6460	—PUBLIC EDUCATION	10660	
850		3260		6660		10910	—DEFENSE PROGRAM
950		3460		6910		11160	
1050		3660		7160		11410	

CARD A9

Annual Household Income Before Taxes

Under $10,000

AMOUNT ACTUALLY PAID IN 1982 FOR WATER QUALITY PROGRAMS

In 1982, households in your income group paid the following amount in local, state and federal taxes and in higher prices for:

All Water Pollution Control Programs   Between $10 and $100

It is uncertain whether annual payments at this level will be enough to reach the fishable and swimmable water quality goals.

CARD A9

Annual Household Income Before Taxes

$10,000–$19,999

AMOUNT ACTUALLY PAID IN 1982 FOR WATER QUALITY PROGRAMS

In 1982, households in your income group paid the following amount in local, state and federal taxes and in higher prices for:

All Water Pollution Control Programs   Between $70 and $150

It is uncertain whether annual payments at this level will be enough to reach the fishable and swimmable water quality goals.

CARD A9

Annual Household Income Before Taxes

$20,000–$29,999

AMOUNT ACTUALLY PAID IN 1982 FOR WATER QUALITY PROGRAMS

In 1982, households in your income group paid the following amount in local, state and federal taxes and in higher prices for:

All Water Pollution Control Programs   Between $175 and $300

It is uncertain whether annual payments at this level will be enough to reach the fishable and swimmable water quality goals.

CARD A9

Annual Household Income Before Taxes

$30,000–$49,999

AMOUNT ACTUALLY PAID IN 1982 FOR WATER QUALITY PROGRAMS

In 1982, households in your income group paid the following amount in local, state and federal taxes and in higher prices for:

All Water Pollution Control Programs   Between $400 and $600

It is uncertain whether annual payments at this level will be enough to reach the fishable and swimmable water quality goals.

CARD A9

Annual Household Income Before Taxes

$50,000 AND OVER

AMOUNT ACTUALLY PAID IN 1982 FOR WATER QUALITY PROGRAMS

In 1982, households in your income group paid the following amount in local, state and federal taxes and in higher prices for:

All Water Pollution Control Programs Between $1,200 and $1,500

It is uncertain whether annual payments at this level will be enough to reach the fishable and swimmable water quality goals.

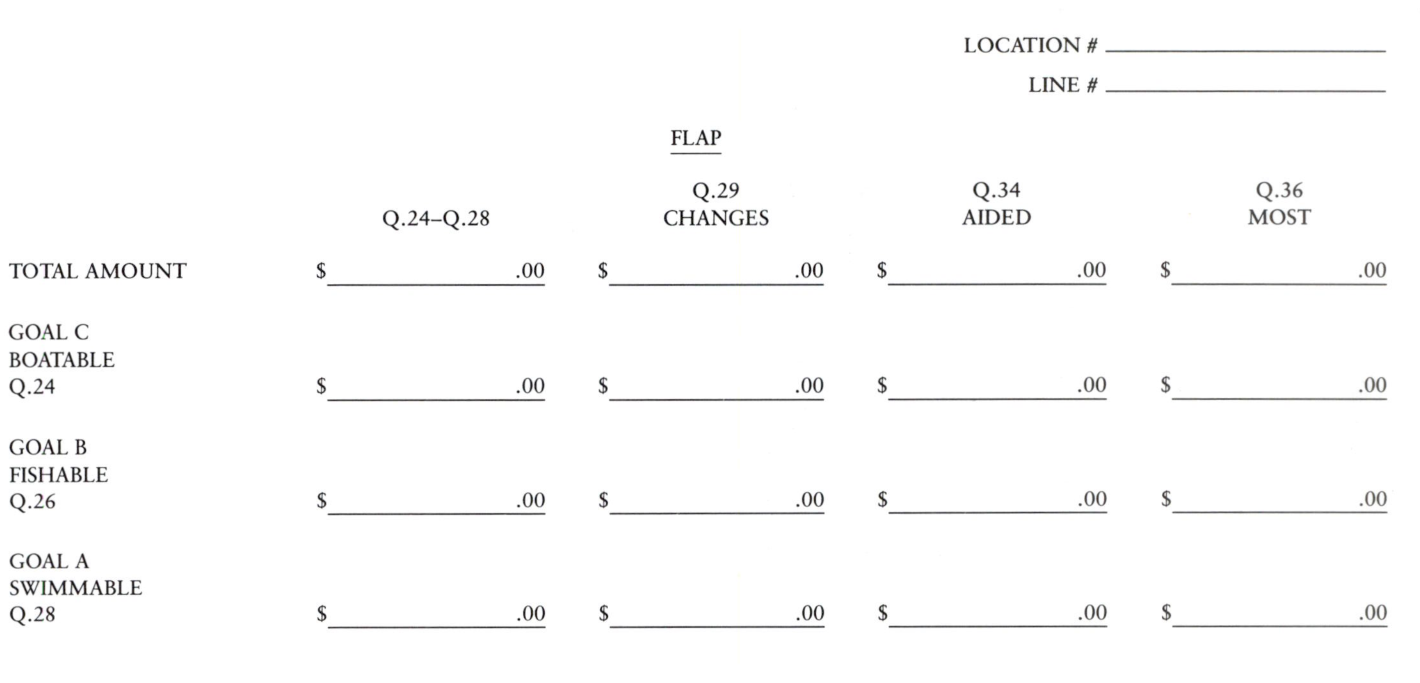

LOCATION # ____________

LINE # ____________

<u>FLAP</u>

	Q.24–Q.28	Q.29 CHANGES	Q.34 AIDED	Q.36 MOST
TOTAL AMOUNT	$ ______ .00	$ ______ .00	$ ______ .00	$ ______ .00
GOAL C BOATABLE Q.24	$ ______ .00	$ ______ .00	$ ______ .00	$ ______ .00
GOAL B FISHABLE Q.26	$ ______ .00	$ ______ .00	$ ______ .00	$ ______ .00
GOAL A SWIMMABLE Q.28	$ ______ .00	$ ______ .00	$ ______ .00	$ ______ .00

INTERVIEWER: THIS FLAP <u>MUST</u> BE ATTACHED TO THE <u>FRONT</u> OF EACH QUESTIONNAIRE!!!

Appendix C

Hypothesis Testing and Experimental Design in Contingent Valuation Surveys

Introduction

Researchers have conducted a large number of experiments to test whether various features of contingent valuation surveys tend to influence the survey results.[1] These experiments typically vary elements of the scenario for equivalent subsamples of respondents, whose aggregate WTP amounts are then assessed for differences.[2] Among the factors tested in the experimental literature are the effects of different types of WTP elicitation methods (the bidding game versus payment cards, for instance), different starting points in bidding games, variations in payment cards, different types of payment vehicles, willingness to pay versus willingness to accept (compensation) formats, different budget constraints, and variations in the information provided to respondents in the survey instrument. Less frequently, CV researchers have been interested in a comparison of the percentage of respondents who are willing to pay different fixed amounts in a discrete-choice format, or in expected willingness to pay as a function of arbitrary starting points.[3] This appen-

[1] Experiments within surveys have frequently been used by survey researchers (for example, Schuman and Presser, 1981). Fienberg and Tanur (1985) examine the similarities of the theories of experimental design and survey sampling. Examples of experiments conducted within the context of contingent valuation surveys include Randall, Ives, and Eastman (1974), Brookshire, Ives, and Schulze (1976), Bishop and Heberlein (1979), Rowe, d'Arge, and Brookshire (1980), Brookshire, Randall, and Stoll (1980), Brookshire and coauthors (1981), Greenley, Walsh, and Young (1981), Mitchell and Carson (1981), Randall, Hoehn, and Tolley (1981), Thayer (1981), Schulze and Brookshire (1983), Desvousges, Smith, and McGivney (1983), Smith, Desvousges, and Freeman (1985), Boyle, Bishop, and Welsh (1985), Roberts, Thompson, and Pawlyk (1985), and Carson and Mitchell (1986). For recent overviews, see Schulze, d'Arge, and Brookshire (1981), Rowe and Chestnut (1983), Cummings and coauthors (1986), and this book.

[2] We will assume throughout this appendix that individuals are randomly assigned to different treatments. This is important and should not be neglected, as it has been in some CV experiments.

[3] Occasionally the hypothesis for an experiment concerns some aspect of a CV survey other than the respondents' actual willingness to pay, such as differences in response rates under different treatments.

dix can be viewed as an extension of chapter 11, which provides a catalogue of potentially testable propositions about the behavior of respondents.

We will first show that the designs of many of the contingent valuation experiments reported in the literature are flawed because their small sample sizes provide insufficient power to reject the null hypothesis that there is no difference in treatment effects. Where sample sizes are too small, a finding of no difference—which is the most common finding in such CV experiments—is relatively meaningless because any difference, to be statistically significant, would have to be very large.[4] In order to assist those who wish to evaluate the power of specific past or proposed experiments, we provide tables that conveniently summarize the relationship between sample size and the size of differences to be detected with a specific level of power and significance. We then turn to a consideration of how the design and analysis of CV experiments might be improved. Obviously a "large enough" increase in sample size, other things being equal, will provide the necessary statistical power for meaningful experiments. Since additional interviews are costly, we examine the factors that influence the power of experiments independent of sample size, in order to identify features which researchers may use to minimize the number of cases needed to achieve a specified ability to detect differences. Particular attention is paid to the role of prior estimates of coefficients of variation in the design of CV experiments.[5] One important finding is that in some cases much more powerful statistical tests can be made by using median rather than mean WTP amounts. We also discuss whether there are advantages to the use of more complex designs, and consider the special case of asking CV questions in a discrete-choice framework. We conclude with recommendations for the design of contingent valuation experiments.

The Power of a Test, and Four Simple Examples

We begin by examining the ability of four typical contingent valuation experiments to discriminate among competing hypotheses. The examples used are taken from Thayer (1981), Desvousges, Smith, and McGivney (1983), Brookshire and coauthors (1981), and Schulze and coauthors (1983). The first two experiments to be considered test the role of a starting point when the iterative bidding game is used as the elicitation method. The third tests whether willingness to pay for a good (air quality) changes if the date at which the good will be available changes. The fourth

[4] This suggests that while one should not put a lot of faith in experiments based on small samples where the null hypothesis was accepted, special note should be taken of those experiments where the null hypothesis was rejected because such a rejection was unlikely even when the alternative hypothesis was true.

[5] The importance of the coefficient of variation for experimental design in CV studies was first recognized by Desvousges, Smith, and McGivney (1983). Although their treatment of the issue was correct, they inadvertently used the standard error of the mean (rather than the standard deviation, which their formula correctly identified) in calculating the coefficients of variation for previous CV studies in their table A-2. This led them to reach considerably more optimistic conclusions about the usefulness of past CV experiments than we do.

examines whether using a mail survey to elicit willingness to pay obtains a different mean estimate than elicitation by personal interview. In the first two cases, the researchers appear to have reached conclusions which were not at variance with the ability of their data to detect differences. In the last two cases, the conclusions reached appear to be widely at odds with the data. In all four examples the primary problem is a lack of statistical power to reject the null hypothesis.

Using two samples of people making day trips, Thayer (1981) compared their mean willingness to pay for a recreation area at Jemez Mountain in New Mexico. One sample received a starting point of \$1 and the other a starting point of \$10. The mean WTP (to preserve the recreation area) X_1, the standard deviation S_1, and the number of observations N_1 for the \$1 starting point sample were \$2.63, \$3.27, and 49, respectively. For the \$10 starting point sample, the equivalent numbers were X_{10} = \$2.44, S_{10} = \$1.85, and $N_{10} = 25$. The formula (Larson, 1982) for calculating Thayer's two-tailed t-test comparing the hypothesis H_0: $X_1 = X_{10}$ against the hypothesis H_1: $X_1 \neq X_{10}$ is given by

$$T = (X_1 - X_{10})/(S_p\sqrt{1/N_1 + 1/N_{10}}), \tag{C-1}$$

where $S_p^2 = [(N_1 - 1)S_1^2 + (N_{10} - 1)S_{10}^2]/[N_1 + N_{10} - 2]$. Performing this calculation yields a value of $T = .269$. If the bids in each of the two samples are considered to be independent normal random variables with equal variances, T has Student's t distribution with $N_1 + N_{10} - 2$ degrees of freedom.[6] The t-value from Thayer's experiment (.269) is insignificant, and the p-value, which is the probability of observing this or a larger value for the t-test, is approximately 0.8. By almost any standard, Thayer's result suggests that there was no starting point bias in this CV survey, and his test for the absence of that bias is perhaps the most convincing in the literature.

A problem arises when one asks how big a difference there would have to be before H_1 would be accepted. One way to answer that question, and probably the best for most readers of CV research reports, is to calculate the confidence intervals for the t-test of the difference of the two means. If a significance level of 5 percent is used, as is typical in most CV studies, the 95 percent confidence interval for the differences between the two means is given by

$$\begin{aligned} &\text{Lower bound: } [X_1 - X_{10} - t_{(.25,\ Np-2)}S_p\sqrt{1/N_1 + 1/N_{10}}] \\ &\text{Upper bound: } [X_1 - X_{10} + t_{(.25,\ Np-2)}S_p\sqrt{1/N_1 + 1/N_{10}}] \end{aligned} \tag{C-2}$$

[6] In terms of the significance level chosen, the t-test has long been known to be quite robust against most violations of these assumptions (Bartlett, 1935; Ito, 1980), but may be very inefficient in terms of power (Efron, 1969). If it is obvious that there are gross violations of these assumptions, the researcher should consider using the nonparametric Wilcoxon test (Lehmann, 1975). The Wilcoxon test may in any case be a desirable alternative to the t-test, since it is almost as efficient as the t-test for the normal distribution, much more efficient for a wide range of other distributions, and because it is based on ranks, much less vulnerable to the effects of outliers. Carson and Mitchell (1986) provide an example of the use of the Wilcoxon test in a case where the distributional assumptions of the t-test were clearly not met.

where $N_p = N_1 + N_{10}$. Calculating this quantity for Thayer's study gives the interval $(-1.19, 1.57)$.[7] This is somewhat disturbing, since it suggests that there would have to be more than a 50 percent difference between the two means before H_0 could be rejected—a large difference indeed.[8]

Desvousges, Smith, and McGivney (DSM) tested for starting point bias in a study of willingness to pay to raise Monongahela River water from boatable to fishable quality (1983, tables 4-10 and 4-11) by comparing the WTP amounts for two samples, which received starting points of \$25 and \$125 respectively. The mean X_{25}, the standard deviation S_{25}, and the number of observations N_{25} of the first sample were \$15.90, \$15.50, and 58, respectively. For the \$125 starting point sample, the equivalent numbers were $X_{125} = \$36.90$, $S_{125} = \$49.50$, and $N_{125} = 48$. The t-test for H_0: $X_{25} = X_{125}$ versus H_1: $X_{25} \neq X_{125}$ has a value of -3.072, which indicates a significant starting point bias. How big would the difference have to be before H_0 is rejected? Again, the test is insensitive. Only the very large differences found allowed DSM to reject H_0. Using equation (C-2), the 95 percent confidence interval for the difference between the two means in DSM is $(-34.5, -7.5)$, showing that X_{125} would have to be at least 47 percent larger than X_{25} in order for H_0 to be rejected. Since the distance between the two starting points (\$125–\$25) in the DSM study is quite large, we would expect there to be a large difference between the two mean WTP values if respondents increased their bids by some percentage of the starting point, as suggested by Rowe, d'Arge, and Brookshire (1980).[9]

In the Thayer and DSM examples just given, the researchers, despite their use of small samples, achieved meaningful results. DSM did so because the difference they found was so large; Thayer did so because it was possible to recast his test as a one-sided alternative.[10] In our next two examples, however, the low power of the tests and the data patterns were such that the conclusions reached by the researchers appear to be unjustified.

[7] Thayer used a 10 percent significance level in his paper. We will argue below that this significance level is much more reasonable than the 5 percent level usually employed in CV experiments, and that in some cases an even higher level should be used.

[8] Thayer's reported conclusion of no starting point bias is not at odds with his data because (1) it is reasonable (and efficient) to formulate the hypothesis in terms of a one-sided test (that is, higher starting points lead to higher WTP responses), and (2) the observed difference is in the direction of higher starting points resulting in lower WTP responses.

[9] Increasing the difference between the starting points is not, in general, a solution to the lack of statistical power (although it should be done when possible). There is a limit to the upper starting point, beyond which the incidence of protest zeros, infinite WTP responses, and nonresponses becomes unacceptably high. DSM found significant problems of this kind with the \$125 starting point, and were probably close to that limit.

[10] Recasting Thayer's test as a one-sided hypothesis (higher starting points cause higher WTP responses) immediately increases its power (although the test is still weak); but perhaps more important, the null hypothesis of no starting point bias would be accepted even if the variance of his WTP responses approached zero, since the larger starting point is associated with the smaller willingness-to-pay amount. Thus Thayer's evidence would never seem to contradict the null hypothesis of no starting point bias.

Our third example is taken from Brookshire and coauthors' (1981) series of experiments on the factors influencing willingness to pay for an increase in visibility in the Los Angeles air basin. They compared the willingness to pay of two very small samples of Montebello respondents. One group was told that the cleanup would take two years, the other that the cleanup would take ten years. There were 10 respondents in the first group, which had a mean of \$19.10 and a standard deviation of \$45.95. The second group had 8 observations, with a mean of \$4.38 and a standard deviation of \$4.30. The *t*-test for the difference between the two means has a value of .895, and the authors, using a two-sided test and an $\alpha = .10$ criteria, concluded that the difference between the means was not significant. Yet the difference between the two means is \$15.52 (more than three times the mean from the second treatment), and in the direction suggested by theory. Because the confidence interval implied by this test is so large (−\$14.76, \$45.80), it was virtually impossible to reject H_0.

The fourth example, from Schulze and coauthors (1983), involves a comparison between the use of personal interviews and of a mail survey to obtain willingness-to-pay information. In this study there were 37 personal interviews, yielding a mean value of \$3.91 and a standard deviation of \$5.65, and 38 returned mail surveys, with a mean value of \$7.85 and a standard deviation of \$17.99. The *t*-test for the difference between the two methods of obtaining willingness-to-pay responses is −1.26. At any reasonable significance level, there is no statistically significant difference between the two treatments; yet there is more than a 100 percent difference between the two means,[11] and again the 95 percent confidence interval for the difference between the means is very large (−10.10, 2.22). In both examples three and four, the problem lies in the use of small samples and large standard deviations relative to the means.

The Power of CV Hypothesis Tests

To show that this problem is generic to contingent valuation experiments, and in order to offer suggestions for the design of future CV experiments, it is necessary to introduce some statistical theory on the power of hypothesis tests. We begin by stating three standard statistical definitions which will apply throughout this appendix:

> Type I Error Probability, denoted by α: probability of rejecting H_0 when H_0 is true;
>
> Type II Error Probability, denoted by β: probability of accepting H_0 when H_1 is true;
>
> Power of a Test, denoted by $1 - \beta$: probability of not accepting H_0 when H_1 is true.

[11] We have suggested in chapter 12 that the mean WTP amount obtained from a mail survey may be biased upward because those who are more interested in the good (and likely to have higher values for it) are those most likely to return the mail questionnaires.

Most of the attention in hypothesis testing in CV experiments has focused on the probability of making a Type I Error. The researcher concentrating on α is saying that $1 - \alpha$ percent of the time he will reject H_1 when H_0 is true.

There is an unfortunate tendency in CV literature to assume that H_0 is true if H_1 is rejected. This tendency is found in all disciplines in spite of constant reminders from statisticians that rejection of H_1 does not imply the truth of H_0.[12] To assume otherwise is an especially dangerous habit when the probability of a Type II Error is large, or, equivalently, when the power of the test used is low, as it was in the examples described above.

The power of a hypothesis test is more difficult to determine than is the probability of making a Type I Error, which is determined automatically by the choice of a significance level (if the data have already been collected). For a given N, the power of a test $(1 - \beta)$ declines inversely with the choice of α, increases as N becomes larger, and increases as S_p becomes smaller. If we are prepared to assume normality and equal variances, the power of the test of the difference between two means in the general two sample cases can be expressed as

$$1 - \beta = \text{Prob}[-t_{(\alpha/2,N_p-2)} \leq d/(S_pN^*) \leq t_{(\alpha/2,N_p-2)} \mid d/S_p] = \text{cl}, \quad \text{(C-3)}$$

or

$$1 - \beta = \text{Prob}[\text{rejecting } H_0 \mid d/S_p] = \text{cl},$$

where, as before, N_p equals the combined sample size $N_1 + N_2$, N^* equals $\sqrt{1/N_1 + 1/N_2}$, and d is a prechosen constant (which represents a possible difference between X_1 and X_2). It should be clear from equation (C-3) that the level of power $(1 - \beta)$ will vary directly with the choice of d. By varying d, a power curve can be traced out for any fixed N_p and S_p. Taking S_p and N_p from the Thayer experiment, in figure C-1 we have traced out power curves where α equals .05, .10, and .20, with d along the horizontal axis and power $(1 - \beta)$ along the vertical axis.

Contingent Valuation Surveys and the Coefficient Variation

Formulas for determining the level of power are most often given in terms of a noncentrality parameter, λ, where

$$\lambda = d/(S_pN^*) \quad \text{(C-4)}$$

The noncentrality parameter is the term between the two inequality signs in equation (C3), and incorporates the two choices (other than the levels of α and β)

[12] For an interesting examination of the widespread lack of statistical power in randomized clinical trials in drug testing, see Freiman and coauthors (1978). Cohen (1962, 1977) discusses some interesting examples from psychology experiments.

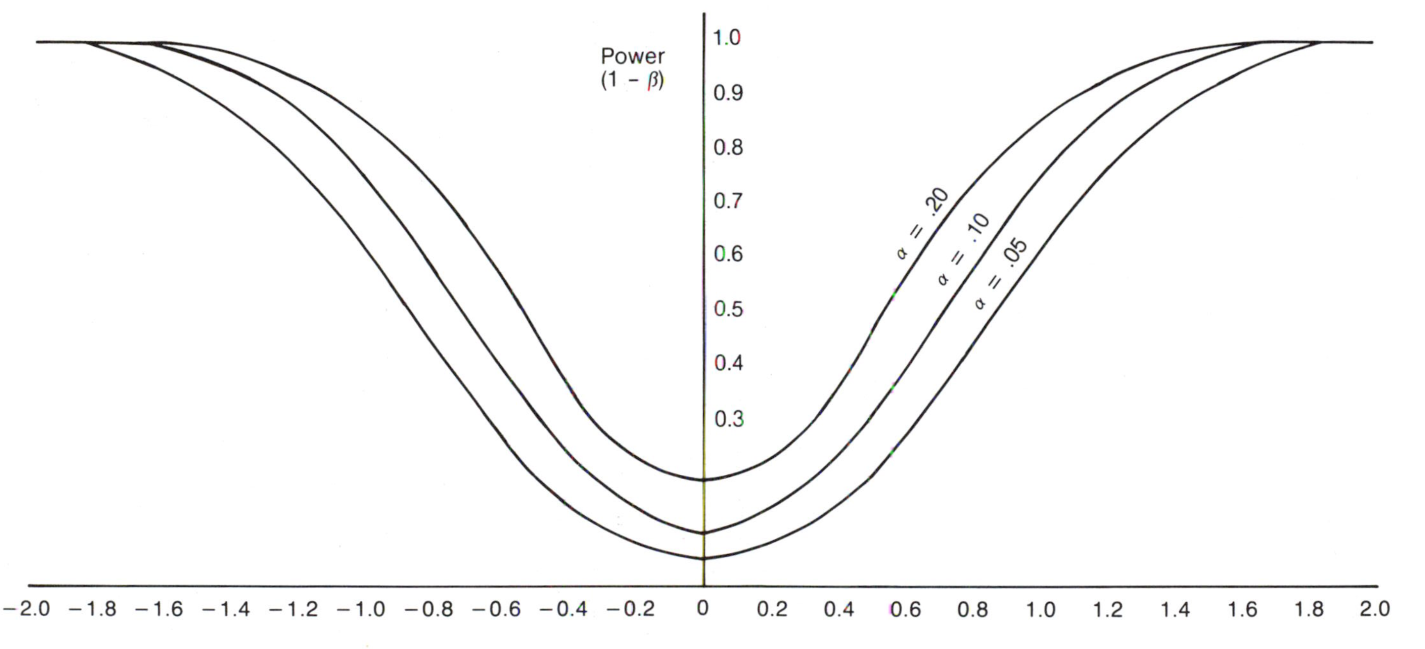

Figure C-1. Power curves for the Thayer starting point experiment

available to the researcher, N^* and d.[13] This parameter is difficult to work with in planning CV experiments because the researcher is unlikely to know S_p before conducting the experiment. The choice of α, which should be the smallest difference in the two means that the researcher wishes to be able to detect, is also troublesome since the researcher frequently does not have good prior information on the probable mean value of either treatment in the experiment.

Typically, however, CV researchers do not think in terms of d, but in terms of some percentage difference between the mean of the first experiment, X_1, and that of the second experiment, X_2. Denoting this percentage of X_1 as ΔX_1, where $\Delta X_1 = d$, and replacing d in equation (C-4) with this term yields

$$\lambda = \frac{\Delta X_1}{S_p\sqrt{1/N_1 + 1/N_2}}, \tag{C-5}$$

or

$$\lambda = \frac{\Delta V^{-1}}{\sqrt{1/N_1 + 1/N_2}}, \tag{C-6}$$

where V is the coefficient of variation S_p/X_1.

The coefficient of variation tends to be much more stable than individual estimates of S_p. Our calculations of the minimum and maximum V of a number of different studies reported in the contingent valuation literature are given in table C-1. In most of these studies the coefficients of variation fall in the range of 1.0 to 3.0. The few that are much larger than 3.0 were calculated from samples which included a number of gross outliers or aberrant cases, or both, which the researcher usually dropped before conducting the data analysis. The few studies having coefficients of variation much smaller than 1.0 tended to be from small samples of homogeneous populations.[14]

[13] If the variances in the two subsamples are approximately equal and sampling costs are approximately equal, then N_p should consist of two equally sized subsamples. In the present discussion we assume this is the case—an assumption that appears to be reasonable in view of past CV experiments. Where that assumption does not hold, it may be optimal to adopt a different sampling plan. For details about sampling plans, see any standard sampling text such as Cochran (1977).

[14] There is nothing inherently good about having a small coefficient of variation. Indeed, a lack of variability in WTP responses is sometimes indicative of a severe problem, as occurs when respondents latch onto a suggestion in the survey that indicates what the "correct" WTP response should be. To take another example, the coefficient of variation for household income in the United States varies from about .6 to 1.0, depending on how the upper-income categories are treated. If willingness to pay for a particular public good is a simple, deterministic, linear function of household income alone, then the coefficient of variation for household income in the United States represents the lower bound for the coefficient of variation for willingness to pay obtained from a national sample of households. If a researcher were to observe a coefficient of variation much smaller than this, she should worry about what went wrong.

Table C-1. Coefficients of Variation (V) from Selected CV Studies

Study	Minimum V	N	Maximum V	N
Randall, Ives, and Eastman (1974)	1.16	526	1.39	526
Brookshire, Ives, and Schulze (1976)	0.47	20	0.74	22
Gramlich (1977)	1.44	165	2.31	165
Cocheba and Langford (1978)	1.31	169	NA	NA
Brookshire, d'Arge, and Schulze (1979)	0.60	4	4.24	9
Rowe, d'Arge, and Brookshire (1980)	0.83	93	2.08	45
Brookshire, Randall, and Stoll (1980)	0.59	9	1.39	10
Greenley, Walsh, and Young (1981)	0.68	85	1.84	25
Mitchell and Carson (1981)	1.29	748	1.45	748
Thayer (1981)	0.59	23	1.24	49
Randall, Hoehn, and Tolley (1981)	0.83	35	4.71	47
Schulze and Brookshire (1983)	0.65	60	3.41	115
Schulze and coauthors (1983)	0.91	35	2.29	39
Brookshire, Eubanks, and Randall (1983)	1.39	65	3.08	245
Desvousges, Smith, and McGivney (1983)	0.61	19	2.85	32
Majid, Sinden, and Randall (1983)	0.98	140	5.18	140
Blomquist (1984)	0.71	65	1.61	163
Loehman (1984)	0.61	9	2.11	83
Carson and Mitchell (1986)	1.01	130	2.16	564
Smith, Desvousges, and Freeman (1985)	1.01	16	1.95	196
Roberts, Thompson, and Pawlyk (1985)	0.85	33	0.98	41
Sorg and coauthors (1985)	3.22	769	3.69	215
Donnelly and coauthors (1985)	1.49	84	6.33	126

Note: The studies selected here are intended to be representative of the range of results reported, and the list is not inclusive. The sample sizes of the various surveys in the table are also representative of those reported in the published CV literature.

Sample Size Tables

A set of tables indicating sample sizes should also be useful to researchers in planning how large a study would be necessary to conduct an experiment on a particular issue.[15] Tables C-2 through C-13 give the total number of observations needed for different levels of α and β in terms of the coefficient of variation S_p/X_1 and in terms of Δ, the difference between X_1 and X_2 expressed as a percentage of X_1,

$$\Delta = \frac{X_1 - X_2}{X_1} \qquad \text{(C-7)}$$

which can be detected. Tables are given for both one-sided and two-sided tests for the following pairs of α and β: [.05, .05], [.05, .10], [.10, .10], [.10, .20],

[15] For those studying the results of CV experiments, these tables should be useful in determining whether a given study is capable of providing information about the issues addressed. Cohen (1977) provides tables in terms of the quantity $d = (X_1 - X_2)/S_p$. Cohen (1977) and Kramer and Thiemann (1987) provide extensive discussions of power calculations in different situations.

[.20, .20], [.25, .50]. To use the tables, the researcher should decide on the desired levels of α and β and on a likely prior estimate of the coefficient of variation V, find the table with that particular α and β, select the column of the likely V, and find the row with the desired Δ. The value in the table is the approximate number of actual WTP responses needed. If the desired levels of α and β are not included in these tables, or if the likely V is outside the range given on a table, the following formula can provide a reasonable approximation:

$$n = 2[Z_{(1-\alpha)} + Z_{(1-\beta)}]^2 \left(\frac{V}{\Delta}\right)^2 \tag{C-8}$$

where $Z_{(\cdot)}$ is the usual standard normal variate. Hodges and Lehmann (1968) provide a straightforward method and tables for interpolation which are more accurate than equation (C-8). SAS macros for doing both power and sample size calculations are also available (Bergstalh, 1984).

A quick perusal of tables C-2 through C-13 will show that the sample sizes required for the traditional $\alpha = .05$ tables are so large as to be unaffordable for most CV researchers, thereby necessitating the selection of some larger value of α. The choice of β depends on how one views making a Type II Error relative to a Type I Error. In most contingent valuation situations we believe that both types of errors are about equally troublesome, although perhaps slightly less weight should be given to the Type II Error. The [α, β] combinations of [.10, .10] and [.10, .20] in tables C-6, C-7, C-8, and C-9 will probably be those most useful to CV researchers.

In these tables the number of observations (N) needed to detect a certain Δ for a given V, α, and β are for *completed usable* WTP responses.[16] Sample and item nonresponse rates will often shrink the number of people in the original sample to a much lower number of usable responses. The researcher should keep this in mind and plan the number of households to be interviewed accordingly.

Ways to Improve the Power of Simple Experiments

Tables C-2 through C-13 show that quite large samples are required for an experiment unless V is small or the desired Δ is large. There are several ways to improve the power characteristics of, or reduce the subsample sizes needed in, simple experiments.

The first is to formulate the hypothesis in terms of a one-sided hypothesis test (instead of the two-sided test which has usually been used). For many CV experiments—such as testing the effect of using a higher starting point, using higher payment card anchors, or using a larger budget constraint—the one-sided test makes sense. For tests involving other CV features, like the effects of providing different descriptions of the good, a two-sided test may still be called for, but even here theory or previous research may suggest the direction of particular treatment effects.

[16] Each of the two subsamples should be one-half of the N given in the tables (assuming equally sized subsamples are desired).

Table C-2. Total Number of Observations Needed for a Two-tailed *t*-Test as a Function of Δ and *V* for $\alpha = .05$ and $\beta = .05$

				Coefficient of variation (*V*)					
Δ[a]	.1	.25	.50	.75	1.00	1.25	1.50	1.75	2.00
.10	32	194	770	1,730	3,076	4,804	6,918	9,416	12,296
.20	8	50	194	434	770	1,202	1,730	2,354	3,706
.30	4	22	86	194	342	534	770	1,048	1,368
.40	2	14	50	110	194	302	434	590	770
.50	2	8	32	70	124	194	278	378	492
.75	2	4	14	32	56	86	124	168	220
1.00	2	2	8	18	32	50	70	96	124

[a]In tables C-2 through C-13, Δ is defined as the difference between the means *x* and *y*, expressed as a percentage (p) of *x* which is detectable with a Type I Error probability of α and a Type II Error probability of β.

Table C-3. Total Number of Observations Needed for a One-tailed *t*-Test as a Function of Δ and *V* for $\alpha = .05$ and $\beta = .05$

				Coefficient of variation (*V*)					
Δ	.1	.25	.50	.75	1.00	1.25	1.50	1.75	2.00
.10	22	136	542	1,220	2,166	3,384	4,874	6,632	8,662
.20	6	34	136	306	542	846	1,220	1,658	2,166
.30	4	16	62	136	242	376	542	738	964
.40	2	10	34	78	136	212	306	416	542
.50	2	6	22	50	88	136	196	266	348
.75	2	4	10	22	40	62	88	118	154
1.00	2	2	6	14	22	34	50	68	88

As an example of the potential research dollars to be saved by the substitution of a one-sided for a two-sided test, consider a situation where $\alpha = .10$, $\beta = .20$, $V = 1.5$, and the percentage difference we are interested in detecting is .2. Using tables C-8 and C-9, the indicated sample size for a two-tailed *t*-test is 964, while a one-tailed *t*-test would require 508 observations. At a rate of $50 for an in-person interview, which is a low estimate for such interviews, the one-tailed test would offer savings of more than $20,000.

When the researcher's interest is primarily in the results of the experiment rather than in making an actual benefit estimate for policy purposes, it is possible to reduce *V* by choosing for the experiment a population which is homogeneous in terms of income, age, and environmental attitudes. There must still be a random assignment of subjects to treatments, however.[17]

Our third suggestion concerns outlier treatment procedures. Because a few outliers can easily increase S_p or distort $\overline{X}_i$, or both, enough to destroy all hope of

[17] The tradeoff is that the more specialized the population, the less possible it is to generalize the results of the experiment to other populations.

Table C-4. Total Number of Observations Needed for a Two-tailed t-Test as a Function of Δ and V for $\alpha = .05$ and $\beta = .10$

	Coefficient of variation (V)								
Δ	.1	.25	.50	.75	1.00	1.25	1.50	1.75	2.00
.10	28	164	652	1,464	2,602	4,062	5,852	7,962	10,402
.20	8	42	164	366	652	1,016	1,464	1,992	2,602
.30	4	20	74	164	290	452	652	886	1,156
.40	2	12	42	92	164	254	366	498	652
.50	2	8	28	60	106	164	236	320	418
.75	2	4	12	28	48	74	106	142	186
1.00	2	2	8	16	28	42	60	80	106

Table C-5. Total Number of Observations Needed for a One-tailed t-Test as a Function of Δ and V for $\alpha = .05$ and $\beta = .10$

	Coefficient of variation (V)								
Δ	.1	.25	.50	.75	1.00	1.25	1.50	1.75	2.00
.10	18	108	430	964	1,714	2,678	3,856	5,248	6,854
.20	6	28	108	242	430	670	964	1,312	1,714
.30	2	12	48	108	192	298	430	584	762
.40	2	8	28	62	108	168	242	328	430
.50	2	6	18	40	70	108	156	210	276
.75	2	2	8	18	32	48	70	94	122
1.00	2	2	6	10	18	28	40	54	70

Table C-6. Total Number of Observations Needed for a Two-tailed t-Test as a Function of Δ and V for $\alpha = .10$ and $\beta = .10$

	Coefficient of variation (V)								
Δ	.10	.25	.50	.75	1.00	1.25	1.50	1.75	2.00
.10	22	136	542	1,220	2,166	3,384	4,874	6,632	8,862
.20	6	34	136	306	542	846	1,220	1,658	2,166
.30	2	16	62	136	242	376	542	738	964
.40	2	10	34	78	136	212	306	416	542
.50	2	6	22	50	88	136	196	266	348
.75	2	2	10	22	40	62	88	118	154
1.00	2	2	6	14	22	34	50	68	88

Table C-7. Total Number of Observations Needed for a One-tailed *t*-Test as a Function of Δ and *V* for $\alpha = .10$ and $\beta = .10$

	Coefficient of variation (*V*)								
Δ	.10	.25	.50	.75	1.00	1.25	1.50	1.75	2.00
.10	14	84	330	740	1,316	2,054	2,958	4,026	5,258
.20	4	22	84	186	330	514	740	1,008	1,316
.30	2	10	38	84	146	230	330	448	586
.40	2	6	22	48	84	130	186	252	330
.50	2	4	14	30	54	84	120	162	212
.75	2	2	6	14	24	37	54	72	94
1.00	2	2	4	8	14	21	30	42	54

Table C-8. Total Number of Observations Needed for a Two-tailed *t*-Test as a Function of Δ and *V* for $\alpha = .10$ and $\beta = .20$

	Coefficient of variation (*V*)								
Δ	.10	.25	.50	.75	1.00	1.25	1.50	1.75	2.00
.10	18	108	430	964	1,714	2,678	3,856	5,248	6,854
.20	6	28	108	242	430	670	964	1,312	1,714
.30	2	12	48	108	192	298	430	584	762
.40	2	8	28	62	108	168	242	328	430
.50	2	6	18	40	70	108	156	210	276
.75	2	2	8	18	32	48	70	94	122
1.00	2	2	6	10	18	28	49	54	70

Table C-9. Total Number of Observations Needed for a One-tailed *t*-Test as a Function of Δ and *V* for $\alpha = .10$ and $\beta = .20$

	Coefficient of variation (*V*)								
Δ	.10	.25	.50	.75	1.00	1.25	1.50	1.75	2.00
.10	10	58	226	508	902	1,410	2,030	2,762	3,606
.20	4	16	58	128	226	354	508	692	902
.30	2	8	26	58	102	158	226	308	402
.40	2	4	16	32	58	90	128	174	226
.50	2	4	10	21	38	58	82	112	146
.75	2	2	6	10	18	26	38	50	66
1.00	2	2	4	6	10	16	22	28	38

Table C-10. Total Number of Observations Needed for a Two-tailed *t*-Test as a Function of Δ and *V* for $\alpha = .20$ and $\beta = .20$

	Coefficient of variation (*V*)								
Δ	.10	.25	.50	.75	1.00	1.25	1.50	1.75	2.00
.10	14	84	330	740	1,316	2,054	2,958	4,026	5,258
.20	4	22	84	186	330	514	740	1,008	1,316
.30	2	10	38	84	148	230	330	448	586
.40	2	6	22	48	84	130	186	252	330
.50	2	4	14	30	54	84	120	162	212
.75	2	2	6	14	24	38	54	72	94
1.00	2	2	4	8	14	22	30	42	54

Table C-11. Total Number of Observations Needed for a One-tailed *t*-Test as a Function of Δ and *V* for $\alpha = .20$ and $\beta = .20$

	Coefficient of variation (*V*)								
Δ	.10	.25	.50	.75	1.00	1.25	1.50	1.75	2.00
.10	6	36	142	320	568	886	1,276	1,736	2,266
.20	2	10	36	80	142	222	320	434	568
.30	2	4	16	36	64	100	142	194	252
.40	2	4	10	20	36	56	80	110	142
.50	2	2	6	14	24	36	52	70	92
.75	2	2	4	6	12	16	24	32	42
1.00	2	2	2	4	6	10	14	18	24

Table C-12. Total Number of Observations Needed for a Two-tailed *t*-Test as a Function of Δ and *V* for $\alpha = .25$ and $\beta = .50$

	Coefficient of variation (*V*)								
Δ	.10	.25	.50	.75	1.00	1.25	1.50	1.75	2.00
.10	8	42	166	396	666	1,042	1,500	2,040	2,664
.20	2	12	42	94	168	262	376	510	666
.30	2	6	20	42	74	116	168	228	296
.40	2	4	12	24	42	66	94	128	168
.50	2	2	8	16	28	42	60	82	108
.75	2	2	4	8	12	20	28	38	48
1.00	2	2	2	4	8	12	16	22	28

Table C-13. Total Number of Observations Needed for a One-tailed *t*-Test as a Function of Δ and V for $\alpha = .25$ and $\beta = .50$

	Coefficient of variation (V)								
Δ	.10	.25	.50	.75	1.00	1.25	1.50	1.75	2.00
.10	2	6	23	52	92	144	206	280	364
.20	2	2	6	14	24	36	52	70	92
.30	2	2	4	6	12	16	24	32	42
.40	2	2	2	4	6	10	14	18	24
.50	2	2	2	4	4	6	10	12	16
.75	2	2	2	2	2	4	4	6	8
1.00	2	2	2	2	2	2	4	4	4

determining whether two experimental treatments have different effects, it is important to either remove the outliers or choose a statistical technique which is insensitive to them (Barnett and Lewis, 1984). If the outliers are removed, the removal procedures should be adopted and justified before the completion of the experiment, and the same procedures should be used on both subsamples. The researcher also should include a list of the dropped outliers in an appendix to her technical report, along with a description of the removal criteria. It is important to recognize that the percentage of outliers deleted from each of the subsamples may provide evidence about the differential effects of the two treatments, since the effects may not necessarily manifest themselves in the mean WTP responses of the two subsamples.[18]

If a researcher chooses not to delete the outliers according to some criteria, it is advisable to report the results of a robust test of the differences between subsamples (Huber, 1981). An analogue of the *t*-test can be based upon one of the robust M-estimators, or upon a robust L-estimator such as the trimmed mean. Inferences based upon an ordinary *t*-test should be made only if these results suggest conclusions similar to those obtained by a robust counterpart. Contingent valuation experiments inevitably generate a number of "bad" observations from respondents who either did not understand the game or who deliberately gave bizarre answers, and as few as a single gross outlier can strongly influence the results of an experiment having a sizable number of observations, and can completely destroy the results of an experiment having a small or moderate number of observations.

Our last suggestion for improving the power characteristics of a given study is to test the difference between the treatment medians, rather than the treatment means. Such a test is particularly advantageous if the distribution of the WTP responses is left-skewed with a long right tail, and resembles the log-normal distribution.[19]

[18] An example is the large number of outliers found by Desvousges, Smith, and McGivney (1983) in their $125 starting point experiment relative to the number they found in their $25 experiment. While the $125 starting point did tend to encourage larger WTP responses, its primary effect appears to have been to encourage a larger number of nonsensical WTP responses. (DSM used procedures suggested by Belsley, Kuh, and Welsch, 1980.)

[19] Willingness-to-pay responses are, of course, constrained to be non-negative.

Letting $X = \ln(Y)$ be distributed normally with mean ξ and standard deviation θ, the random variable Y is then said to possess the log-normal distribution with mean

$$\mu = \mathrm{EXP}[\xi + \theta^2/2], \tag{C-9}$$

and standard deviation

$$\sigma = \mathrm{EXP}[\xi + \theta^2/2]\ \mathrm{EXP}[\theta^2 - 1]^{1/2} \tag{C-10}$$

The coefficient of variation for X is given by

$$V_x = \theta/\xi \tag{C-11}$$

while the coefficient of variation for Y is given by

$$V_Y = \mathrm{EXP}[\theta^2 - 1]^{1/2} \tag{C-12}$$

It can be shown that

$$E(Y) \neq E(\mathrm{EXP}[X]) \tag{C-13}$$

but that the

$$E(Y) = \mathrm{EXP}[E(X) + \mathrm{VAR}(X)/2] = \mathrm{EXP}[\xi + \theta^2/2] \tag{C-14}$$

As a result of the symmetry of the normal distribution, however,

$$M(Y) = M(\mathrm{EXP}[X]) = \mathrm{EXP}[E(Y)] = \mathrm{EXP}[\xi] \tag{C-15}$$

where $M(\cdot)$ is the median operator. Thus, tests about the mean of X are equivalent to tests about the median of Y.[20]

In order for it to be advantageous to test differences between the medians of experiments rather than the means, we must show that $V_Y > V_X$, at least in the range of values for ξ and θ that is likely to be relevant for the goods being valued in CV surveys. This region is where

$$[\mathrm{EXP}(\theta^2) - 1]^{1/2} > \theta/q \tag{C-16}$$

[20] The fact that one is making an inference about the median of the original log-normal distribution after taking logs has been noted by Goldberger (1968) in the Cobb-Douglas context. Stynes, Peterson, and Rosenthal (1986) discuss the use of log transformations in travel cost analysis. The properties of the log-normal distribution are discussed in Johnson and Kotz (1970) and Aitchinson and Brown (1957). See Koopmans, Owens, and Rosenblatt (1964) for a nice discussion of the coefficients of variation for the normal and log-normal distributions, including confidence intervals.

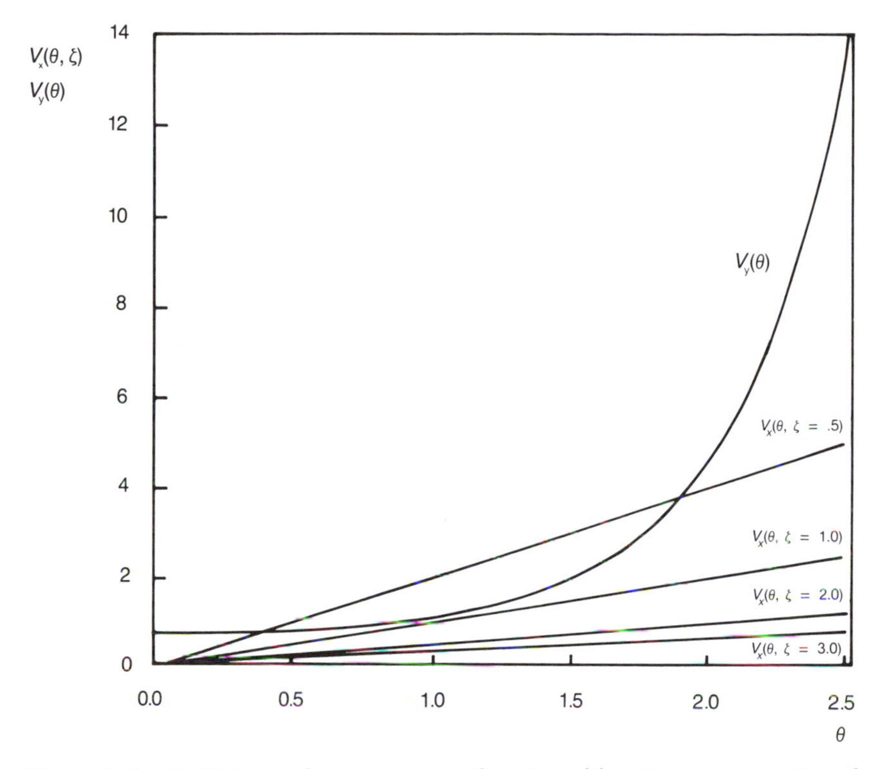

Figure C-2. Coefficients of variation as a function of location parameter ξ and scale parameter θ

We will first narrow down the likely range of ξ and θ. Note that θ must always be positive, and we have yet to observe a CV survey where ξ was not larger than .5, usually substantially so. Since V_Y is not a function of ξ, the likelihood that the inequality will hold increases as ξ increases, as θ must be positive. Figure C-2 shows θ from 0.0 to 2.5 on the horizontal axis, and V_Y and V_X on the vertical axis. Curves for $V_Y(\theta)$ are given for ξ = .5, 1.0, 2.0, and 3.0.[21]

To illustrate the greater efficiency of medians over means for a given power level, we use the data for total willingness to pay for water quality from Carson and Mitchell (1986). In that study the unadjusted mean WTP amount for national water quality improvements is $275, with a standard deviation of $601, which yields a coefficient of variation for this variable of 2.2. An examination of the WTP amounts shows a distinct left-skewed distribution with a thick right tail.[22] Taking the log of these WTP responses yields a mean value of 4.545, with a

[21] This gives values of Y from roughly 1 to 25.

[22] The distribution looks very log-normal except for a spike at zero, particularly if one takes into account some grouping effects at round numbers like $10, $25, and $100. Specific techniques are available to take account of such self-grouping (see Tallis and Young, 1962; Hasselblad, Stead, and Galke, 1980).

standard deviation of 1.753, giving a coefficient of variation of .39.[23] Using these parameters, we want to determine what sample size is required to detect a difference of 20 percent, using a two-tailed test with $\alpha = .05$ and $\beta = .10$. According to table C-4, more than 2,000 observations would be necessary for a test of the difference between the means of two different treatments. In contrast, a comparable test of the difference between the medians of two treatments would require only about 100 observations.

The logarithmic transformation is one of a larger group of variance stabilizing transformations that may be useful when the variance is a function of the mean. A transformation $Z = \sigma(Y)$ is said to be variance stabilizing for Y if σ_Z is approximately independent of μ_Z, while σ_Y is a function of μ_Y. It is generally desirable to transform the original distribution to normality. If the transformation used is monotonic, then the median observation is the same for both distributions.[24]

More Complicated Experiments

Two types of more complicated experimental designs should also be considered: (1) more general forms of analysis of variance (ANOVA), and (2) response surface designs.[25]

The simplest extension of the *t*-test is to include more than two treatments of the same factor in a one-way ANOVA layout. Denote the factor by T and its k levels by $\tau_1, \tau_2, \ldots, \tau_k$. The linear model can be written

$$y_{ik} = \tau_1 + \tau_2 + \ldots + \tau_k + \epsilon_{ik}, \quad \text{(C-17)}$$

[23] A problem occurs here because the log of zero is undefined, and taking the log of a very small positive number can have severely distorting effects. We follow Bartlett's suggestion (1947) that 1 should be added to the variable, if it does not materially alter the statistics from the original distribution. It does not in this case: the mean changes from $275 to $276. This data set also demonstrates the large effect that a single outlier can have (in this case, it was verified that the outlier came from a respondent whose income was extremely large). Dropping that one observation causes a reduction in the coefficient of variation for the untransformed total WTP amount from 2.2 to 1.6. It should be noted that the change in the coefficient of variation of the logged form of WTP with and without this outlier is small, only about .002.

[24] Various transformations and the distributions for which they are appropriate are discussed by Bartlett (1947), Box, Hunter, and Hunter (1978), Emerson (1983), and McCullagh and Nelder (1983).

[25] The *t*-test is the simplest form of ANOVA. The commonly used regression estimation with dummy variables for the treatments is also a form of ANOVA, if only the dummy variables appear. If quantitative regressors also appear it is a form of analysis of covariance. We note that there is a disturbing trend among researchers to report such regression equations without providing the simple basic statistics on WTP by treatment which are necessary for the reader to make an informed judgment about the tests. Extremely large differences between treatments are often present, even though the reported *t*-statistics for the coefficients may be insignificant. A good basic reference on ANOVA experiments is Box, Hunter, and Hunter (1978). For a theoretical discussion of power in ANOVA designs see Scheffe (1959), and for a more advanced text on possible experimental designs and their properties see Cochran and Cox (1957). Box, Hunter, and Hunter also cover response surface models. A

where the y_{ik} is the WTP response and $i = 1, 2, \ldots, n_k$. Since a particular respondent receives only one of the treatments, all of the estimated $\tilde{\tau}_k$ for that respondent are identically zero, except for the treatment he received. The null hypothesis H_0 is usually expressed as $\tau_k = \tau_k' \ \forall\ k, k'$, and the alternative hypothesis H_1 as $\tau_k \neq \tau_k'$ for one or more pairs of k and k'. Being tested is whether the estimated $\tilde{\tau}_j$ as a group are significantly different from zero. The analysis is a simple extension of that given above, except that the F-test is substituted for the t-test, and that if H_0 is rejected further analysis would be needed to determine which differences between the means are responsible for the rejection.[26] Sample size tables in terms of the $\underset{k,k'}{\text{MAX}} \mid \tilde{\tau}_k - \tilde{\tau}_k' \mid / \tilde{\sigma}$ and the level of power are given in Katsenbaum, Hoel, and Bowman (1970) and in Harter and Owen (1975).

A more general ANOVA design may be desirable if it is possible to take advantage of "blocking" variables that reduce the overall unexplained variance. These variables allow the effects of the treatments to be estimated with more precision, and reduce the probability that the observed effects are simply a result of random variations in treatment response. For the case with one blocking variable (two-way ANOVA classification layout), the linear model is

$$y_{ijk} = \gamma_1 + \ldots + \gamma_j + \tau_1 + \ldots + \tau_2 + \epsilon_{ijk}, \qquad \text{(C-18)}$$

where the blocks are represented by γ_j. This model is usually reparameterized and estimated with a constant term (grand mean) and restrictions that the $\Sigma_j \gamma$ and the $\Sigma_k \tilde{\tau}_k$ are both equal to zero. The model yields the straightforward interpretation that the expected value of y_{ijk} is the grand mean plus the block effect plus the treatment effect. We have implicitly assumed here that there are no interaction effects between the treatments and blocks, but this is not a necessary assumption.

To illustrate how these ANOVA designs can be used, we use the data from Mitchell and Carson's (1981) payment card experiment, in which they randomly assigned respondents to three different payment card treatments.[27] Their results,

---

more advanced discussion of optimal ANOVA and response surface designs is provided by Silvey (1980).

[26] It is possible to use multiple comparison tests and develop simultaneous confidence intervals (Miller, 1981) to make inferences about the significance of the differences between pairs of means. Tukey's least significance difference test seems to work well in practice for paired comparisons. This test and others are discussed in Miller (1981). The 1985 *SAS User's Guide: Statistics* (SAS Institute) contains a somewhat more recent bibliography, as well as brief descriptions of the characteristics and performances of various tests that are easily available to the researcher.

[27] Each payment card consisted of an array of dollar amounts from $0 to a high amount. Certain dollar amounts (benchmarks) were identified as the annual amount respondents in households of specified income levels pay for public goods such as police and fire protection and national defense. The first payment card version (A), which used four benchmarks, was a base or control treatment; the second (B) used payment card benchmarks set 20 percent higher than those of the control; and the third (C) added an additional benchmark in the range where a sizable number of responses were expected to occur. The three subsamples were of approximately equal size, and had a grand mean WTP of $259, with a standard deviation of $308 for fishable quality water.

Table C-14. Analysis of Variance: Treatment Effects Only

Source of variance	Sum of squares	df	Mean square	F
Treatment	105,360	2	52,680	0.55
Error	63,635,071	670	94,978	—
Corrected total	63,740,431	672	—	—

Table C-15. Analysis of Variance: with Income Groups as a Blocking Variable

Source of variance	Sum of squares	df	Mean square	F
Blocks (income)	1,492,543	3	4,975,248	68.02
Treatment	26,961	2	13,481	0.18
Error	48,788,017	667	73,145	—
Corrected total	63,740,431	672	—	—

Table C-16. Analysis of Variance: with Interaction between Blocks and Treatment

Source of variance	Sum of squares	df	Mean square	F
Blocks (income)	14,295,453	3	4,765,151	68.13
Treatment	26,961	2	13,481	0.18
Blocks*treatment	517,785	6	86,298	1.18
Error	48,270,232	661	73,026	—
Corrected total	63,740,431	672	—	—

based on paired *t*-tests, show that the three payment card treatments are not significantly different from each other. The ANOVA results in table C-14 indicate the same conclusion.

In addition to the three payment card treatments, Mitchell and Carson used different levels of payment card benchmarks for four different income groups,[28] and it may be advantageous, in terms of efficiency, to take into account the four income groups as a blocking variable. The results of rerunning the ANOVA with two factors—the different payment card treatments and the four income categories—are given in table C-15. Note that while the different payment card treatments still have a small and insignificant effect, there is a large reduction in the unexplained variance which will allow us to draw much tighter confidence intervals for the differences between various pairs of treatment means.[29]

The sum of squares for the four income categories is quite large, and it is natural to ask if the effects of the different payment card treatments are the same for each

[28] The different benchmark levels were designed to take into account the actual levels paid for the goods in taxes and higher prices by households in four income groups.

[29] It is possible to use a blocking approach with the two-treatment *t*-test scheme simply by randomly assigning those two treatments to individuals in each block and performing an analysis of variance based on two factors.

of the four income categories. This can be tested by allowing an interaction between the two factors. The results are shown in table C-16. Again, we conclude that the different payment card treatments have no effect on willingness to pay.

It is instructive to go further and look at the simultaneous confidence intervals for the differences between the three pairs of treatment means. These are given in table C-17 for each of the three ANOVA models and for $\alpha = .05, .10, .20$.[30] The effect of reducing the significance level from .05 to .10 is a sizable decrease in the width of the confidence intervals. The reasons for using a blocking variable can be clearly seen by comparing the width of the confidence intervals for the ANOVA in tables C-14 and C-15.[31]

A large number of special designs, such as Latin squares, Youden squares, and balanced incomplete blocks, are available for use in different situations (Cochran and Cox, 1957; Box, Hunter, and Hunter, 1978). We call attention to three caveats to the use of such designs: (1) power is lost fairly rapidly as the number of observations in each of the cells becomes small, so further blocking is advisable only when it results in a fairly large reduction in variance; (2) it becomes ever harder to make the treatments random as the number of blocking variables increases, especially under the field conditions in which CV surveys are usually conducted; and (3) nonresponses will almost always result in an unbalanced design, even if the initial design was balanced, and all but simple unbalanced designs are very hard to work with.

Another type of ANOVA design which has possible applications to contingent valuation experiments is the 2^n factorial design. A 2^n factorial experiment explicitly allows for an interaction between n different factors.[32] A number of ingenious schemes have been proposed to reduce the necessary sample size when the number of factors is between 3 and 8, and especially when it is possible to assume that some of the higher-order interaction effects are zero. The researcher should note, however, that the large coefficients of variation typical of CV studies call for quite large sample sizes when factorial designs are used. Nevertheless, the 2^n factorial design is probably the best one for CV experimenters who wish to test a large

[30] The Tukey-Kramer least significance difference test (Kramer, 1956) considers two means $\bar{y}_j$ and $\bar{y}_i$ to be significantly different if

$$|\bar{y}_i - \bar{y}_j| / (s(1/n_i + 1/n_j)/2)^{1/2} \geq q(\alpha, k, v),$$

where s is the standard deviation of the error from the ANOVA, and $q\ (\alpha, k, v)$ is the α-level critical value of the Studentized range of the distribution of k (the number of different means) independent normal random variables with v degrees of freedom.

[31] The confidence interval for the difference between treatments (A) and (B) can be significantly narrowed further by using a one-tailed test.

[32] Interactions will usually be present in CV experiments. If there is no interaction between the two factors, then one of the factors may be treated as a block when analyzing the effects of the other factor. A relatively common practice in CV experiments has been to test numerous explanatory factors by a series of t-tests. This practice implicitly assumes that there is no interaction between the factors—a dangerous assumption, especially when the total sample size is small or the cell sizes (of usable responses) are not equal, or both.

Table C-17. Tukey Simultaneous Upper and Lower Confidence Limits

Comparison	SLL(.05)	SLL(.1)	SLL(.2)	Actual	SUL(.2)	SUL(.1)	SUL(.05)
A–B (IIIa)	−53.62	−45.14	−35.32	14.27	68.86	73.68	82.16
A–C (IIIa)	−37.67	−29.15	−19.27	30.61	80.49	90.37	98.90
B–C (IIIa)	−85.29	−76.69	−66.70	−16.34	34.02	44.00	52.61
A–B (IIIb)	−45.31	−37.87	−29.25	14.27	57.79	66.41	73.80
A–C (IIIb)	−29.31	−21.84	−19.27	30.61	80.49	83.06	90.54
B–C (IIIb)	−76.85	−69.30	−60.54	−16.34	27.85	36.61	44.17
A–B (IIIc)	−45.26	−37.83	−29.25	14.27	57.75	66.37	73.80
A–C (IIIc)	−29.27	−21.79	−13.12	30.61	74.35	83.02	90.49
B–C (IIIc)	−76.80	−69.26	−60.50	−16.34	27.82	36.57	44.12

SLL = Simultaneous Lower Limit; SUL = Simultaneous Upper Limit

number of factors when it is also possible to use a blocking variable to reduce the variance within treatment groups.

Response surface designs are desirable when the stimulus given to subjects naturally takes a continuous form, such as the starting point in a bidding game.[33] The first question to ask when considering a response surface design is whether or not the functional form $Y = f(X, \theta)$ is known except for the vector of parameters θ. The next question to ask is whether $f(X)$ is linear or can be written in some form which is linear in the parameters. If so, and if the researcher is prepared to make some additional assumptions such as a constant (or known) variance function, the results from the optimal design literature (Silvey, 1980) are fairly easy to use.

This can be illustrated by considering the case of estimating or testing for starting point bias. The most commonly used specification in the literature is

$$Y_i = \psi_0 + \psi_1 x_i + \xi_i \tag{C-19}$$

where x_i is the starting point received by the *i*th respondent. The typical objective is to minimize some form of the variance-covariance matrix of the estimate of the ψ vector, usually the determinant (D-optimality), given a fixed total sample size, N. There is a unique solution if the range of possible x_is is restricted, as it is in the starting point case,[34] which is to take half of the x_i to be zero and the other half of the x_i to be the largest possible starting point. The general solution for the k-degree polynomial is to use $k + 1$ design points, with two of those being the two end-points of the allowable range of x_i.

Typically, however, there are two or more competing functional forms, or the response function is completely unknown. In some instances it may be possible to nest competing functional forms within a more general one and reduce the number of observations needed to distinguish between them.[35] If so, certain design points may be most powerful in distinguishing between the two models. This topic, as well as the non-nested case, is covered in more detail by Cox (1961), Box and Hill (1967), and Atkinson and Cox (1974). Sequential design and experimentation are called for in instances where the researcher explores the implications of different sized x_is. The situation in which nothing is known about the response surface almost always requires some sort of sequential design whereby the researcher chooses a small number of x_is, obtains responses and analyzes them, and then chooses a new set of x_is in the direction where there appears to be the largest

[33] The only published study to employ this type of design that we know of is Boyle, Bishop, and Welsh's (1985) starting point experiment, in which a large number of starting points between $10 and $120 was used in valuing a Wisconsin river. There has been some work on response surface experimental design in connection with income maintenance and time-of-day electricity pricing experiments. See Conlisk (1973), Ferber and Hirsch (1982), Hausman and Wise (1985), and, in particular, the special issue of the *Journal of Econometrics* published in 1979, and edited by Aigner and Morris.

[34] The lower limit for x_i is obviously zero. There will be some amount beyond which x_i is no longer plausible as a place to start the bidding game.

[35] See Carson, Casterline, and Mitchell (1985) for a discussion of different possible functional forms for starting point bias.

informational gain. Box, Hunter, and Hunter (1978) provide some easy-to-follow guidelines for doing this.

Optimal design of response surface models becomes mathematically very complicated for models having nonlinear parameters. Different optimality criteria can result in quite different designs. There is usually no alternative to the use of some type of sequential design, as the information matrix is almost always a function of the mean responses for the values of the x_i chosen. Thus, in contrast to the simple linear case, variance is not independent of the expected response.

When pretests are undertaken to improve the wording of a questionnaire, contingent valuation surveys are well suited to the sequential approach to response surface exploration in either the classical or the Bayesian context. It should be noted that obtaining an estimate of the response surface to a stimulus is a necessary step in going from a simple finding that the stimulus has a significant effect on the WTP response to the development of constructive correction methods and to a deeper understanding of how and why the stimulus affects WTP values.

Discrete Choice/Quantal Response Experiments

There are other problems that arise when the response variable from a CV experiment takes a discrete form. Discrete choice analysis is used when the respondent is asked whether or not he is willing to pay a single specified amount, such as $15, or when the variable of interest is whether the respondent gave a usable response or a nonusable response (for example, a protest zero). The researcher wishes to determine whether the percentage of respondents willing to pay a specified amount is greater under one treatment than another, or whether the percentage giving protest zeros is greater under one of the treatments. Hypothesis testing in situations such as these calls for variants of the *t*-test which have some particular properties worth noting.

The binomial distribution, which is appropriate is these situations unless the underlying population is small, is particularly easy to work with, and the normal approximation to it is quite good when n is large. Let r_1 be the number of respondents who are willing to pay the specified amount under treatment 1, and r_2 be the number of respondents who are willing to pay the same amount under treatment 2, where the sample size of those playing the game under the first treatment is n_1 and under the second treatment is n_2. The percentage who will agree to the payment under treatment 1—$\hat{p}_1$—can be estimated by r_1/n_1, while $\hat{p}_2$ can be estimated in a similar manner.[36] The variance of the difference $\hat{p}_1 - \hat{p}_2$ is given by

$$\frac{\hat{p}_1(1-\hat{p}_1) + \hat{p}_2(1-\hat{p}_2)}{n_1}, \qquad \text{(C-20)}$$

where we have assumed $n_1 = n_2$ for convenience. In this case, the researcher may have a clear idea of how large a difference he is interested in detecting, and it is

[36] The coefficient of variation (V) for the binomial case can be written as $\sqrt{(1-\hat{p})/n\hat{p}}$. Note that if p is small, it will take very large samples to obtain accurate estimates of p.

easy to set equation (C-20) equal to the desired variance and solve for the required n_1 (given a set of prior estimates of p_1 and p_2). For the case where H_0 is $p_1 = p_2$, simply set

$$\frac{2p_1(1-p_1)}{\sigma^2} = n_1, \tag{C-21}$$

where σ is the desired standard deviation. Extensive tables of the appropriate sample sizes for different α, β, and p_1s and p_2s are given in Fleiss (1981).

There has been a growing interest in the use of a discrete choice (take-it-or-leave-it) format for CV surveys that ask respondents whether or not they are willing to pay a specified amount for some amenity, and then repeat the process for different (equivalent) groups using different amounts (see Bishop and Heberlein, 1979; Bishop and coauthors, 1984; Sellar, Stoll, and Chavas, 1985; Bishop and Heberlein, 1986; Carson, Hanemann, and Mitchell, 1986; Cummings and coauthors, 1986; Cameron and James, 1987; Cameron, forthcoming). The discrete-choice elicitation format is considered easier for respondents to answer than other elicitation formats that use the payment card or open-ended question. It also is easy to use in a mail survey. The percentage who are willing to pay the specified amount is then considered to be a function of the specified amount. This can be viewed as a special type of response surface regression, and has a long history in the medical literature, where it is known as a bioassay (Finney, 1978).

The basic procedure here is to choose k willingness to pay amounts, and ask k different random samples of n_k respondents if they are willing to pay that amount. The percentage who agree is calculated at each of the k values of willingness to pay. By making an appropriate transformation of the percentage WTP (usually the logistic or inverse normal [probit]), a curve can be drawn through the k points, yielding a function of WTP.[37]

One potentially very troublesome difficulty with a discrete-choice elicitation format is the possibility of a non-zero response background.[38] There is always going to be some percentage of respondents who in essence refuse to play the game, and do so by saying no to an offered WTP amount, no matter what the amount (or who, in the WTA case, refuse to accept any amount of compensation). These respondents are easily identified in the standard CV survey format because they either give protest zeros or unreasonably large (often infinite) amounts. In the discrete choice format they cannot be distinguished from those who give genuine

[37] For a discussion of these procedures, see Cox (1970), Finney (1978), and McCullagh and Nelder (1983). Since these are generally nonlinear functions, it is not usually possible to arrive at an optimal placement of the k different "dose" points (Silvey, 1980). There are, however, good approximate solutions available if the researcher has a fairly good idea of the percentage accepting at each dollar amount. These designs are discussed by Cox (1970), Cochran (1973), and Finney (1978). Ford and Silvey (1980) propose an interesting sequential approach to estimating a logistic response curve. Tsutakawa (1980) considers the case where the researcher desires to estimate an extreme quantal. See also Abdelbasit and Plackett (1981, 1983), and Whittemore (1981).

[38] Discussed in more detail by Carson and Mitchell (1983), and Hanemann (1984b).

"no" answers. Their effect is most obviously seen in one of the tails (depending on whether the variable is WTP or WTA), when increases in the amounts do not cause the acceptance percentage to converge to the axis. The researcher can either directly attempt to take account of this behavior, through the use of maximum-likelihood techniques (Hasselblad, Stead, and Creason, 1980), or the use of a robust quantal estimation technique (Miller and Halpern, 1980).[39] Hanemann (1984c) provides results which suggest that the median is more robust than the mean.

The discussion above has taken a bioassay approach. However, it is also possible to take a utility theoretic approach using discrete-choice observations (Hanemann, 1984a, 1984b, 1984c). Often economic theory will suggest what sort of variables (in addition to the WTP amount) should influence a respondent's choice. Use of these variables can be viewed as a form of blocking because it reduces the variance from random assignment of the respondents to the different treatment (that is, WTP amount) groups. The key problem with this approach is that it is often hard to specify a particular functional form for the indirect utility function. Cameron and James (1987) postulate a distribution for WTP, rather than the indirect utility function as Hanemann does. While such a specification may be easier to test empirically, it should be clear that doing so implies a particular functional form for the indirect utility function.

A weak postulate of utility theory, based on revealed preferences, can be used to gain efficiency in discrete-choice CV experimental designs. This approach takes advantage of the fact that when someone reveals willingness to pay, say $10, that respondent gives the researcher information about what she would do if asked to pay an amount less that $10. For an application of this procedure, see Carson, Hanemann, and Mitchell's (1986) study of California referendum voting for water quality improvements.

Concluding Remarks

In this appendix we have focused on the design of experiments to test for response effects in contingent valuation surveys. The use of CV surveys for that purpose is distinct from their use to estimate benefits for policy purposes. We believe that both purposes cannot usually be achieved with a single survey sample unless the experimental portions of the survey instrument come *after* that portion used to make the policy estimates. Irrespective of the purpose of a study, whenever experiments are conducted it is necessary to randomly assign respondents to the treatments in order to make statistical inferences about the hypotheses. If a study is solely intended to examine response effects it is acceptable to use a less-than-random sample, or a sample deliberately designed to increase the homogeneity of the respondents and thus reduce response variance, as long as the assignments to the treatments are random. On the other hand, if the primary purpose is to estimate benefits for policy purposes, the key is a good sampling plan (and method of imputing nonresponses) which provides an unbiased estimate of the likely benefits, along with an appropriate confidence interval for that estimate.

[39] See also Pregibon (1981, 1982).

These two purposes can often be complementary. A CV survey instrument which is going to be used for policy purposes needs extensive pretesting. Pretests, provided they have large enough samples, may be used to conduct experiments to determine the effects of changing various features of the CV scenario.

Our discussion of experimental design has touched on many issues that must be taken into account by the researcher who wishes to test hypotheses about response effects. It may be useful to conclude by summarizing these factors in the form of recommendations for researchers who are planning contingent valuation experiments.

1. Before carrying out an experiment, determine the desired levels of α, β, and Δ, and form an estimate of V (or alternatively of σ and d). A conservative assumption would be that V will be approximately 2. In no case, in a CV experiment, should it be assumed that V is likely to be less than 1 without firm evidence to this effect (such as information from a large pretest). Next, make the necessary sample size calculation. If sufficient funds are not available for the chosen sample size, the researcher should reconsider her objectives. In many cases, the researcher (and proposal reviewers) should recommend that the experiment not be carried out. A test having no power is often worse than no test because of the potential for misinterpreting the results.
2. When determining the sample size required for the experiment, expect a large number of unusable responses.
3. Either adopt criteria for removing outliers, or use a robust estimation technique which is relatively insensitive to these observations. For reasons described at length in this book, any CV survey will generate a significant amount of what could be best described as noise. In general, the mean of the respondents' WTP amounts will not be a good estimate of the population's mean willingness to pay for the amenity.
4. Report confidence intervals along with the standard *t*-test (or ANOVA) results for a hypothesis test. This is perhaps the easiest and clearest way for a researcher to communicate the level of power of the experiment to readers.
5. If the distribution of WTP results has a distinct non-normal shape, consider using a nonparametric test based on ranks or a test performed on transformation of the data. If the data have a log-normal shape, consider testing the difference between the medians of the treatments, as opposed to the means. The necessary sample size for means tests may be unaffordable, whereas the size necessary for median tests may be easily affordable.
6. Consider the possibility of using blocking variables, such as income categories, to greatly increase the power of tests by random assignment of the different treatments within blocks.
7. Be particularly aware of the sample size requirements necessary for a 2^n and other factorial designs, and of the necessity of including interaction terms in the analysis if this type of design is used. CV researchers (ourselves included) tend to try to estimate more effects than the sample size warrants.

8. Consider using a sequential set of experiments when estimating the response surface for a change in a continuous CV stimulus.
9. Be aware that while the discrete-choice CV framework (such as the take-it-or-leave-it approach) is easy to work with and has a number of advantages, not all yes or no answers are necessarily valid. Such nonplaying behavior is difficult to detect and correct.
10. Always keep in mind that good contingent valuation experiments need good CV survey instruments. If respondents do not understand what is being asked of them, a test of different forms of the same basic instrument is meaningless.

Bibliography

Abdelbasit, K. M., and R. L. Plackett. 1981. "Experimental Design for Categorized Data," *International Statistical Review* vol. 49, pp. 111–126.

Abdelbasit, K. M., and R. L. Plackett. 1983. "Experimental Design for Binary Data," *Journal of the American Statistical Association* vol. 78, pp. 90–98.

Abelson, Robert P., and Ariel Levi. 1985. "Decision Making and Decision Theory," in Gardner Lindzey and Elliot Aronson, eds., *The Handbook of Social Psychology,* vol. 1 (3d ed., New York, L. Erlbaum Associates).

Acton, Jan P. 1973. "Evaluating Public Programs to Save Lives: The Case of Heart Attacks," Research Report R-73-02, Rand Corporation, Santa Monica, Calif.

Adams, F. G., and L. R. Klein. 1972. "Anticipations Variables in Macro-Economic Models," in Burkhard Strumpel, James N. Morgan, and Ernest Zahn, eds., *Human Behavior in Economic Affairs: Essays in Honor of George Katona* (San Francisco, Jossey-Bass).

Adelman, Irma, and Zvi Griliches. 1961. "On an Index of Quality Change," *Journal of the American Statistical Association* vol. 56, pp. 531–548.

Aigner, Dennis J., and C. N. Morris, eds. 1979. "Experimental Design in Econometrics," *Journal of Econometrics* vol. 11, no. 1.

Aitchinson, J., and J. A. C. Brown. 1957. *The Lognormal Distribution with Special Reference to Its Uses in Economics* (London, Cambridge University Press).

Ajzen, Icek, and Martin Fishbein. 1977. "Attitude-Behavior Relations: A Theoretical Analysis and Review of Empirical Research," *Psychological Bulletin* vol. 84, no. 5, pp. 888–918.

Ajzen, Icek, and George L. Peterson. 1986. "Contingent Value Measurement: The Price of Everything and the Value of Nothing?" Paper presented at the National Workshop on Integrating Economic and Psychological Knowledge in Valuations of Public Amenity Resources, Fort Collins, Colo., May.

Akerlof, George A. 1983. "Loyalty Filters," *American Economic Review* vol. 73, pp. 54–63.

Akerlof, George A., and William T. Dickens. 1982. "The Economic Consequences of Cognitive Dissonance," *American Economic Review* vol. 72, pp. 307–319.

Akin, J., G. Fields, and W. Neenan. 1973. "A Socioeconomic Explanation for the Demand for Public Goods," *Public Finance Quarterly* vol. 1, pp. 168–189.

Alwin, Duane F. 1977. "Making Errors in Surveys, an Overview," *Sociological Methods & Research* vol. 6, no. 2, pp. 131–150.

Amemiya, Takeshi. 1973. "Regression Analysis When the Dependent Variable Is Truncated Normal," *Econometrica* vol. 41, pp. 997–1016.

Amemiya, Takeshi. 1981. "Qualitative Response Models: A Survey," *Journal of Economic Literature* vol. 19, pp. 1483–1536.

Amemiya, Takeshi. 1985. *Advanced Econometrics* (Cambridge, Mass., Harvard University Press).

American Psychological Association. 1974. *Standards for Educational and Psychological Tests* (Washington, D.C.).

Anderson, Andy B., Alexander Basilevsky, and Derek P. J. Hum. 1983. "Missing Data," in Peter H. Rossi, James D. Wright, and Andy B. Anderson, eds., *Handbook of Survey Research* (New York, Academic Press).

Anderson, Eric, and Dean Devereaux. 1986. "Testing for Two Kinds of Bias in a Contingent Valuation Survey of Anglers Using an Artificial Reef." Paper presented at the Eastern Economic Association Meeting, Philadelphia, March.

Anderson, R. J. 1981. "A Note on Option Value and the Expected Value of Consumer Surplus," *Journal of Environmental Economics and Management* vol. 8, pp. 187–191.

Anderson, Ronald, Judith Kasper, and Martin R. Frankel, eds. 1979. *Total Survey Error: Applications to Improve Health Surveys* (San Francisco, Jossey-Bass).

Anderson, R. W. 1980. "Some Theory of Inverse Demand for Applied Demand Analysis," *European Economic Review* vol. 14, pp. 281–290.

Arndt, J., and E. Crane. 1975. "Response Bias, Yea-Saying, and the Double Negative," *Journal of Marketing Research* vol. 12, pp. 218–220.

Arrow, Kenneth J. 1951. *Social Choice and Individual Values* (New York, John Wiley).

Arrow, Kenneth J. 1982. "Risk Perceptions in Psychology and Economics," *Economic Inquiry* vol. 20, pp. 1–19.

Arrow, Kenneth J. 1986. "Comments," in Ronald G. Cummings, David S. Brookshire, and William D. Schulze, eds., *Valuing Environmental Goods* (Totawa, N. J., Rowman and Allanheld).

Arrow, Kenneth J., and Anthony C. Fisher. 1974. "Environmental Preservation, Uncertainty, and Irreversibility," *Quarterly Journal of Economics* vol. 88, pp. 313–319.

Atkinson, A. C., and D. R. Cox. 1974. "Planning Experiments for Discriminating Between Models," with discussion, *Journal of the Royal Statistical Society,* series B, vol. 36, pp. 321–348.

Atkinson, Anthony B., and Joseph E. Stiglitz. 1980. *Lectures on Public Economics* (New York, McGraw-Hill).

Attiyeh, Richard, and Robert F. Engle. 1979. "Testing Some Propositions About Proposition 13," *National Tax Journal* vol. 32, no. 2, pp. 131–146.

Axelrod, Robert M. 1984. *The Evolution of Cooperation* (New York, Basic Books).

Bailar, J. C., and D. A. Bailar. 1978. "Comparison of Two Procedures for Imputing Missing Survey Values," in *Proceedings of the Section of Survey Research Methods* (Washington, D.C., American Statistical Association) pp. 462–467.

Banford, Nancy D., Jack L. Knetsch, and Gary A. Mauser. 1977. "Compensating and Equivalent Variation Measures of Consumer's Surplus: Future Survey

Results," Department of Economics and Commerce, Simon Fraser University, Vancouver.

Barnett, Vic, and Toby Lewis. 1984. *Outliers in Statistical Data* (2d ed., Chichester, England, John Wiley).

Barr, James L., and Otto A. Davis. 1966. "An Elementary Political and Economic Theory of the Expenditures of Local Government," *Southern Economic Journal* vol. 33, no. 2, pp. 149–165.

Bartlett, M. S. 1935. "The Effect of Non-Normality on the t Distribution," in *Proceedings of the Cambridge Philosophical Society* vol. 31, pp. 223–231.

Bartlett, M. S. 1947. "The Use of Transformations," *Biometrics* vol. 3, pp. 39–52.

Baumol, William J. 1986. *Superfairness: Application and Theory* (Cambridge, Mass., MIT Press).

Baumol, William J., and Oates, Wallace E. 1979. *Economics, Environmental Policy, and the Quality of Life* (Englewood Cliffs, N.J., Prentice-Hall).

Beardsley, W. G. 1971. "Economic Value of Recreation Benefits Determined by Three Methods," U.S. Forest Service Research Notes, RM-176 (Colorado Springs, Rocky Mountain Experiment Station).

Becker, Gary S. 1965. "A Theory of the Allocation of Time," *Economic Journal* vol. 75, pp. 493–517.

Becker, Gary S. 1976. *The Economic Approach to Human Behavior* (Chicago, University of Chicago Press).

Becker, Gary S. 1981. *A Treatise on the Family* (Cambridge, Mass., Harvard University Press).

Beeghley, Leonard, 1986. "Social Class and Political Participation: A Review and an Explanation," *Sociological Focus* vol. 1, no. 3, pp. 496–513.

Beggs, S., S. Cardell, and Jerry A. Hausman. 1981. "Assessing the Potential Demand for Electric Cars," *Journal of Econometrics* vol. 16, pp. 1–19.

Bell, F. W., and V. R. Leaworthy. 1985. "An Economic Analysis of the Importance of Saltwater Beaches in Florida," Sea Grant Project no. R/C C-P-12, Department of Economics, Florida State University, Tallahassee.

Bell, Ralph. 1984. "Item Nonresponse in Telephone Surveys: An Analysis of Who Fails to Report Income," *Social Science Quarterly* vol. 65, no. 1, pp. 207–215.

Belsley, David A., Edwin Kuh, and Roy E. Welsch. 1980. *Regression Diagnostics: Identifying Influential Data and Sources of Collinearity* (New York, John Wiley and Sons).

Belson, W. A. 1968. "Respondent Understanding of Survey Questions," *Polls* vol. 3, pp. 1–13.

Bentkover, Judith D., Vincent T. Covello, and Jeryl Mumpower. 1985. *Benefits Assessment: The State of the Art* (Hingham, Mass., Kluwer Academic Publishers).

Berdie, Douglas R., and John F. Anderson. 1976. "Mail Questionnaire Response Rates: Updating Outmoded Thinking," *Journal of Marketing Research* vol. 40, no. 1, pp. 71–73.

Berg, George G., and H. David Maillie, eds. 1981. *Measurement of Risks* (New York, Plenum).

Berger, Gretchen J. 1984. "Application and Assessment of the Contingent Valuation Method for Federal Hazardous Waste Policy in the Washington, D.C. Area" (Ph.D. dissertation, University of New Mexico, Albuquerque).

Bergland, Olvar. 1985. "Exact Welfare Analysis with Rationing." Paper presented at the winter meeting of the Econometric Society, New York.

Bergson, Abram. 1938. "A Reformulation of Certain Aspects of Welfare Economics," *Quarterly Journal of Economics* vol. 52, pp. 310–334.

Bergstalh, E. 1984. "SAS Macros for Sample Size and Power Calculations," in *SUGI 84: Proceedings of the 9th Annual SAS Users Group International Conference* (Cary, N.C., SAS Institute) pp. 633–638.

Bergstrom, John C., B. L. Dillman, and John R. Stoll. 1985. "Public Environmental Amenity Benefits of Private Land: The Case of Prime Agricultural Land," *Southern Journal of Agricultural Economics* vol. 17, no. 1.

Bergstrom, John C., and John R. Stoll. 1985. "Cognitive Decision Process, Information, and Contingent Valuation." Paper presented at the annual summer meeting of the American Agricultural Economics Association, Ames, Iowa, August.

Bergstrom, Theodore C., and Richard C. Cornes. 1983. "Independence of Allocation Efficiency from Distribution in the Theory of Public Goods," *Econometrica* vol. 51, pp. 1753–1765.

Bergstrom, Theodore C., and Richard P. Goodman. 1973. "Private Demands for Public Goods," *American Economic Review* vol. 63, no. 3, pp. 280–296.

Bergstrom, Theodore C., Daniel L. Rubinfeld, and Perry Shapiro. 1982. "Micro-Based Estimates of Demand Functions for Local School Expenditures," *Econometrica* vol. 50, no. 5, pp. 1183–1205.

Berry, D. 1974. "Open Space Values: A Household Survey of Two Philadelphia Parks," Discussion Paper Series no. 76, Regional Science Research Institute, Philadelphia.

Bettman, James R. 1979. *An Information Processing Theory of Consumer Choice* (Reading, Mass., Addison-Wesley).

Bickel, Peter, and Kjell Doksum. 1977. *Mathematical Statistics: Ideas and Concepts* (San Francisco, Holden-Day).

Binkley, Clark S., and W. Michael Hanemann. 1978. "The Recreation Benefits of Water Quality Improvement: Analysis of Day Trips in an Urban Setting," Report to the U.S. Environmental Protection Agency (Washington, D.C.).

Bishop, George F., 1981. "Survey Research," chap. 21 in Dan D. Nimmo and Keith R. Sanders, eds., *Handbook of Political Communication* (Beverly Hills, Calif., and London, Sage).

Bishop, George F. 1985. "Context Effects on Self-Perceptions of Interest in Government and Public Affairs," Project Report to the National Science Foundation (Cincinnati, Ohio, Behavioral Sciences Laboratory, Institute for Policy Research, University of Cincinnati).

Bishop, George F., David L. Hamilton, and John B. McConahay. 1980. "Attitudes and Nonattitudes in the Belief Systems of Mass Publics," *Journal of Social Psychology* vol. 110, pp. 53–64.

Bishop, George F., Alfred J. Tuchfarber, and Robert W. Oldendick. 1986. "Opinions on Fictitious Issues: The Pressure to Answer Survey Questions," *Public Opinion Quarterly* vol. 50, no. 2, pp. 240–250.

Bishop, John, and Charles J. Cicchetti. 1975. "Some Institutional and Conceptual Thoughts on the Measurement of Indirect and Intangible Benefits and Costs," in Henry M. Peskin and Eugene P. Seskin, eds., *Cost-Benefit Analysis and Water Pollution Policy* (Washington, D.C., Urban Institute).

Bishop, Richard C. 1982. "Option Value: An Exposition and Extension," *Land Economics* vol. 58, pp. 1–15.

Bishop, Richard C., and Kevin J. Boyle. 1985. "The Economic Value of Illinois Beach State Nature Preserve," report to the Illinois Department of Conservation (Madison, Wis., Heberlein and Baumgartner Research Services).

Bishop, Richard C., and Thomas A. Heberlein. 1979. "Measuring Values of Extra-Market Goods: Are Indirect Measures Biased?" *American Journal of Agricultural Economics* vol. 61, no. 5, pp. 926–930.

Bishop, Richard C., and Thomas A. Heberlein. 1980. "Simulated Markets, Hypothetical Markets, and Travel Cost Analysis: Alternative Methods of Estimating Outdoor Recreation Demand," Staff Paper Series no. 187, Department of Agricultural Economics, University of Wisconsin.

Bishop, Richard C., and Thomas A. Heberlein. 1984. "Contingent Valuation Methods and Ecosystem Damage from Acid Rain," Staff Paper Series no. 217, Department of Agricultural Economics, University of Wisconsin.

Bishop, Richard C., and Thomas A. Heberlein. 1985. "Progress Report of the 1984 Sandhill Study," preliminary report to the Wisconsin Department of Natural Resources.

Bishop, Richard C., and Thomas A. Heberlein. 1986. "Does Contingent Valuation Work?" in Ronald G. Cummings, David S. Brookshire, and William D. Schulze, eds., *Valuing Environmental Goods* (Totawa, N.J., Rowman and Allanheld).

Bishop, Richard C., Thomas A. Heberlein, and Mary Jo Kealy. 1983. "Hypothetical Bias in Contingent Valuation: Results from a Simulated Market," *Natural Resources Journal* vol. 23, no. 3, pp. 619–633.

Bishop, Richard C., Thomas A. Heberlein, Michael P. Welsh, and Robert A. Baumgartner. 1984. "Does Contingent Valuation Work? A Report on the Sandhill Study." Paper presented at the Joint Meeting of the Association of Environmental and Resource Economists and the American Economics Association, Cornell University, Ithaca, N.Y., August.

Black, Duncan. 1958. *The Theory of Elections and Committees* (Cambridge, Cambridge University Press).

Blackorby, Charles, Daniel Primont, and R. Robert Russell. 1978. *Duality, Separability, and Functional Structure: Theory and Economic Applications* (New York, North-Holland).

Blank, Frederick M., David S. Brookshire, Thomas D. Crocker, Ralph C. d'Arge, R. L. Horst, and Robert D. Rowe. 1978. "Valuation of Aesthetic Preferences: A Case Study of the Economic Value of Visibility," report to the Electric Power Research Institute (Resource and Environmental Economics Laboratory, University of Wyoming).

Bliss, Christopher, and Barry Nalebuff. 1984. "Dragon-Slaying and Ballroom Dancing: The Private Supply of a Public Good," *Journal of Public Economics* vol. 25, pp. 1–12.

Blomquist, Glenn C. 1982. "Estimating the Value of Life and Safety: Recent Developments," in M. W. Jones-Lee, ed., *The Value of Life and Safety* (Amsterdam, North-Holland).

Blomquist, Glenn C. 1983. "Measurement of the Benefits of Water Quality Improvements," in George S. Tolley, Dan Yaron, and Glenn C. Blomquist, eds., *Environmental Policy: Water Quality* (Cambridge, Mass., Ballinger).

Blomquist, Glenn C. 1984. "Measuring the Benefits of Public Goods Provision Using Implicit and Contingent Markets," working paper, College of Business and Economics, University of Kentucky.

Bloomgarden, Kathy. 1982. "Toward Responsible Growth: Economic and Environmental Concern in the Balance" (Stamford, Conn., The Continental Group).

Bockstael, Nancy E., and Kenneth E. McConnell. 1980a. "Calculating Equivalent and Compensating Variation for Natural Resource Facilities," *Land Economics* vol. 56, no. 1, pp. 56–62.

Bockstael, Nancy E., and Kenneth E. McConnell. 1980b. "Measuring the Worth of Natural Resource Facilities: Reply," *Land Economics* vol. 56, no. 4, pp. 487–490.

Bockstael, Nancy E., and Kenneth E. McConnell. 1983. "Welfare Measurement in the Household Production Framework," *American Economic Review* vol. 73, no. 4, pp. 806–814.

Bockstael, Nancy E., W. Michael Hanemann, and Ivar E. Strand. 1985. "Measuring the Benefits of Water Quality Improvements Using Recreation Demand Models," report to the Economic Analysis Division, U.S. Environmental Protection Agency, Washington, D.C.

Bohm, Peter. 1972. "Estimating Demand for Public Goods: An Experiment," *European Economic Review* vol. 3, pp. 111–130.

Bohm, Peter. 1975. "Option Demand and Consumer Surplus: Comment," *American Economic Review* vol. 65, pp. 733–736.

Bohm, Peter. 1977. "Estimating Access Value," in Lowdon Wingo and Alan Evans, eds., *Public Economics and the Quality of Life* (Baltimore, The Johns Hopkins University Press for Resources for the Future).

Bohm, Peter. 1979. "Estimating Willingness to Pay: Why and How?" *Scandinavian Journal of Economics* vol. 81, no. 2, pp. 142–153.

Bohm, Peter. 1984. "Revealing Demand for an Actual Public Good," *Journal of Public Economics* vol. 24, pp. 135–151.

Bohm, Peter, and Clifford S. Russell. 1985. "Comparative Analysis of Alternative Policy Instruments," in Allen V. Kneese and James L. Sweeney, eds., *Handbook of Natural Resource and Energy Economics,* vol. 1 (Amsterdam, North-Holland).

Bohrnstedt, George W. 1983. "Measurement," in Peter H. Rossi, James D. Wright, and Andy B. Anderson, eds., *Handbook of Survey Research* (New York, Academic Press).

Bok, Sissela. 1978. *Lying: Moral Choice in Public and Private Life* (New York, Random House).

Bolnick, Bruce R. 1976. "Collective Goods Provision Through Community Development," *Economic Development and Cultural Change* vol. 25, no. 1, pp. 137–150.

Borcherding, Thomas E., and Robert T. Deacon. 1972. "The Demand for the Services of Non-Federal Governments," *American Economic Review* vol. 62, no. 5, pp. 891–901.

Bowen, Howard R. 1943. "The Interpretation of Voting in the Allocation of Economic Resources," *Quarterly Journal of Economics* vol. 58, pp. 27–48.

Box, G. E. P., and W. J. Hill. 1967. "Discriminating Among Mechanistic Models," *Technometrics* vol. 9, pp. 57–71.

Box, G. E. P., William G. Hunter, and J. Stuart Hunter. 1978. *Statistics for Experimenters* (New York, John Wiley).

Boyle, Kevin J., and Richard C. Bishop. 1984a. "A Comparison of Contingent Valuation Techniques," Staff Paper Series no. 222, Department of Agricultural Economics, University of Wisconsin, Madison.

Boyle, Kevin J., and Richard C. Bishop. 1984b. "Economic Benefits Associated with Boating and Canoeing on the Lower Wisconsin River," Economic Issues no. 84, Department of Agricultural Economics, University of Wisconsin, Madison.

Boyle, Kevin J., and Richard C. Bishop. 1985. "The Total Value of Wildlife Resources: Conceptual and Empirical Issues." Paper presented at the Association of Environmental and Resource Economics Workshop on Recreation Demand Modeling, Boulder, Colo., May.

Boyle, Kevin J., Richard C. Bishop, and Michael P. Welsh. 1985. "Starting Point Bias in Contingent Valuation Surveys," *Land Economics* vol. 61, pp. 188–194.

Bradburn, Norman M. 1982. "Question-Wording Effect in Surveys," in Robin M. Hogarth, ed., *Question Framing and Response Consistency* (San Francisco, Jossey-Bass).

Bradburn, Norman M. 1983. "Measurement: Theory and Techniques," in Peter H. Rossi, James D. Wright, and Andy B. Anderson, eds., *Handbook of Survey Research* (New York, Academic Press).

Bradburn, Norman M., and S. Sudman. 1979. *Improving Interview Method and Questionnaire Design* (San Francisco, Jossey-Bass).

Bradford, David F. 1970. "Benefit-Cost Analysis and Demand Curves for Public Goods," *Kyklos* vol. 23, pp. 775–791.

Brannon, R., G. Cyphers, S. Hess, S. Hesselbart, R. Keane, H. Schuman, T. Vaccaro, and D. Wright. 1973. "Attitude and Action: A Field Experiment Joined to a General Population Survey," *American Sociological Review* vol. 38, pp. 625–636.

Bredemeier, Harry C., and Richard M. Stephenson. 1962. *The Analysis of Social Systems* (New York, Holt, Rinehart and Winston).

Breiman, L., and D. Freeman. 1983. "How Many Variables Should Be Entered in a Regression Equation," *Journal of the American Statistical Association* vol. 78, pp. 131–136.

Breiman, L., J. H. Friedman, R. A. Olshen, and C. Stone. 1984. *Classification and Regression Trees* (Belmont, Calif., Wadsworth).

Brennan, Geoffrey, and James Buchanan. 1984. "Voter Choice," *American Behavioral Scientist* vol. 28, no. 2, pp. 185–201.

Broadway, Robin W., and Neil Bruce. 1984. *Welfare Economics* (Oxford, Basil Blackwell).

Brody, Charles J. 1986. "Things Are Rarely Black and White: Admitting Gray into the Converse Model of Attitude Stability," *American Journal of Sociology* vol. 92, no. 3, pp. 657–677.

Brookshire, David S., Don L. Coursey, and Karen M. Radosevich. 1986. "Market Methods of Benefits, Some Future Results." Paper presented at the Workshop on Integrating Psychology and Economics in Valuing Public Amenity Resources, Estes Park, Colo., May.

Brookshire, David S., Don L. Coursey, and William D. Schulze. 1986. "Experiments in the Solicitation of Public and Private Values: An Overview," in L. Green and J. Kasel, eds., *Advances in Behavioral Economics* (Greenwich, Conn., JAI Press).

Brookshire, David S., and Thomas D. Crocker. 1981. "The Advantages of Contingent Valuation Methods for Benefit-Cost Analysis," *Public Choice* vol. 36, pp. 235–252.

Brookshire, David S., Ralph C. d'Arge, and William D. Schulze. 1979. "Experiments in Valuing Non-marketed Goods: A Case Study of Alternative Benefit Measures of Air Pollution Control in the South Coast Air Basin of Southern California," in *Methods Development for Assessing Tradeoffs in Environmental Management,* vol. 2, EPA-60076-79-0016 (Washington, D.C., NTIS).

Brookshire, David S., Ralph C. d'Arge, William D. Schulze, and Mark A. Thayer. 1981. "Experiments in Valuing Public Goods," in V. Kerry Smith, ed., *Advances in Applied Microeconomics* (Greenwich, Conn., JAI Press).

Brookshire, David S., and Larry S. Eubanks. 1978. "Contingent Valuation and Revealing Actual Demand for Public Environmental Commodities," manuscript, University of Wyoming.

Brookshire, David S., Larry S. Eubanks, and Alan Randall. 1983. "Estimating Option Price and Existence Values for Wildlife Resources," *Land Economics* vol. 59, no. 1, pp. 1–15.

Brookshire, David S., Larry S. Eubanks, and Cindy F. Sorg. 1986. "Existence Values and Normative Economics: Implications for Valuing Water Resources," *Water Resources Research* vol. 22, no. 11, pp. 1509–1518.

Brookshire, David S., Berry C. Ives, and William D. Schulze. 1976. "The Valuation of Aesthetic Preferences," *Journal of Environmental Economics and Management* vol. 3, no. 4, pp. 325–346.

Brookshire, David S., and Alan Randall. 1978. "Public Policy Alternatives, Public Goods, and Contingent Valuation Mechanisms." Paper presented at the Western Economic Association Meeting, Honolulu, Hawaii.

Brookshire, David S., Alan Randall, and John R. Stoll. 1980. "Valuing Increments and Decrements in Natural Resource Service Flows," *American Journal of Agricultural Economics* vol. 62, no. 3, pp. 478–488.

Brookshire, David S., William D. Schulze, and Mark A. Thayer. 1985. "Some Unusual Aspects of Valuing a Unique Natural Resource," manuscript, University of Wyoming.

Brookshire, David S., Mark A. Thayer, William P. Schulze, and Ralph C. d'Arge. 1982. "Valuing Public Goods: A Comparison of Survey and Hedonic Approaches," *American Economic Review* vol. 72, no. 1, pp. 165–176.

Brookshire, David S., Mark A. Thayer, John Tschirhart, and William D. Schulze.

1985. "A Test of the Expected Utility Model: Evidence from Earthquake Risks," *Journal of Political Economy* vol. 93, no. 2, pp. 369–389.

Brown, Gardner Mallard, Jr., and Henry O. Pollakowski. 1977. "Economic Valuation of a Shoreline," *Review of Economics and Statistics* vol. 59, pp. 272–278.

Brown, G., J. J. Charbonneau, and M. J. Hay. 1978. "Estimating Values of Wildlife: Analysis of the 1975 Hunting and Fishing Survey," Working Paper no. 7, U.S. Fish and Wildlife Service Division of the Program Plans (Washington, D.C.).

Brown, James N., and Harvey S. Rosen. 1982. "On the Estimation of Structural Hedonic Price Models," *Econometrica* vol. 50, no. 3, pp. 765–768.

Brown, Thomas C. 1984. "The Concept of Value in Resource Allocation," *Land Economics* vol. 60, no. 3, pp. 231–246.

Brubaker, Earl. 1975. "Free Ride, Free Revelation or Golden Rule?" *Journal of Law and Economics* vol. 18, no. 1, pp. 147–159.

Brubaker, Earl. 1982. "Sixty-Eight Percent Free Revelation and Thirty-Two Percent Free Ride? Demand Disclosures Under Varying Conditions of Exclusion," in V. L. Smith, ed., *Research in Experimental Economics,* vol. 2 (Greenwich, Conn., JAI Press).

Buchanan, James M. 1954. "Individual Choice in Voting and the Market," *Journal of Political Economy* vol. 62, pp. 334–344.

Buchanan, James M. 1975. "Public Finance and Public Choice," *National Tax Journal* vol. 28, no. 4, pp. 383–394.

Buchanan, James M., and W. C. Stubblebine. 1962. "Externality," *Economica* vol. 29, pp. 371–384.

Burness, H. S., Ronald G. Cummings, A. F. Mehr, and M. S. Walbert. 1983. "Valuing Policies Which Reduce Environmental Risk," *Natural Resources Journal* vol. 23, pp. 675–682.

Calabresi, Guido, and A. Douglas Melamed. 1972. "Property Rules, Liability Rules, and Inalienability: One View of the Cathedral," *Harvard Law Review* vol. 85, no. 6, pp. 1089–1128.

Callicott, J. Baird. 1986. "On the Intrinsic Value of Nonhuman Species," in Bryan G. Norton, ed., *The Preservation of Species* (Princeton, Princeton University Press).

Cameron, Trudy Ann. Forthcoming. "A New Paradigm for Valuing Non-Market Goods Using Referendum Data: Maximum Likelihood Estimation by Censored Logistic Regression," *Journal of Environmental Economics and Management.*

Cameron, Trudy Ann, and Daniel D. Huppert. 1987. "Non-Market Resource Valuation: Assessment of Value Elicitation by 'Payment Card' versus 'Referendum' Methods." Paper presented at the Western Economic Association meetings, Vancouver, July.

Cameron, Trudy Ann, and M. D. James. 1987. "Efficient Estimation Methods for Use with 'Closed-Ended' Contingent Valuation Survey Data," *Review of Economics and Statistics* vol. 69, pp. 269–276.

Canary, Daniel J., and David R. Seibold. 1983. *Attitudes and Behavior: An Annotated Bibliography* (New York, Praeger).

Cannell, C. F., S. A. Lawson, and D. L. Hausser. 1975. *A Technique for Evaluating Interviewer Performance* (Ann Arbor, Institute for Social Research, University of Michigan).

Cannell, C. F., P. V. Miller, and L. Oksenberg. 1981. "Research on Interviewing Techniques," in S. Leinhardt, ed., *Sociological Methodology, 1981* (San Francisco, Jossey-Bass).

Carmines, Edward G., and Richard A. Zeller. 1979. *Reliability and Validity Assessment* (Beverly Hills, Calif., Sage).

Carroll, James L., and James R. Nest. 1982. "Moral Development," in Benjamin B. Wolman, ed., *Handbook of Developmental Psychology* (Englewood Cliffs, N.J., Prentice-Hall).

Carson, Richard T. 1984. "Compensating for Missing and Invalid Data in Contingent Valuation Surveys," in *Proceedings of the Survey Research Section of the American Statistical Association* (Washington, D.C., American Statistical Association).

Carson, Richard T. 1986a. "Notes on Option Value and Contingent Valuation," Discussion Paper QE86-03, Resources for the Future, Washington, D.C.

Carson, Richard T. 1986b. "A Comparison of Methods for Imputing Missing Values in Survey Data," manuscript, University of California, San Diego.

Carson, Richard T. 1986c. "The Use of Dichotomous Choice Formats for Contingent Valuation: Current Research Issues and Concerns." Paper presented at the U.S. Department of Agriculture Conference on Research Issues in Resource Decisions Involving Marketed and Nonmarketed Goods, San Diego, February.

Carson, Richard T., Gary L. Casterline, and Robert Cameron Mitchell. 1985. "A Note on Testing and Correcting for Starting Point Bias in Contingent Valuation Surveys," Discussion Paper QE85-11, Resources for the Future, Washington, D.C.

Carson, Richard T., and W. E. Foster. 1984. "A Theory of Auctions from the Auctioneer's Perspective." Paper presented at the Econometric Society meetings, Stanford University, Palo Alto, August.

Carson, Richard T., Theodore Graham-Tomasi, Charles F. Rund, and William W. Wade. 1986. "Problems with the MPCA Contingent Valuation Survey Instrument and Survey Results: Exhibit 324," report prepared for Northern States Power by Dames and Moore, San Francisco.

Carson, Richard T., W. Michael Hanemann, and Robert Cameron Mitchell. 1986. "Determining the Demand for Public Goods by Simulating Referendums at Different Tax Prices," manuscript, University of California, San Diego.

Carson, Richard T., Mark Machina, and John Horowitz. 1987. *Discounting Mortality Risks: Final Technical Report to the U.S. Environmental Protection Agency* (La Jolla, Calif., University of California, San Diego).

Carson, Richard T., and Robert Cameron Mitchell. 1983. "A Reestimation of Bishop and Heberlein's Simulated Market-Hypothetical Markets-Travel Cost Results Under Alternative Assumptions," Discussion Paper D-107, Resources for the Future, Washington, D.C.

Carson, Richard T., and Robert Cameron Mitchell. 1986. "The Value of Clean

Water: The Public's Willingness to Pay for Boatable, Fishable, and Swimmable Quality Water," Discussion Paper QE85-08, rev., Resources for the Future, Washington, D.C.

Carson, Richard T., and Robert Cameron Mitchell. 1987. "Economic Value of Reliable Water Supplies for Residential Water Users in the State Water Project Service Area," report prepared for the Metropolitan Water District of Southern California.

Cassady, R. 1967. *Auctions and Auctioneering* (Berkeley, University of California Press).

Caulkins, Peter P., Richard C. Bishop, and Nicolaas W. Bouwes, Sr. 1986. "The Travel Cost Model for Lake Recreation: A Comparison of Two Methods of Incorporating Site Quality and Substitution Effects," *American Journal of Agricultural Economics* vol. 68, no. 2, pp. 291–297.

Cesario, Frank. 1976. "Value of Time in Recreation Benefit Studies," *Land Economics* vol. 55, pp. 32–41.

Chamberlain, Gary, and Michael Rothschild. 1981. "A Note on the Probability of Casting a Decisive Vote," *Journal of Economic Theory* vol. 25, no. 1, pp. 152–162.

Charbonneau, John, and Michael J. Hay. 1978. "Determinants and Economic Values of Hunting and Fishing." Paper presented at the 43d North American Wildlife and Natural Resources Conference, Phoenix.

Chavas, Jean-Paul, Richard C. Bishop, and Kathleen Segerson. 1986. "Ex Ante Consumer Welfare Evaluation in Cost-Benefit Analysis," *Journal of Environmental Economics and Management* vol. 13, no. 3, pp. 255–268.

Christainsen, Gregory B. 1982. "Evidence for Determining the Optimal Mechanism for Providing Collective Goods," *American Economist* vol. 26, no. 1, pp. 57–61.

Cicchetti, Charles J., Anthony C. Fisher, and V. Kerry Smith. 1976. "An Econometric Evaluation of a Generalized Consumer Surplus Measure: The Mineral King Controversy," *Econometrica* vol. 44, pp. 1269–1276.

Cicchetti, Charles J., and A. Myrick Freeman III. 1971. "Option Demand and Consumer's Surplus: Further Comment," *Quarterly Journal of Economics* vol. 85, no. 3, pp. 528–539.

Cicchetti, Charles J., and V. Kerry Smith. 1973. "Congestion, Quality Deterioration, and Optimal Use: Wilderness Recreation in the Spanish Peaks Primitive Area," *Social Science Research* vol. 2, pp. 15–30.

Cicchetti, Charles J., and V. Kerry Smith. 1976a. *The Cost of Congestion* (Cambridge, Mass., Ballinger).

Cicchetti, Charles J., and V. Kerry Smith. 1976b. "The Measurement of Individual Congestion Costs: An Economic Application to a Wilderness Area," in S. A. Lin, ed., *Theory and Measurement of Externalities* (New York, Academic Press).

Ciriacy-Wantrup, S. V. 1947. "Capital Returns from Soil-Conservation Practices," *Journal of Farm Economics* vol. 29, pp. 1181–1196.

Ciriacy-Wantrup, S. V. 1952. *Resource Conservation: Economics and Policies* (Berkeley, University of California Press).

Citrin, Jack. 1979. "Do People Want Something for Nothing: Public Opinion on Taxes and Government Spending," *National Tax Journal* vol. 32, no. 2, pp. 113–129.

Clarke, Edward H. 1971. "Multipart Pricing of Public Goods," *Public Choice* vol. 11, pp. 19–33.

Clarke, Edward H. 1975. "Experimenting with Public Goods Pricing: A Comment," *Public Choice* vol. 23, pp. 49–53.

Clarke, Edward H. 1980. *Demand Revelation and the Provision of Public Goods* (Cambridge, Mass., Ballinger).

Clawson, Marion. 1959. "Methods of Measuring the Demand for and the Value of Outdoor Recreation," Reprint no. 10, Resources for the Future, Washington, D.C.

Clawson, Marion, and Jack Knetsch. 1966. *Economics of Outdoor Recreation* (Baltimore, The Johns Hopkins University Press for Resources for the Future).

Coase, Ronald H. 1960. "The Problem of Social Cost," *Journal of Law and Economics* vol. 3, pp. 1–44.

Coase, Ronald H. 1974. "The Lighthouse in Economics," *Journal of Law and Economics* vol. 17, pp. 357–376.

Cocheba, D. J., and W. A. Langford. 1978. "Wildlife Valuation: The Collective Good Aspect of Hunting," *Land Economics* vol. 54, pp. 490–504.

Cochran, William G. 1973. "Experiments for Nonlinear Functions," *Journal of the American Statistical Association* vol. 68, no. 344, pp. 771–781.

Cochran, William G. 1977. *Sampling Techniques* (3d ed., New York, Wiley).

Cochran, William G., and G. M. Cox. 1957. *Experimental Designs* (2d ed., New York, Wiley).

Cohen, Jacob. 1962. "The Statistical Power of Abnormal-Social Psychological Research," *Journal of Abnormal and Social Psychology* vol. 65, pp. 145–153.

Cohen, Jacob. 1977. *Statistical Power Analysis for the Behavioral Sciences* (rev. ed., New York, Academic Press).

Coleman, James S. "Social Theory, Social Research, and a Theory of Action," *American Journal of Sociology* vol. 91, pp. 1309–1335.

Conlisk, John. 1973. "Choice of Response Function in Designing Subsidy Experiments," *Econometrica* vol. 41, pp. 643–656.

Conrad, Jon M. 1980. "Quasi-Option Value and the Expected Value of Information," *Quarterly Journal of Economics* vol. 85, pp. 813–820.

Conrad, Jon M., and David LeBlanc. 1979. "The Supply of Development Rights: Results from a Survey in Hadley, Massachusetts," *Land Economics* vol. 55, no. 2, pp. 269–276.

Converse, Jean M., and Stanley Presser. 1986. *Survey Questions: Handcrafting the Standardized Questionnaire* (Beverly Hills, Calif., Sage).

Converse, Philip E. 1964. "The Nature of Belief Systems in Mass Publics," in David E. Apter, ed., *Ideology and Discontent* (New York, Free Press) pp. 206–261.

Converse, Philip E. 1970. "Attitudes and Non-Attitudes: Continuation of a Dialogue," in E. R. Tufte, ed., *The Quantitative Analysis of Social Problems* (Reading, Mass., Addison-Wesley).

Converse, Philip E. 1974. "Comment: The Status of Nonattitudes," *American Political Science Review* vol. 68, no. 2. pp. 650–660.

Conway, James M., and Richard T. Carson. 1984. "The Demand for State-Level Assistance: An Examination of a 1980 California Referendum," manuscript, University of California, Berkeley.

Coppinger, V. M., V. L. Smith, and J. A. Titus. 1980. "Incentives and Behavior in English, Dutch, and Sealed Bid Auctions," *Economic Inquiry* vol. 18, pp. 1–22.

Cornes, Richard, and Todd Sandler. 1984. "Easy Riders, Joint Production, and Public Goods," *Economic Journal* vol. 94. pp. 580–598.

Cornes, Richard, and Todd Sandler. 1986. *The Theory of Externalities, Public Goods, and Club Goods* (New York, Cambridge University Press).

Cottrell, Leonard S., Jr., and Sylvia Eberhart. 1948. *American Opinion on World Affairs in the Atomic Age* (Princeton, Princeton University Press).

Couch, A., and K. Keniston. 1960. "Yeasayers and Naysayers: Agreeing Response Set as a Personality Variable," *Journal of Abnormal and Social Psychology* vol. 60, pp. 151–174.

Coursey, Don L., John Hovis, and William D. Schulze. 1987. "The Disparity Between Willingness to Accept and Willingness to Pay Measures of Value," *Quarterly Journal of Economics* vol. 102, pp. 679–690.

Coursey, Don L., and William D. Schulze. 1986. "The Application of Laboratory Experimental Economics to the Contingent Valuation of Public Goods," *Public Choice* vol. 49, no. 1, pp. 47–68.

Covello, Vincent, W. Gary Flamm, Joseph V. Rodricks, and Robert G. Tardiff, eds. 1983. *The Analysis of Actual Versus Perceived Risks* (New York, Plenum).

Cox, D. R. 1961. "Test of Separate Families of Hypotheses," in *Proceedings of the Fourth Berkeley Symposium on Mathematical Statistics and Probability,* vol. 1 (Berkeley, University of California Press) pp. 105–123.

Cox, D. R. 1970. *Analysis of Binary Data* (London, Methuen).

Craig, C. Samuel, and John M. McCann. 1978. "Item Nonresponse in Mail Surveys: Extent and Correlates," *Journal of Marketing Research* vol. 15, pp. 285–289.

Cramer, J. S. 1964. "Efficient Grouping, Regression, and Correlation in Engel Curve Analysis," *Journal of the American Statistical Association* vol. 59, pp. 233–250.

Cramer, J. S. 1987. "Mean and Variance of R^2 in Small and Moderate Samples," *Journal of Econometrics* vol. 35, pp 253–266.

Crenson, Matthew A. 1971. *The Un-politics of Air Pollution: A Study of Non-decisonmaking in the Cities* (Baltimore, The Johns Hopkins University Press).

Crespi, I. 1971. "What Kinds of Attitude Measures Are Predictive of Behavior?" *Public Opinion Quarterly* vol. 35, pp. 327–334.

Crocker, T. D. 1984. "On the Value of the Condition of a Forest Stock," manuscript, Department of Economics, University of Wyoming.

Cronin, Francis J. 1982. "Valuing Nonmarket Goods Through Contingent Markets," report to the U.S. Environmental Protection Agency (Richland, Wash., Pacific Northwest Laboratory and Battelle Memorial Institute).

Cullis, John, and Alan Lewis. 1985. "Some Hypotheses and Evidence on Tax Knowledge and Preferences," *Journal of Economic Psychology* vol. 6, pp. 271–287.

Cummings, Ronald G., David S. Brookshire, and William D. Schulze, eds. 1986. *Valuing Environmental Goods: A State of the Arts Assessment of the Contingent Method* (Totowa, N.J., Rowman and Allanheld).

Cummings, Ronald G., Louis Anthony Cox, Jr., and A. Myrick Freeman III. 1984. "General Methods for Benefits Assessment," in Arthur D. Little Co., comp., *Evaluation of the State-of-the-Art in Benefits Assessment Methods for Public Policy Purposes,* report to the Division of Policy Research and Analysis, National Science Foundation (Cambridge, Mass.)

Cummings, Ronald G., William D. Schulze, Shelby D. Gerking, and David S. Brookshire. 1986. "Measuring the Elasticity of Substitution of Wages for Municipal Infrastructure: A Comparison of the Survey and Wage Hedonic Approaches," *Journal of Environmental Economics and Management* vol. 13, no. 3, pp. 269–276.

Curtin, Richard T. 1982. "Indicators of Consumer Behavior: The University of Michigan Surveys of Consumers," *Public Opinion Quarterly* vol. 46, no. 3, pp. 340–352.

Curtis, T. D., and E. W. Shows. 1982. "Economic and Social Benefits of Artificial Beach Nourishment Civil Works at Delray Beach," report to the Department of Natural Resources, Division of Beaches and Shores, STAR Grant no. 81-046, Department of Economics, University of South Florida.

Curtis, T. D., and E. W. Shows. 1984. "A Comparative Study of Social Economic Benefits of Artificial Beach Nourishment Civil Works in Northeast Florida," report to the Department of Natural Resources, Division of Beaches and Shores, STAR Grant, Department of Economics, University of South Florida.

d'Arge, Ralph C. 1985. "Water Quality Benefits: Analysis of the Lakes at Okoboji, Iowa," draft report for the U.S. Environmental Protection Agency.

d'Arge, Ralph C., William D. Schulze, and David S. Brookshire. 1980. "Benefit-Cost Valuation of Long Term Future Effects: The Case of CO_2." Paper presented at the Resources for the Future/National Climate Program Office Workshop, Fort Lauderdale, Fla.

Darling, Arthur H. 1973. "Measuring Benefits Generated by Urban Water Parks," *Land Economics* vol. 49, no. 1, pp. 22–34.

Daubert, John T., and Robert A. Young. 1981. "Recreational Demands for Maintaining Instream Flows: A Contingent Valuation Approach," *American Journal of Agricultural Economics* vol. 63, no. 4, pp. 666–676.

David, Martin, Roderick J. A. Little, Michael E. Samuhel, and Robert K. Triest. 1986. "Alternative Methods for CPS Income Imputation," *Journal of the American Statistical Association* vol. 81, no. 393, pp. 29–41.

Davidson, Andrew R., and Diane M. Morrison. 1982. "Social Psychological Models of Decision Making," in *Choice Models for Buyer Behavior,* Supplement 1 to *Research in Marketing,* pp. 91–112.

Davis, Robert K. 1963a. "Recreation Planning as an Economic Problem," *Natural Resources Journal* vol. 3, no. 2, pp. 239–249.

Davis, Robert K. 1963b. "The Value of Outdoor Recreation: An Economic Study of the Maine Woods" (Ph.D. dissertation, Harvard University).

Davis, Robert K. 1964. "The Value of Big Game Hunting in a Private Forest," in *Transactions of the 29th North American Wildlife and Natural Resources Conference* (Washington, D.C., Wildlife Management Institute).

Davis, Robert K. 1980. "Analysis of the Survey to Determine the Effects of Water Quality on Participation in Recreation." Davis to John Parsons, National Park Service, July 28, 1980.

Deacon, Robert, and P. Shapiro. 1975. "Private Preference for Collective Goods Revealed Through Voting on Referenda," *American Economic Review* vol. 65, no. 5, pp. 943–955.

Deaton, Angus, and John Muellbauer. 1980. *Economics and Consumer Behavior* (New York, Cambridge University Press).

Deaton, Brady J., Larry C. Morgan, and Kurt R. Anschel. 1982. "The Influence of Psychic Costs on Rural-Urban Migration," *American Journal of Agricultural Economics* vol. 64, pp. 177–187.

Debreu, G. 1959. *Theory of Value* (New York, Wiley).

DeMaio, Theresa J. 1984. "Social Desirability and Survey Measurement: A Review," in Charles F. Turner and Elizabeth Martin, eds., *Surveying Subjective Phenomena,* vol. 2 (New York, Russell Sage Foundation).

Dempster, A. P., N. M. Laird, and D. B. Rubin. 1977. "Maximum Likelihood from Incomplete Data Via the EM Algorithm," *Journal of the Royal Statistical Society* vol. 39, pp. 1–38.

Dennis, Steve, and R. W. Hodgson. 1984. "Starting Point Bias in Contingent Valuation Methods: An Experiment." Paper presented at the U.S. Department of Agriculture W-133 Annual Meeting, Las Vegas, February.

Denzau, Arthur T., and Robert P. Parks. 1983. "Existence of Voting-Market Equilibria," *Journal of Economic Theory* vol. 30, pp. 243–265.

Department of the Interior. 1985. "Proposed Rule for Natural Resource Damage Assessments under the Comprehensive Environmental Response, Compensation, and Liability Act of 1980" (CERCLA), *Federal Register* vol. 50, no. 245 (December 20).

Department of the Interior. 1986. "Final Rule for Natural Resource Damage Assessments under the Comprehensive Environmental Response, Compensation, and Liability Act of 1980" (CERCLA), *Federal Register* vol. 51, no. 148 (August 1) pp. 27674–27753.

Desvousges, William H., V. Kerry Smith, Diane H. Brown, and D. Kirk Pate. 1984. "The Role of Focus Groups in Designing a Contingent Valuation Survey to Measure the Benefits of Hazardous Waste Management Regulations," draft report for the U.S. Environmental Protection Agency (Research Triangle Park, N.C., Research Triangle Institute).

Desvousges, William H., V. Kerry Smith, and Matthew P. McGivney. 1983. "A Comparison of Alternative Approaches for Estimating Recreation and Related Benefits of Water Quality Improvements," EPA-230-05-83-001 (Washington, D.C., Office of Policy Analysis, U.S. Environmental Protection Agency).

Devine, D. Grant, and Bruce W. Marion. 1979. "The Influence of Consumer Price

Information on Retail Pricing and Consumer Behavior," *American Journal of Agricultural Economics* vol. 61, pp. 228–237.

Deyak, Timothy A., and V. Kerry Smith. 1978. "Congestion and Participation in Outdoor Recreation: A Household Production Function Approach," *Journal of Environmental Economics and Management* vol. 5, no. 1, pp. 63–80.

Dickie, Mark, Ann Fisher, and Shelby Gerking. 1987. "Market Transactions and Hypothetical Demand Data: A Comparative Study," *Journal of the American Statistical Association* vol. 82, no. 397, pp. 69–75.

Dijkstra, Wil, and Johannes van der Zouwen. 1982. Introduction, in Wil Dijkstra and Johannes van der Zouwen, eds., *Response Behavior in the Survey-Interview* (New York, Academic Press).

Dillman, Don A. 1978. *Mail and Telephone Surveys—The Total Design Method* (New York, Wiley).

Dillman, Don A. 1983. "Mail and Other Self-Administered Questionnaires," in Peter H. Rossi, James D. Wright, and Andy B. Anderson, eds., *Handbook of Survey Research* (New York, Academic Press).

Donnelly, Dennis M., John B. Loomis, Cindy F. Sorg, and Louis J. Nelson. 1985. "Net Economic Value of Recreational Steelhead Fishing in Idaho," Resource Bulletin RM-9, Rocky Mountain Forest and Range Experiment Station, U.S. Forest Service, Fort Collins, Colo.

Dorfman, Robert. 1977. "Incidence of the Benefits and Costs of Environmental Programs," *American Economic Review Papers and Proceedings* vol. 67, pp. 333–340.

Downs, A. 1957. *An Economic Theory of Democracy* (New York, Harper and Row).

Dreze, J., and D. De la Vallee Poussin. 1971. "A Tatonnement Process for Public Goods," *Review of Economic Studies* vol. 38, pp. 133–150.

Driver, Beverly L., Thomas L. Brown, and William R. Burch, Jr. 1986. "Toward More Comprehensive and Integrated Valuations of Public Amenity Goods and Services." Paper presented at the Workshop on Integrating Psychology and Economics in Valuation of Amenity Resources, Estes Park, Colo., May.

Duffield, J. 1984. "Travel Cost and Contingent Valuation: A Comparative Analysis," in V. Kerry Smith, ed., *Advances in Applied Microeconomics,* vol. 3 (Greenwich, Conn., JAI Press) pp. 67–87.

DuMouchel, William H., and Greg J. Duncan. 1983. "Using Sample Survey Weights in Multiple Regression Analysis of Stratified Samples," *Journal of the American Statistical Association* vol. 78, no. 383, pp. 535–543.

Dwyer, John F., John R. Kelly, and Michael D. Bowes. 1977. *Improved Procedures for Valuation of the Contribution of Recreation to National Economic Development* (Urbana-Champaign, Ill., Water Resources Center, University of Illinois).

Eagleton Institute of Politics. 1984. "Toxic Wastes: Panic Lessens but Problem Remains a Major Concern," press release dated 11 March, issued at Rutgers University, New Brunswick, N.J.

Eastman, Clyde, Alan Randall, and Peggy L. Hoffer. 1974. "How Much to Abate Pollution," *Public Opinion Quarterly* vol. 38, pp. 575–584.

Eastman, Clyde, Alan Randall, and Peggy L. Hoffer. 1978. "A Socioeconomic Analysis of Environmental Concern: Case of the Four Corners Electric Power

Complex," Bulletin 626, Agricultural Experiment Station, University of New Mexico, Albuquerque.

Economic Analysis, Inc. and Applied Science Associates. 1986. "Measuring Damages to Coastal and Marine Natural Resources: Concepts and Data Relevant for CERCLA Type A Damage Assessments," 2 vols., prepared for the CERCLA 301 Project, U.S. Department of the Interior, Washington, D.C.

ECO Northwest. 1984. "Economic Valuation of Potential Loss of Fish Populations in the Swan River Drainage," report prepared for the Montana Department of Fish, Wildlife, and Parks.

Edwards, Steven F. 1984. "An Analysis of the Non-Marketed Benefits of Protecting the Salt Water Pond Quality in Southern Rhode Island: An Application of the Hedonic Price and Contingent Valuation Techniques" (Ph.D. dissertation, University of Rhode Island).

Edwards, Steven F. 1985. "Genuine Altruism and Intrinsic Value: Implications for Contingent Valuation Research on Existence Values," manuscript, Woods Hole (Mass.) Oceanographic Institution.

Edwards, Steven F., and Glen D. Anderson. 1987. "Overlooked Biases in Contingent Valuation Surveys: Some Considerations," *Land Economics* vol. 62, no. 2, pp. 168–178.

Efron, Brad. 1969. "Student's t-Test Under Symmetry Conditions," *Journal of the American Statistical Association* vol. 64, pp. 1278–1303.

Emerson, John D. 1983. "Mathematical Aspects of Transformation," in David C. Goaglin, Frederick Mosteller, and John W. Tukey, eds., *Understanding Robust and Exploratory Data Analysis* (New York, Wiley).

Enelow, James M., and Melvin J. Hinich. 1984. *The Spatial Theory of Voting: An Introduction* (Cambridge, Cambridge University Press).

Environmental Defense Fund. 1984. *The Tuolumne River: Preservation or Development? An Economic Assessment* (Berkeley, Calif., Environmental Defense Fund).

Erskine, Hazel. 1971. "The Polls: Pollution and Its Costs," *Public Opinion Quarterly* vol. 35, pp. 120–135.

Etzioni, Amitai. 1968. *The Active Society* (New York, Free Press).

Etzioni, Amitai. 1985. "Opening the Preferences: A Socio-Economic Research Agenda," *Journal of Behavioral Economics* vol. 14, pp. 183–205.

Evans, John S. 1984. "Theoretically Optimal Metrics and Their Surrogates," *Journal of Environmental Economics and Management* vol. 10, pp. 1–10.

Evans, Roy C., and Frederick H. DeB. Harris. 1982. "A Bayesian Analysis of the Free Rider Meta Game," *Southern Economic Journal* vol. 49, pp. 137–149.

Fazio, Russell H., Jeaw-mei Chen, Elizabeth C. McDonel, and Steven J. Sherman. 1982. "Attitude Accessibility, Attitude-Behavior Consistency, and the Strength of the Object-Evaluation Association," *Journal of Experimental Social Psychology* vol. 18, pp. 339–357.

Fazio, Russell H., and Mark P. Zanna. 1981. "Direct Experience and Attitude-Behavior Consistency," in Leonard Berkowitz, ed., *Advances in Experimental Social Psychology,* vol. 14 (New York, Academic Press).

Federal Energy Administration. 1977. "The Surveys of Public Attitudes and Response to Federal Energy Policy," data prepared by Opinion Research Corp. (Ann Arbor, Inter-University Consortium for Political and Social Research).

Fee, J. 1979. "Symbols and Attitudes: How People Think About Politics" (Ph.D. dissertation, University of Chicago).

Feenberg, Daniel, and Edwin S. Mills. 1980. *Measuring the Benefits of Water Pollution Abatement* (New York, Academic Press).

Ferber, Robert, and Werner Z. Hirsch. 1982. *Social Experimentation and Economic Policy* (New York, Cambridge University Press).

Ferris, James M. 1983. "Demand for Public Spending: An Attitudinal Approach," *Public Choice* vol. 40, no. 2, pp. 135–154.

Fields, James M., and Howard Schuman. 1976. "Public Beliefs About the Beliefs of the Public," *Public Opinion Quarterly* vol. 40, pp. 427–449.

Fienberg, Stephen E., and Judith M. Tanur. 1985. "A Long and Honorable Tradition: Intertwining Concepts and Constructs in Experimental Design and Sample Surveys." Paper presented at the International Statistical Institute Meeting, Amsterdam.

Findlater, P. A., and J. A. Sinden. 1982. "Estimation of Recreation Benefits from Measured Utility Functions," *American Journal of Agricultural Economics* vol. 64, pp. 102–109.

Finney, David. 1978. *Statistical Methods in Biological Assay* (3d ed., New York, Macmillan).

Fischel, William A. 1979. "Determinants of Voting on Environmental Quality: A Study of a New Hampshire Pulp Mill Referendum," *Journal of Environmental Economics and Management* vol. 6, pp. 107–118.

Fischer, David Hackett. 1970. *Historians' Fallacies* (New York, Harper and Row).

Fischhoff, Baruch, Paul Slovic, and Sarah Lichtenstein. 1980. "Knowing What You Want: Measuring Labile Values," in Thomas S. Wallsten, ed., *Cognitive Processes in Choice and Decision Behavior* (Hillsdale, N.J., Lawrence Erlbaum Associates).

Fishbein, Martin, and Icek Ajzen. 1975. *Belief, Attitude, Intention and Behavior: An Introduction to Theory and Research* (Reading, Mass., Addison-Wesley).

Fisher, Ann, Gary H. McClelland, and William D. Schulze. 1986. "Measures of Willingness to Pay versus Willingness to Accept: Evidence, Explanations, and Potential Reconciliation." Paper presented at the Workshop on Integrating Psychology and Economics in Valuation of Amenity Resources, Estes Park, Colo., May.

Fisher, Ann, and Robert Raucher. 1984. "Intrinsic Benefits of Improved Water Quality: Conceptual and Empirical Perspectives," in V. Kerry Smith, ed., *Advances in Applied Economics* (Greenwich, Conn., JAI Press).

Fisher, Anthony C. 1981. *Resources and Environmental Economics* (New York, Cambridge University Press).

Fisher, Anthony C., and W. Michael Hanemann. 1985a. "Endangered Species and the Economics of Irreversible Damage," in D. O. Hall, N. Myers, and N. S. Margaris, eds., *Economics of Ecosystem Management* (Dordrecht, Netherlands, Kluwer Academic Publishers Group).

Fisher, Anthony C., and W. Michael Hanemann. 1985b. "Valuing Pollution Controls: The Hysteresis Phenomenon in Aquatic Ecosystems," Giannini Foundation Working Paper no. 361, University of California, Berkeley.

Fisher, Anthony C., and W. Michael Hanemann. 1986. "Option Value and the Extinction of Species," in V. Kerry Smith, ed., *Advances in Applied Microeconomics,* vol. 4 (Greenwich, Conn., JAI Press).

Fisher, Anthony C., and W. Michael Hanemann. 1987. "Quasi-Option Value: Some Misconceptions Dispelled," *Journal of Environmental Economics and Management* vol. 14, pp. 183–190.

Fleiss, Joseph L. 1981. *Statistical Methods for Rates and Proportions* (2d ed., New York, Wiley).

Ford, I., and S. D. Silvey. 1980. "A Sequentially Constructed Design for Estimating a Non-Linear Parametric Function," *Biometrika* vol. 67, pp. 381–388.

Fort, Rodney, and Jon B. Christianson. 1981. "Determinants of Public Services Provision in Rural Communities: Evidence from Voting on Hospital Referenda," *American Journal of Agricultural Economics* vol. 63, no. 2, pp. 228–236.

Foster, John, J. Halstead, and T. H. Stevens. 1982. "Measuring the Non-Market Value of Agricultural Land: A Case Study," Research Bulletin no. 672, Massachusetts Agricultural Experiment Station, College of Food and Natural Resources, University of Massachusetts, Amherst.

Foxall, Gordon. 1984. "Evidence for Attitudinal-Behavioral Consistency: Implications for Consumer Research Paradigms," *Journal of Economic Psychology* vol. 5, pp. 71–92.

Frankel, M. 1979. "Opportunity and the Valuation of Life," preliminary report for the Department of Economics, University of Illinois, Urbana-Champaign.

Frankel, Martin. 1983. "Sampling Theory," in Peter H. Rossi, James D. Wright, and Andy B. Anderson, eds., *Handbook of Survey Research* (New York, Academic Press).

Frankena, William K. 1973. *Ethics* (2d ed., Englewood Cliffs, N.J., Prentice-Hall).

Freeman, A. Myrick III. 1979a. "The Benefits of Air and Water Pollution Control: A Review and Synthesis of Recent Estimates," report to the Council on Environmental Quality (Washington, D.C., Council on Environmental Quality).

Freeman, A. Myrick III. 1979b. *The Benefits of Environmental Improvement: Theory and Practice* (Baltimore, The Johns Hopkins University Press for Resources for the Future).

Freeman, A. Myrick III. 1979c. "Hedonic Prices, Property Values and Measuring Environmental Benefits: A Survey of the Issues," *Scandinavian Journal of Economics* vol. 81, pp. 154–173.

Freeman, A. Myrick III. 1981. "Notes on Defining and Measuring Existence Values," manuscript, Department of Economics, Bowdoin College.

Freeman, A. Myrick III. 1982. *Air and Water Pollution Control: A Benefit Cost Assessment* (New York, Wiley).

Freeman, A. Myrick III. 1984a. "The Quasi-Option Value of Irreversible Development," *Journal of Environmental Economics* vol. 11, pp. 292–295.

Freeman, A. Myrick III. 1984b. "The Size and Sign of Option Value," *Land Economics* vol. 60, pp. 1–13.

Freeman, A. Myrick III. 1986. "On Assessing the State of the Arts of the Contingent Valuation Method of Valuing Environmental Changes," in Ronald G. Cummings, David S. Brookshire, and William D. Schulze, eds., *Valuing Environmental Goods* (Totawa, N. J., Rowman and Allanheld).

Freiman, Jennie A., Thomas C. Chalmers, Harry Smith, Jr., and Roy R. Kuebler. 1978. "The Importance of Beta, the Type II Error and Sample Size in the Design and Interpretation of the Randomized Control Trial," *New England Journal of Medicine* vol. 299, pp. 690–694.

Freudenburg, William R., and Rodney K. Baxter. 1985. "Nuclear Reactions: Public Attitudes and Policies Toward Nuclear Power," *Policy Studies Review* vol. 5, no. 1, pp. 96–110.

Frey, James H. 1983. *Survey Research by Telephone* (Beverly Hills, Calif., Sage).

Friedman, Lee S. 1984. *Microeconomic Policy Analysis* (New York, McGraw-Hill).

Fromm, Gary. 1968. "Comment on T. C. Schelling's Paper, 'The Life You Save May Be Your Own,' " in S. B. Chape, ed., *Problems in Public Expenditure Analysis* (Washington, D.C., Brookings Institution).

Furubotn, Eirik, and Svetozar Pejorvich. 1972. "Property Rights and Economic Theory: A Survey of Recent Literature," *Journal of Economic Literature* vol. 10, no. 4, pp. 1137–1162.

Gallagher, David R., and V. Kerry Smith. 1985. "Measuring Values for Environmental Resources Under Uncertainty," *Journal of Environmental Economics and Management* vol. 12, pp. 132–143.

Garbacz, Christopher, and Mark A. Thayer. 1983. "An Experiment in Valuing Senior Companion Program Services," *Journal of Human Resources* vol. 18, pp. 147–153.

Georgescu-Roegen, N. 1958. "Threshhold in Choice and the Theory of Demand," *Econometrica* vol. 26, pp. 157–168.

Gibbard, A. 1973. "Manipulation of Voting Schemes: A General Result," *Econometrica* vol. 41, pp. 587–601.

Gibson, Betty Blecha. 1980. "Estimating Demand Elasticities for Public Goods from Survey Data," *American Economic Review* vol. 70, no. 5, pp. 1069–1076.

Gillroy, John M., and Robert Y. Shapiro. 1986. "The Polls: Environmental Protection," *Public Opinion Quarterly* vol. 50, no. 2, pp. 270–279.

Goldberger, Arthur S. 1968. "The Interpretation and Estimation of Cobb-Douglas Functions," *Econometrica* vol. 35, pp. 464–472.

Goldman, Steven. 1978. "Gift Equilibria and Pareto Optimality," *Journal of Economic Theory* vol. 18, pp. 368–370.

Goldsmith, Barbara J., and Company. 1986. "Comments in Response to Proposed Rule on Natural Resource Damage Assessments," report to the U.S. Department of the Interior, submitted by Allied-Signal, CIBA-Geigy, and the General Electric Company.

Golob, Thomas F., Abraham D. Horowitz, and Martin Wachs. 1979. "Attitude-Behavior Relationships in Travel-Demand Modelling," in David A. Heusher and Peter R. Stolpfer, eds., *Behavioral Travel Modelling* (London, Croom Helm).

Gordon, Irene M., and Jack L. Knetsch. 1979. "Consumer's Surplus Measures and the Evaluation of Resources," *Land Economics* vol. 55, pp. 1–10.

Gove, Walter R. 1982. "Systematic Response Bias and Characteristics of the Respondent," in Wil Dijkstra and Johannes van der Zouwen, eds., *Response Behavior in the Survey-Interview* (New York, Academic Press).

Graham, Daniel A. 1981. "Cost-Benefit Analysis Under Uncertainty," *American Economic Review* vol. 71, pp. 715–725.

Graham-Tomasi, Theodore. 1986. "Irreversibility and Uncertainty in State-Level Acid Deposition Policy Formulation," manuscript, School of Natural Resources, University of Michigan.

Graham-Tomasi, Theodore, and Frank Wen. 1987. "Option Value and the Bias for Ignoring Uncertainty," manuscript, School of Natural Resources, University of Michigan.

Gramlich, Edward, and Daniel Rubinfeld. 1982. "Using Micro Data to Estimate Public Spending Demand Functions and Test the Tiebout and Median Voter Hypotheses," *Journal of Political Economy* vol. 90, pp. 336–360.

Gramlich, Frederick W. 1975. "Estimating the Net Benefits of Improvements in Charles River Quality" (Ph.D. dissertation, Harvard University).

Gramlich, Frederick W. 1977. "The Demand for Clean Water: The Case of the Charles River," *National Tax Journal* vol. 30, no. 2, pp. 183–194.

Green, Jerry R., and Jean-Jacques Laffont. 1978. "A Sampling Approach to the Free Rider Problem," in Agnar Sandmo, ed., *Essays in Public Economics* (Lexington, Mass., Lexington Books).

Green, Jerry R., and Jean-Jacques Laffont. 1979. *Incentives in Public Decision Making* (Amsterdam, North-Holland).

Green, P. E., and V. Srinivasan. 1978. "Conjoint Analysis in Consumer Research: Issues and Outlook," *Journal of Consumer Research* vol. 3, pp. 103–123.

Greenley, Douglas A., Richard G. Walsh, and Robert A. Young. 1981. "Option Value: Empirical Evidence from a Case Study of Recreation and Water Quality," *Quarterly Journal of Economics* vol. 96. no. 4, pp. 657–672.

Greenley, Douglas A., Richard G. Walsh, and Robert A. Young. 1982. *Economic Benefits of Improved Water Quality: Public Perceptions of Option and Preservation Values* (Boulder, Colo., Westview Press).

Greenley, Douglas A., Richard G. Walsh, and Robert A. Young. 1985. "Option Value: Empirical Evidence from a Case Study of Recreation and Water Quality: Reply," *Quarterly Journal of Economics* vol. 100, no. 1, pp. 294–299.

Gregory Research. 1982. "The Economic Value of the British Columbia Provincial Museum," report to the Friends of the British Columbia Provincial Museum, Vancouver.

Gregory, Robin. 1982. *Valuing Non-Market Goods: An Analysis of Alternative Approaches* (Ph.D. dissertation, University of British Columbia).

Gregory, Robin. 1986. "Interpreting Measures of Economic Loss: Evidence from Contingent Valuation and Experimental Studies," *Journal of Environmental Economics and Management* vol. 13, pp. 325–337.

Gregory, Robin, and Richard C. Bishop. 1986. "Willingness to Pay or Compensation Demanded." Paper presented at the Workshop on Integrating Psychology and Economics in Valuing Public Amenity Resources, Estes Park, Colo., May.

Gregory, Robin, and Lita Furby. 1985. "Auctions, Experiments, and Contingent Valuation," draft manuscript, Decision Research, Eugene, Ore.

Grether, D. M., and C. R. Plott. 1979. "Economic Theory of Choice and the Preference Reversal Phenomenon," *American Economic Review* vol. 69, no. 4, pp. 623–638.

Griliches, Zvi, ed. 1971. *Price Indexes and Quality Change* (Cambridge, Mass., Harvard University Press).

Groves, Robert M., and R. Kahn. 1979. *Comparing Telephone and Personal Interview Surveys* (New York, Academic Press).

Groves, Robert M., and Lou J. Magilavy. 1986. "Measuring and Explaining Interviewer Effects in Centralized Telephone Surveys," *Public Opinion Quarterly* vol. 50, no. 2, pp. 251–266.

Groves, Theodore. 1973. "Incentive in Teams," *Econometrica* vol. 41, pp. 617–631.

Groves, Theodore, and John O. Ledyard. 1977. "Optimal Allocation of Public Goods: A Solution to the Free Rider Problem," *Econometrica* vol. 45, no. 4, pp. 783–809.

Groves, Theodore, and John O. Ledyard. 1987. "Incentive Compatibility since 1972," in T. Groves, R. Radner, and S. Reiter, eds., *Information, Incentives, and Economic Mechanisms* (Minneapolis, University of Minnesota Press).

Groves, Theodore, and Martin Loeb. 1975. "Incentives and Public Inputs," *Journal of Public Economics* vol. 4, pp. 211–226.

Gruson, Lindsey. 1986. "Widespread Illiteracy Burdens the Nation," *New York Times,* 22 July.

Guttman, J. 1978. "Understanding Collective Action: Matching Behavior," *American Economic Review Papers and Proceedings* vol. 68, pp. 251–255.

Hageman, Rhonda. 1985. "Valuing Marine Mammal Populations: Benefit Valuations in a Multi-Species Ecosystem," administrative report no. LJ-85-22, Southwest Fisheries Center, National Marine Fisheries Service, La Jolla, Calif.

Halstead, John M. 1984. "Measuring the Nonmarket Value of Massachusetts Agricultural Land: A Case Study," *Northeastern Journal of Agriculture and Resource Economics* vol. 14, pp. 12–19.

Halvorsen, Robert, and Henry O. Pollakowski. 1981. "Choice of Functional Form for Hedonic Price Functions," *Journal of Urban Economics* vol. 10, no. 1, pp. 37–49.

Hammack, Judd, and Gardner Mallard Brown, Jr. 1974. *Waterfowl and Wetlands: Toward Bioeconomic Analysis* (Baltimore, The Johns Hopkins University Press for Resources for the Future).

Hammerton, M., M. W. Jones-Lee, and V. Abbott, 1982. "The Consistency and Coherence of Attitudes to Physical Risk. Some Empirical Evidence," *Journal of Transport Economics and Policy,* May, pp. 181–199.

Hammitt, James K. 1986. "Estimating Consumer Willingness to Pay to Reduce Food-Borne Risks," R-3447-EPA, report to the U.S. Environmental Protection Agency (Santa Monica, Calif., Rand Corp.).

Hanemann, W. Michael. 1978. "A Methodological and Empirical Study of the Recreation Benefits from Water Quality Improvement" (Ph.D. dissertation, Harvard University).

Hanemann, W. Michael. 1980. "Measuring the Worth of Natural Resource Facilities: Comment," *Land Economics* vol. 56, no. 4, pp. 482–486.

Hanemann, W. Michael. 1981. "Some Further Results on Exact Consumer Surplus," Giannini Foundation Working Paper no. 190, University of California, Berkeley.

Hanemann, W. Michael. 1982. "Quality and Demand Analysis," in Gordon C. Rausser, ed., *New Directions in Econometric Modeling and Forecasting in U.S. Agriculture* (New York, North-Holland).

Hanemann, W. Michael. 1983a. "Marginal Welfare Measures for Discrete Choice Models," *Economics Letters* vol. 13, pp. 129–136.

Hanemann, W. Michael. 1983b. "Welfare Evaluation with Simulated and Hypothetical Market Data: Bishop and Heberlein Revisited," Giannini Foundation Working Paper no. 276, University of California, Berkeley.

Hanemann, W. Michael. 1984a. "Discrete/Continuous Models of Consumer Demand," *Econometrica* vol. 52, no. 3, pp. 541–561.

Hanemann, W. Michael. 1984b. "Statistical Issues in the Discrete-Response Contingent Valuation Studies," *Northeastern Journal of Agriculture and Resource Economics* vol. 14, pp. 5–12.

Hanemann, W. Michael. 1984c. "Welfare Evaluations in Contingent Valuation Experiments with Discrete Responses," *American Journal of Agricultural Economics* vol. 66, pp. 332–341.

Hanemann, W. Michael. 1984d. "On Reconciling Two Concepts of Option Value," Giannini Foundation Working Paper no. 295, University of California, Berkeley.

Hanemann, W. Michael. 1986a. "Willingness to Pay and Willingness to Accept: How Much Can They Differ?" draft manuscript, Department of Agricultural and Resource Economics, University of California, Berkeley.

Hanemann, W. Michael. 1986b. "Implications from Biometrics for the Design of Dichotomous Choice Contingent Valuation Markets." Paper presented at the U.S. Department of Agriculture Conference of Research Issues in Resource Decisions Involving Marketed and Nonmarketed Goods, San Diego, February.

Hanemann, W. Michael. 1986c. "Weak Complementarity Revisited." Working paper, University of California, Berkeley.

Hanemann, W. Michael. Forthcoming. "Information and the Concept of Option Value," *Journal of Environmental Economics and Management.*

Hansen, Christopher. 1977. "A Report on the Value of Wildlife," Intermountain Regional Office, U.S. Forest Service, Ogden, Utah.

Hansen, Morris A., William G. Madow, and Benjamin J. Tepping. 1983. "An Evaluation of Model-Dependent and Probability-Sampling Inference in Sample Surveys," with discussion, *Journal of the American Statistical Association* vol. 78, pp. 776–807.

Hansen, William J., and Allan S. Mills. 1983. "Different Results with Two Contingent Valuation Measures." Paper presented to the Research Methodology and Statistics Session of the National Recreation and Parks Association Research Symposium, Kansas City, October.

Hardie, Ian, and Ivar E. Strand. 1979. "Measurement of Economic Benefits for Potential Public Goods," *American Journal of Agricultural Economics* vol. 61, no. 2, pp. 311–317.

Hardin, Russell. 1982. *Collective Action* (Baltimore, The Johns Hopkins University Press for Resources for the Future).

Hargreaves, George, John D. Claxton, and Frederick H. Siller. 1976. "New Product Evaluation: Electric Vehicles for Commercial Applications," *Journal of Marketing* vol. 40, no. 1, pp. 74–77.

Harris, B. S. 1984. "Contingent Valuation of Water Pollution Control," *Journal of Environmental Management* vol. 19, pp. 199–208.

Harris, B. S., and A. D. Meister. 1981. "A Report on the Use of a Travel Cost Demand Model for Recreation Analysis in New Zealand: An Evaluation of Lake Tutira," Discussion Paper in Natural Resource Economics no. 4, Department of Agricultural Economics and Farm Management, Massey University, Palmerston North, New Zealand.

Harris, Charles C., B. L. Driver, and W. J. McLaughlin. c. 1985. "Assessing Contingent Valuation Methods from a Psychological Perspective," draft manuscript, Department of Wildland Recreation Management, University of Idaho, Moscow.

Harris, Charles C., Howard E. A. Tinsley, and Dennis M. Donnelly. 1986. "Research Methods for Amenity Resource Valuation: Issues and Recommendations." Paper presented at the Workshop on Integrating Psychology and Economics in Valuation of Amenity Resources, Estes Park, Colo., May.

Harris, Louis, and Associates. 1969. "Study Number 1939," done for the National Wildlife Federation. Louis Harris Data Center, University of North Carolina, Chapel Hill.

Harrison, David, and Daniel L. Rubinfeld. 1978. "The Distribution of Benefits from Improvements in Urban Air Quality," *Journal of Environmental Economics and Management* vol. 5, pp. 313–332.

Harsanyi, John C. 1978. "Bayesian Decision Theory and Utilitarian Ethics," *American Economic Review* vol. 68, no. 2, pp. 223–228, 231–232.

Harstad, Ronald M., and Michael Marrese. 1982. "Behavioral Explanations of Efficient Public Good Allocations," *Journal of Public Economics* vol. 19, pp. 367–383.

Harter, H. L. and D. B. Owen, eds. 1975. *Selected Tables in Mathematical Statistics,* vol. 3 (Providence, R. I., American Mathematical Society).

Hasselblad, Victor, Andrew G. Stead, and J. P. Creason. 1980. "Multiple Probit with Non-Zero Background," *Biometrics* vol. 36, pp. 659–663.

Hasselblad, Victor, Andrew G. Stead, and Warren Galke. 1980. "Analysis of Coarsely Grouped Data from the Lognormal Distribution," *Journal of the American Statistical Association* vol. 75, no. 272, pp. 771–778.

Hausman, Jerry A. 1979. "Individual Discount Rates and the Purchase and Utilization of Energy-Using Durables," *Bell Journal of Economics* vol. 10, pp. 33–54.

Hausman, Jerry A. 1981. "Exact Consumer Surplus and Dead Weight Loss," *American Economic Review* vol. 71, pp. 662–676.

Hausman, Jerry A., and David Wise, eds. 1985. *Social Experimentation* (Chicago, University of Chicago Press for the National Bureau of Economic Research).

Haveman, Robert H., and Burton Weisbrod. 1975. "The Concept of Benefits in Cost-Benefit Analysis: With Emphasis on Water Pollution Control Activities," in Henry M. Peskin and Eugene P. Seskin, eds., *Cost-Benefit Analysis and Water Pollution Policy* (Washington, D.C., Urban Institute).

Hay, Michael J., and Kenneth E. McConnell. 1979. "An Analysis of Participation in Nonconsumptive Wildlife Recreation," *Land Economics* vol. 55, pp. 460–471.

Hayes, Michael T. 1981. *Lobbyists and Legislators: A Theory of Political Markets* (New Brunswick, N. J., Rutgers University Press).

Heberlein, Thomas A. 1986. "Measuring Resource Values: The Reliability and Validity of Dichotomous Contingent Valuation Measures." Paper presented at the American Sociological Association Meeting, New York, August.

Heberlein, Thomas A., and Robert Baumgartner. 1978. "Factors Affecting Response Rates to Mailed Questionnaires: A Quantitative Analysis of the Published Literature," *American Sociological Review* vol. 43, no. 4, pp. 447–462.

Heberlein, Thomas A., and Richard C. Bishop. 1986. "Assessing the Validity of Contingent Valuation: Three Field Experiments," *Science of the Total Environment* vol. 56, pp. 99–107.

Heberlein, Thomas A., and J. S. Black. 1976. "Attitudinal Specificity and the Prediction of Behavior in a Field Setting," *Journal of Personality and Social Psychology* vol. 33, pp. 434–479.

Heckman, J. J. 1976. "The Common Structure of Statistical Models of Truncation, Sample Selection, and Limited Dependent Variables and a Simple Estimation for Such Models," *Annals of Economic and Social Measurement* vol. 5, pp. 475–492.

Heckman, J. J. 1979. "Sample Selection Bias as a Specification Error," *Econometrica* vol. 47, pp. 153–161.

Hedlund, Ronald D., and H. Paul Friesema. 1972. "Representatives' Perceptions of Constituency Opinion," *Journal of Politics* vol. 34, pp. 730–752.

Heintz, H. T., A. Hershaft, and G. C. Horak. 1976. *National Damages of Air and Water Pollution* (Rockville, Md., Enviro Control Inc.).

Henry, Elaude. 1974. "Option Values in the Economics of Irreplaceable Assets," *Review of Economic Studies* vol. 64, pp. 89–104.

Hensher, D. A., J. K. Stanley, and P. McLeod. 1975. "Usefulness of Attitudinal Measures in Investigating the Choice of Travel Mode," *International Journal of Transport Economics* vol. 2, pp. 51–75.

Hicks, John R. 1939. "The Foundations of Welfare Economics," *Economic Journal* vol. 49, pp. 696–700, 711–712.

Hicks, John R. 1941. "The Rehabilitation of Consumer's Surplus," *Review of Economics Studies* vol. 8, pp. 108–116.

Hicks, John R. 1943. "The Four Consumer Surpluses," *Review of Economic Studies* vol. 11, pp. 31–41.

Hicks, John R. 1956. *A Revision of Demand Theory* (Oxford, Clarendon Press).

Hill, Richard J. 1981. "Attitudes and Behavior," in Morris Rosenberg and Ralph H. Turner, eds., *Social Psychology: Sociological Perspectives* (New York, Basic Books).

Hirsch, Werner Z. 1979. *Law and Economics: An Introductory Analysis* (New York, Academic Press).

Hockley, G. C., and G. Harbour. 1983. "Revealed Preference Between Public Expenditures and Taxation Cuts: Public Section Choice," *Journal of Public Economics* vol. 22, no. 3, pp. 387–391.

Hodges, J. L., and E. L. Lehmann. 1968. "A Compact Table for the Power of the t-Test," *Annals of Mathematical Statistics* vol. 39, pp. 1629–1637.

Hoehn, John P. 1983. "The Benefits-Costs Evaluation of Multi-Part Public Policy: A Theoretical Framework and Critique of Estimation Methods" (Ph.D. dissertation, University of Kentucky).

Hoehn, John P., and Alan Randall. 1982. "Aggregation and Disaggregation of Program Benefits in a Complex Policy Environment: A Theoretical Framework and Critique of Estimation Methods." Paper presented at the American Agricultural Economics Association summer meetings, Logan, Utah.

Hoehn, John P., and Alan Randall. 1983. "Incentives and Performance in Contingent Policy Valuation." Paper presented at the American Agricultural Economics Association summer meetings, Purdue University.

Hoehn, John P., and Alan Randall. 1985a. "A Satisfactory Benefit Cost Indicator from Contingent Valuation," Staff Paper 85-4, Department of Agricultural Economics, Michigan State University.

Hoehn, John P., and Alan Randall. 1985b. "Demand Based and Contingent Valuation: An Empirical Comparison." Paper presented at the Annual American Agricultural Economics Association meeting, Ames, Iowa, August.

Hoehn, John P., and Alan Randall. 1986. "Too Many Proposals Pass the Benefit-Cost Test," Staff Paper no. 86-22, Department of Agricultural Economics, Michigan State University.

Hoehn, John P., and Alan Randall. 1987. "A Satisfactory Benefit Cost Indicator from Contingent Valuation," *Journal of Environmental Economics and Management* vol. 14, no. 3, pp. 226–247.

Hoehn, John P., and Cindy F. Sorg. 1986. "Toward a Satisfactory Model of Contingent Valuation Behavior." Paper presented at the Workshop on Integrating Economic and Psychological Knowledge in Valuations of Public Amenity Resources, Estes Park, Colo., May.

Holloway, Robert T. 1967. "An Experiment on Consumer Dissonance," *Journal of Marketing* vol. 31, pp. 39–43.

Hori, H. 1975. "Revealed Preference for Public Goods," *American Economic Review* vol. 65, pp. 978–991.

Horvarth, J. C. 1974. "Detailed Analysis: Survey of Wildlife Recreation," manuscript, Georgia State University.

Houthakker, Henry, and L. D. Taylor. 1970. *Consumer Demand in the United States 1929–1970* (2d ed., Cambridge, Mass., Harvard University Press).

Hovis, John, Don C. Coursey, and William D. Schulze, 1983. "A Comparison of Alternative Valuation Mechanisms for Non-Marketed Commodities," manuscript, Department of Economics, University of Wyoming, Laramie.

Huber, Peter J. 1981. *Robust Statistics* (New York, Wiley).

Hurwicz, L. 1972. "On Informationally Decentralized Systems," in R. Radner and C. B. McGuire, eds., *Decisions and Organization* (Amsterdam, North-Holland).

Isaac, R. Mark, Kenneth F. McCue, and Charles R. Plott. 1985. "Public Goods Provision in an Experimental Environment," *Journal of Public Economics* vol. 26, no. 1, pp. 51–74.

Isaac, R. Mark, James H. Walker, and Susan H. Thomas. 1984. "Divergent Evidence on Free Riding: An Experimental Examination of Possible Explanations," *Public Choice* vol. 43, pp. 113–149.

Ito, P. K. 1980. "Robustness of ANOVA and MANOVA Test Procedures," in P. R. Krishnaiah, ed., *Handbook of Statistics,* vol. 1 (Amsterdam: North-Holland).

Jabine, Thomas B., Miron L. Straf, Judith M. Tanur, and Roger Tourangeau, eds. 1984. *Cognitive Aspects of Survey Methodology: Building a Bridge Between Disciplines* (Washington, D.C., National Research Council/National Academy Press).

Jackman, M. R. 1973. "Education and Prejudice or Education and Response-set?" *American Sociological Review* vol. 38, pp. 327–339.

Jackson, John E. 1983. "Measuring the Demand for Environmental Quality with Survey Data," *Journal of Politics* vol. 45, pp. 335–350.

Johansen, L. 1963. "Some Notes on the Lindahl Theory of Determination of Public Expenditures," *International Economic Review* vol. 4, pp. 346–358.

Johansen, L. 1977. "The Theory of Public Goods: Misplaced Emphasis," *Journal of Public Economics* vol. 7, pp. 147–152.

Johansen, L. 1982. "On the Status of the Nash Type of Noncooperative Equilibrium Theory," *Scandinavian Journal of Economics* vol. 84, pp. 421–441.

Johnson, Norman L., and Samuel Kotz. 1970. *Continuous Univariate Distributions,* vol. 1 (New York: Wiley).

Johnson, Rebecca, Bo Shelby, and Neil Bregenzer. 1986. "Comparison of Contingent Valuation Method Results: Dichotomous Choice versus Open-Ended Response," manuscript, Oregon State University.

Jones, Frank D. 1975. "A Survey Technique to Measure Demand Under Various Pricing Strategies," *Journal of Marketing* vol. 39, no. 3, pp. 75–77.

Jones-Lee, M. W. 1976. *The Value of Life: An Economic Analysis* (Chicago, University of Chicago Press).

Jones-Lee, M. W., M. Hammerton, and R. R. Phillips. 1985. "The Value of Safety: Results from a National Survey," *Economic Journal* vol. 95, pp. 49–72.

Judd, Charles M., and J. A. Krosnick. 1981. "Attitude Centrality, Organization, and Measurement," *Journal of Personality and Social Psychology* vol. 42, pp. 436–447.

Judd, Charles M., and Michael A. Milburn. 1980. "The Structure of Attitude Systems in the General Public: Comparisons of a Structural Equation Model," *American Sociological Review* vol. 45, pp. 627–643.

Judge, George G., William E. Griffiths, R. Carter Hill, and Tsoung-Chao Lee. 1980. *The Theory and Practice of Econometrics* (New York, Wiley).

Just, Richard E., Darrell L. Hueth, and Andrew Schmitz. 1982. *Applied Welfare Economics and Public Policy* (Englewood Cliffs, N.J., Prentice-Hall).

Juster, F. Thomas. 1966. "Consumer Buying and Purchase Probability. An Experiment in Survey Design," *American Statistical Association Journal* vol. 61, pp. 658–696.

Juster, F. Thomas. 1969. "Consumer Anticipations and Models of Durable Goods Demand," in Jacob Mincer, ed., *Economic Forecasts and Expectations* (New York, National Bureau of Economic Research).

Kahneman, Daniel. 1986. "Comments," in Ronald G. Cummings, David S. Brookshire, and William D. Schulze, eds., *Valuing Environmental Goods* (Totawa, N.J., Rowman and Allanheld).

Kahneman, Daniel, Jack L. Knetsch, and Richard Thaler. 1986. "Fairness as a Constraint on Profit Seeking: Entitlements in the Market," *American Economic Review* vol. 76, no. 4, pp. 728–741.

Kahneman, Daniel, Paul Slovic, and Amos Tversky, eds. 1982. *Judgement Under Uncertainty: Heurestics and Biases* (New York, Cambridge University Press).

Kahneman, Daniel, and Amos Tversky. 1979. "Prospect Theory: An Analysis of Decisions Under Risk," *Econometrica* vol. 47, no. 2, pp. 263–291.

Kahneman, Daniel, and Amos Tversky. 1982. "The Psychology of Preferences," *Scientific American* vol. 246, no. 1, pp. 160–173.

Kaldor, N. 1939. "Welfare Propositions of Economics and Interpersonal Comparisons of Utility," *Economic Journal* vol. 49, pp. 549–551.

Kalton, Graham. 1983. *Compensating for Missing Survey Data* (Ann Arbor, Survey Research Center, University of Michigan).

Kalton, Graham, and Howard Schuman. 1982. "The Effect of the Question on Survey Responses: A Review," *Journal of the Royal Statistical Society,* series A, vol. 145, pp. 42–73.

Kamieniecki, Sheldon. 1980. *Public Representation in Environmental Policy-making: The Case of Water Quality Management* (Boulder, Colo., Westview Press).

Kanuk, Leslie, and Conrad Berenson. 1975. "Mail Surveys and Response Rates: A Literature Review," *Journal of Marketing Research* vol. 12, pp. 440–453.

Katona, George. 1975. *Psychological Economics* (New York, Elsevier).

Katsenbaum, M. A., D. G. Hoel, and K. O. Bowman. 1970. "Sample Size Requirements for One-Way Analysis of Variance," *Biometrika* vol. 57, pp. 421–430.

Kealy, Mary Jo, John F. Dovidio, and Mark L. Rockel. 1986. "On Assessing the Magnitude of Bias in Contingent Values." Paper, Department of Economics, Colgate University.

Keene, Karlyn. 1984. "What Do We Know About the Public's Attitude on Progressivity?" *National Tax Journal* vol. 36, pp. 371–376.

Keeney, Ralph L., and Howard Raiffa. 1976. *Decisions with Multiple Objectives: Preferences and Value Tradeoffs* (New York, Wiley).

Kelly, S., and T. W. Mirer. 1974. "The Simple Act of Voting," *American Political Science Review* vol. 68, pp. 572–591.

Kelman, Steven. 1981. *What Price Incentives—Economists and the Environment* (Boston, Auburn House).

Kendall, M. G., and A. Stuart. 1973. *The Advanced Theory of Statistics,* vol. 2 (New York, Hafner).

Kim, Oliver, and Mark Walker. 1984. "The Free Rider Problem: Experimental Evidence," *Public Choice* vol. 43, pp. 3–24.

Kinder, Donald R., and David O. Sears. 1985. "Public Opinion and Political Action," in Gardner Lindzey and Elliot Aronson, eds., *Handbook of Social Psychology,* vol. 2 (New York, Random House).

Kirsch, Irwin S., and Ann Jungeblut. 1986. *Literacy: Profiles of America's Young Adults,* National Assessment of Education Progress Report no. 16-Pl-02 (Princeton, Educational Testing Service).

Kish, L. 1965. *Survey Sampling* (New York, Wiley).

Kleindorfer, Paul, and Howard Kunreuther. 1986. "Ex Ante and Ex Post Problems: Economic and Psychological Considerations." Paper presented at the Workshop on Integrating Economic and Psychological Knowledge in Valuations of Public Amenity Resources, Estes Park, Colo., May.

Kneese, Allen V. 1984. *Measuring the Benefits of Clean Air and Water* (Washington, D.C., Resources for the Future).

Kneese, Allen V., and William D. Schulze. 1985. "Ethics and Environmental Economics," in Allen V. Kneese and James L. Sweeney, eds., *Handbook of Natural Resource and Energy Economics,* vol. 1 (Amsterdam, North-Holland).

Knetsch, Jack L. 1983. *Property Rights and Compensation: Compulsory Acquisition and Other Losses* (Toronto, Butterworth).

Knetsch, Jack L. 1985. "Values, Biases and Entitlements," *Annals of Regional Science* vol. 19, pp. 1–9.

Knetsch, Jack L., and Robert K. Davis. 1966. "Comparisons of Methods for Recreation Evaluation," in Allen V. Kneese and Stephen C. Smith, eds., *Water Research* (Baltimore, The Johns Hopkins University Press for Resources for the Future).

Knetsch, Jack L., and J. A. Sinden. 1984. "Willing to Pay and Compensation Demanded: Experimental Evidence of an Unexpected Disparity in Measures of Value," *Quarterly Journal of Economics* vol. 94, no. 3, pp. 507–521.

Kohut, Andrew. 1983. "Illinois Politics Confound the Polls," *Public Opinion* vol. 5, no. 6, pp. 42–43.

Koopmans, L. H., D. B. Owens, and J. I. Rosenblatt. 1964. "Confidence Intervals for the Coefficient of Variation for the Normal and Log Normal Distributions," *Biometrika* vol. 51, pp. 25–32.

Kopp, Raymond J., and Paul R. Portney. 1985. "Valuing the Outputs of Environmental Programs: A Scoping Study," report prepared for the Electric Power Research Institute (Washington, D.C., Resources for the Future).

Kraemer, Helena, and Sue Thiemann. 1987. *How Many Subjects: Statistical Power Analysis in Research* (Beverly Hills, Calif., Sage).

Kragt, Alphons van de, John M. Orbell, and Robyn M. Daubes. 1982. "The Minimal Contribution Set as a Solution to Public Goods," *American Political Science Review* vol. 77, pp. 112–122.

Kramer, C. Y. 1956. "Extensions of Multiple Range Test to Group Means with Unequal Numbers of Replications," *Biometrics* vol. 12, pp. 307–310.

Krasker, William S., Edwin Kuh, and Roy E. Welsch. 1983. "Estimation for Dirty Data and Flawed Models," in Zvi Griliches and Michael D. Intrilligator, eds., *Handbook of Econometrics,* vol. 1 (Amsterdam, North-Holland).

Krebs, D. 1970. "Altruism: An Examination of the Concept and a Review of the Literature," *Psychological Bulletin* vol. 73, pp. 258–302.

Kreps, D. M., P. Milgram, J. Roberts, and R. Wilson. 1982. "Rational Cooperation in Finitely Repeated Prisoner's Dilemma," *Journal of Economic Theory* vol. 27, no. 2, pp. 245–252.

Krutilla, John V. 1960. *Sequencing and Timing in River Basin Development* (Washington, D.C., Resources for the Future).

Krutilla, John V. 1967. "Conservation Reconsidered," *American Economic Review* vol. 57, pp. 787–796.

Krutilla, John V., Charles J. Cicchetti, A. Myrick Freeman, III, and Clifford S.

Russell. 1972. "Observations on the Economics of Irreplaceable Assets," in Allen V. Kneese and Blair T. Bower, eds., *Environmental Quality Analysis: Theory and Method in the Social Sciences* (Baltimore, The Johns Hopkins University Press for Resources for the Future).

Krutilla, John V., and Anthony C. Fisher. 1975. *The Economics of Natural Environments: Studies in the Valuation of Commodity and Amenity Resources* (Baltimore, The Johns Hopkins University Press for Resources for the Future).

Kunreuther, Howard. 1976. "Limited Knowledge and Insurance Protection," *Public Policy* vol. 24, pp. 227–261.

Kurz, Mordecai. 1974. "Experimental Approach to the Determination of the Demand for Public Goods," *Journal of Public Economics* vol. 3, pp. 329–348.

Kurz, Mordecai. 1978. "Altruism as an Outcome of Social Interaction," *American Economic Review* vol. 68, no. 2, pp. 216–222.

Ladd, Everett C., and G. Donald Ferree. 1981. "Were the Pollsters Really Wrong?" *Public Opinion* vol. 3, no. 6, pp. 13–20.

Laffont, Jean-Jacques. 1979. *Aggregation and Revelation of Preferences* (Amsterdam, North-Holland).

Lah, David. 1985. "Work Valuation of the California Conservation Corps," manuscript, U.S. Department of Labor, Washington, D.C.

Lancaster, K. 1966. "A New Approach to Consumer Theory," *Journal of Political Economy* vol. 74, pp. 132–157.

Lane, Robert E. 1986. "Market Justice, Political Justice," *American Political Science Review* vol. 80, pp. 383–402.

Langkford, R. Hamilton. 1983. "Exact Consumer's Surplus for Changes in Imposed Quantities," working paper, Department of Economics, State University of New York, Albany.

Langkford, R. Hamilton. 1985. "Preferences of Citizens for Public Expenditures on Elementary and Secondary Education," *Journal of Econometrics* vol. 27, pp. 1–20.

Lansing, John B., and James B. Morgan. 1971. *Economic Survey Methods* (Ann Arbor, Institute for Social Research, University of Michigan).

LaPage, W. F. 1968. "The Role of Fees in Camper's Decisions," USDA Forest Service Research Paper no. 188, Northeastern Forest Experiment Station, Upper Darby, Pa.

La Palombara, Joseph G. 1950. *The Initiative and Referendum in Oregon: 1938–1948* (Corvallis, Oregon State College Press).

LaPiere, R. T. 1934. "Attitudes vs. Actions," *Social Forces* vol. 13, pp. 230–237.

Lareau, Thomas J., and Douglas A. Rae. 1985. "Valuing Diesel Odor Reductions: Results from a Philadelphia Survey," draft manuscript, U.S. Environmental Protection Agency, Washington, D.C.

Larson, Harold J. 1982. *Introduction to Probability Theory and Statistical Inference* (3d ed., New York, Wiley).

Lazarsfeld, P. F., B. Berelson, and H. Gaudet. 1948. *The People's Choice* (New York, Columbia University Press).

Lehmann, E. L. 1975. *Non-Parametrics: Statistical Methods Based on Ranks* (San Francisco, Holden-Day).

Lemert, James B. 1986. "Picking the Winners: Politician vs. Voter Predictions of Two Controversial Ballot Measures," *Public Opinion Quarterly* vol. 50, pp. 208–221.

Lessler, Judith T. 1984. "Measurement Error in Surveys," in Charles F. Turner and Elizabeth Martin, eds., *Surveying Subjective Phenomena,* vol. 2 (New York, Russell Sage Foundation).

Lichtenstein, Sarah, and Paul Slovic. 1973. "Response-Induced Reversals of Preference in Gambling: An Extended Replication in Las Vegas," *Journal of Experimental Psychology* vol. 101, pp. 16–20.

Likert, Rensis. 1951. "The Sample Interview Survey as a Tool of Research and Policy Formation," in D. Lerner and D. Laswell, eds., *The Policy Sciences* (Stanford, Stanford University Press).

Lillard, Lee, James P. Smith, and Finis Welch. 1986. "What Do We Really Know About Wages? The Importance of Nonreporting and Census Imputation," *Journal of Political Economy* vol. 94, pp. 489–506.

Lind, Robert, ed. 1982. *Discounting for Time and Risk in Energy Policy* (Baltimore, The Johns Hopkins University Press for Resources for the Future).

Lindblom, Charles E. 1977. *Politics and Markets: The World's Political Economic Systems* (New York, Basic Books).

Linden, Fabian. 1982. "The Consumer as Forecaster," *Public Opinion Quarterly* vol. 46, no. 3, pp. 353–360.

Arthur D. Little Company. 1984. *Evaluation of the State-of-the-Art in Benefits Assessment Methods for Public Policy Purposes,* report to the Division of Policy Research and Analysis, National Science Foundation (Cambridge, Mass.).

Little, Roderick J. A. 1985. "A Note About Models for Selectivity Bias," *Econometrica* vol. 53, no. 6, pp. 1469–1474.

Loehman, Edna T. 1984. "Willingness to Pay for Air Quality: A Comparison of Two Methods," Staff Paper 84-18, Department of Agricultural Economics, Purdue University.

Loehman, Edna T., S. Berg, A. Arroyo, R. Hedinger, J. Schwartz, M. Shaw, R. Fahien, V. De, D. Rio, W. Rossley, and A. Green. 1979. "Distributional Analysis of Regional Benefits and Cost of Air Quality Control," *Journal of Environmental Economics and Management* vol. 6, pp. 222–243.

Loehman, Edna T., David Boldt, and Kathleen Chaikin. 1981. "Measuring the Benefits of Air Quality Improvements in the San Francisco Bay Area," final report (Menlo Park, Calif., SRI International).

Loehman, Edna T., and Vo Hu De. 1982. "Application of Stochastic Choice Modeling to Policy Analysis of Public Goods: A Case Study of Air Quality Improvements," *Review of Economics and Statistics* vol. 64, no. 3, pp. 474–480.

Loomis, John B. 1987a. "An Economic Evaluation of the Public Trust Resources of Mono Lake," Institute of Ecology Report no. 30, College of Agriculture and Environmental Sciences, University of California, Davis.

Loomis, John B. 1987b. "Test-Retest Reliability of the Contingent Valuation Method," manuscript, Department of Agricultural Economics, University of California, Davis.

Lucas, Robert E. B. 1977. "Hedonic Wage Equations and Psychic Wages in the

Returns to Schooling," *American Economic Review* vol. 67, no. 4, pp. 549–558.

Luce, R. D. 1956. "Semiorder and a Theory of Utility Discrimination," *Econometrica* vol. 24, pp. 178–191.

Luce, R. D., and H. Raiffa. 1957. *Games and Decisions* (New York, Wiley).

Lynn, Frances M. 1986. "The Interplay of Science and Values in Assessing and Regulating Environmental Risks," *Science, Technology, and Human Values* vol. 11, no. 2, pp. 40–50.

Machina, Mark. 1982. "Expected Utility Analysis Without the Interdependence Axiom," *Econometrica* vol. 50, no. 2, pp. 277–323.

Machina, Mark. 1983. "The Economic Theory of Individual Behavior Toward Risk: Theory, Evidence, and New Directions," Technical Report no. 433, Institute for Mathematical Studies in the Social Sciences, Stanford University.

Machina, Mark. 1984. "Temporal Risk and the Nature of Induced Preferences," *Journal of Economic Theory* vol. 31, pp. 199–231.

Madariaga, Bruce, and Kenneth E. McConnell. 1985. "Some Implications of Existence Value." Paper presented at the Association of Environmental and Resource Economics Workshop on Recreation Demand Modeling, Boulder, Colo., May.

Maddala, G. S. 1983. *Limited Dependent Variables and Qualitative Variables in Econometrics* (New York, Cambridge University Press).

Maddox, James G. 1960. "Private and Social Cost of Moving People Out of Agriculture," *American Economic Review* vol. 50, pp. 392–412.

Madow, W. G., H. Nisselson, and I. Olkin. 1983. See Panel on Incomplete Data, 1983.

Magleby, David B. 1984. *Direct Legislation: Voting on Ballot Propositions in the United States* (Baltimore, The Johns Hopkins University Press).

Majid, I., J. A. Sinden, and Alan Randall. 1983. "Benefit Evaluation of Increments to Existing Systems of Public Facilities," *Land Economics* vol. 59, pp. 377–392.

Maler, Karl-Goran. 1974. *Environmental Economics: A Theoretical Inquiry* (Baltimore, The Johns Hopkins University Press for Resources for the Future).

Maler, Karl-Goran. 1985. "Welfare Economics and the Environment," in Allen V. Kneese and James L. Sweeney, eds., *Handbook of Natural Resource and Energy Economics,* vol. 1 (Amsterdam, North-Holland).

Maler, Karl-Goran. n.d. "Some Thoughts on the Distinction Between User and Non-User Values of an Environmental Resource," manuscript, Stockholm School of Economics.

Malinvaud, E. 1971. "A Planning Approach to the Public Goods Production," *Swedish Journal of Economics* vol. 73, pp. 96–112.

Malinvaud, E. 1972. *Lectures on Microeconomic Theory* (Amsterdam, North-Holland).

Margolis, Howard. 1982. *Selfishness, Altruism, and Rationality: A Theory of Social Choice* (New York, Cambridge University Press).

Martin, Elizabeth. 1983. "Surveys as Social Indicators: Problems in Monitoring Trends," in Peter H. Rossi, James D. Wright, and Andy B. Anderson, eds., *Handbook of Survey Research* (New York, Academic Press).

Martinez-Vazquez, J. 1981. "Selfishness Versus Public 'Regardingness' in Voting Behavior," *Journal of Public Economics* vol. 15, pp. 349–361.

Marwell, Gerald, and Ruth E. Ames. 1981. "Economists Free Ride, Does Anyone Else? Experiments on the Provision of Public Goods," pt. 4, *Journal of Public Economics* vol. 15, pp. 295–310.

Mathews, S. B., and Gardner Mallard Brown, Jr. 1970. "Economic Evaluation of the 1967 Sport Salmon Fisheries of Washington," Washington Department of Fisheries Technical Report no. 2, Olympia, Washington.

McCloskey, Donald N. 1983. "The Rhetoric of Economics," *Journal of Economic Literature* vol. 21, no. 2, pp. 481–517.

McConnell, Kenneth E. 1977. "Congestion and Willingness to Pay: A Study of Beach Use," *Land Economics* vol. 53, no. 2, pp. 185–195.

McConnell, Kenneth E. 1983. "Existence and Bequest Value," in Robert P. Rowe and Lauraine G. Chestnut, eds., *Managing Air Quality and Scenic Resources at National Parks and Wilderness Areas* (Boulder, Colo., Westview Press).

McConnell, Kenneth E. 1985. "The Economics of Outdoor Recreation," in Allen V. Kneese and James L. Sweeney, eds., *Handbook of Natural Resource and Energy Economics*, vol. 2 (Amsterdam, North-Holland).

McConnell, Kenneth E., and Ivan E. Strand. 1981. "Measuring the Cost of Time in Recreation Demand Analysis: An Application to Sport Fishing," *American Journal of Agricultural Economics* vol. 63, pp. 153–156.

McCullagh, P., and J. Nelder. 1983. *Generalized Linear Models* (London, Chapman and Hall).

McFadden, Daniel. 1974. "Conditional Logit Analysis of Qualitative Choice Behavior," in P. Zarembka, ed., *Frontiers in Econometrics* (New York, Academic Press).

McKean, John R., and Robert R. Keller. 1982. "The Shaping of Tastes, Pareto Efficiency and Economic Policy," *Journal of Behavioral Economics* vol. 12, no. 1, pp. 23–41.

McKenzie, G. W. 1983. *Measuring Economic Welfare: New Methods* (Cambridge, Cambridge University Press).

McMillan, John. 1979a. "The Free-Rider Problem: A Survey," *Economic Record* vol. 55, pp. 95–107.

McMillan, John. 1979b. "Individual Incentives in the Supply of Public Inputs," *Journal of Public Economics* vol. 12, pp. 87–98.

Mead, R., and R. N. Curnow. 1983. *Statistical Methods in Agriculture and Experimental Biology* (London, Chapman and Hall).

Mendelsohn, Robert. 1986. "Damage Assessments Under Proposed CERCLA Regulations: A Critical Analysis," appendix to Barbara J. Goldsmith and Company, "Comments in Response to Proposed Rule on Natural Resource Damage Assessments," report to the U.S. Department of the Interior submitted by Allied-Signal, CIBA-Geigy, and the General Electric Co.

Mendelsohn, Robert, and Gardner Mallard Brown, Jr. 1983. "Revealed Preference Approaches to Valuing Outdoor Recreation," *Natural Resources Journal* vol. 23, no. 3, pp. 607–618.

Mendelsohn, Robert, and William J. Strang. 1984. "Cost-Benefit Analysis Under Uncertainty: Comment," *American Economic Review* vol. 74, no. 5, pp. 1096–1099.

Meyer, Phillip A. 1974a. "Recreation and Preservation Values Associated with Salmon of the Fraser River," Information Report series no. PAC/IN-74-1,

Environmental Canada Fisheries and Marine Service, Southern Operations Branch, Vancouver.

Meyer, Phillip A. 1974b. "A Comparison of Direct Questioning Methods for Obtaining Dollar Values for Public Recreation and Preservation," Technical Report series 110, no. PAC 17-75-6, Environment Canada, Vancouver.

Meyer, Phillip A. 1980. "Recreational/Aesthetic Values Associated with Selected Groupings of Fish and Wildlife in California's Central Valley," report to the U. S. Fish and Wildlife Service, Sacramento, Calif.

Michalson, Edgar L., and Robert L. Smathers. 1985. "Comparative Estimates of Outdoor Recreation Benefits in the Sawtooth National Park Area, Idaho," Agricultural Economics series no. 249, University of Idaho.

Miles, George A. 1967. "Water Based Recreation in Nevada," report B-14, College of Agriculture, University of Nevada, Reno.

Milgram, Paul R., and Robert J. Weber. 1982. "A Theory of Auctions and Competitive Bidding," *Econometrica* vol. 50, pp. 1089–1123.

Miller, G. A. 1956. "The Magical Number Seven, Plus or Minus Two; Some Limits on our Capacity for Processing Information," *Psychology Review* vol. 53, no. 2, pp. 81–97.

Miller, J. R., and F. Lad. 1984. "Flexibility, Learning, and Irreversibility in Environmental Decisions: A Bayesian Approach," *Journal of Environmental Economics and Management* vol. 11, pp. 161–172.

Miller, Peter V., and Robert M. Groves. 1985. "Matching Survey Responses to Records: An Exploration of Validity in Victimization Reporting," *Public Opinion Quarterly* vol. 49, no. 3, pp. 366–380.

Miller, Ruppert G. 1981. *Simultaneous Statistical Inference* (2d ed., New York, Springer-Verlag).

Miller, Ruppert G., and Jerry W. Halpern. 1980. "Robust Estimators for Quantal Bioassay," *Biometrika* vol. 67, pp. 103–110.

Milleron, T. C. 1972. "Theory of Value with Public Goods: A Survey Article," *Journal of Economic Theory* vol. 5, pp. 419–477.

Mills, Allan S. 1986. "Survey Sampling Issues in Empirical Research." Paper presented at the U.S. Department of Agriculture Conference on Research Issues in Resource Decisions Involving Marketed and Nonmarketed Goods, San Diego, February.

Mills, A. S., Nick E. Mathews, and J. P. Titre. 1982. "Participants' Willingness-to-Pay and Expressed Preferences for River Floating and Camping in Texas." Paper presented at the 85th Annual Meeting of the Texas Academy of Science, San Angelo, Tex.

Milon, J. Walter. 1986. "Strategic Incentives Revisited: Valuation Contexts and Specification Comparability for Local Public Goods." Paper presented at the Annual Meeting of the Association of Environmental and Resource Economists, New Orleans.

Mirrlees, J. A. 1986. "The Theory of Optimal Taxation," in Kenneth J. Arrow and Michael D. Intrilligator, eds., *Handbook of Mathematical Economics*, vol. 3 (Amsterdam, North-Holland).

Mishan, E. J. 1971. "Evaluation of Life and Limb: A Theoretical Approach," *Journal of Political Economy* vol. 90, pp. 827–853.

Mishan, E. J. 1976. *Cost-Benefit Analysis* (2d ed., New York, Praeger).

Mitchell, Robert Cameron. 1979a. "National Environmental Lobbies and the Apparent Illogic of Collective Action," in Clifford S. Russell, ed., *Collective Decision Making* (Baltimore, The Johns Hopkins University Press for Resources for the Future).

Mitchell, Robert Cameron. 1979b. "Silent Spring/Solid Majorities," *Public Opinion* vol. 2, no. 4, pp. 16–20, 55.

Mitchell, Robert Cameron. 1980a. "Polling on Nuclear Power: A Critique of the Polls After Three Mile Island," in Albert H. Cantril, ed., *Polling on the Issues* (Washington, D.C., Seven Locks Press).

Mitchell, Robert Cameron. 1980b. *Public Opinion on Environmental Issues: Results of a National Opinion Survey* (Washington, D.C., Council on Environmental Quality).

Mitchell, Robert Cameron. 1982. "A Note on the Use of the Contingent Valuation Approach to Value Public Services in Developing Nations," report to the World Bank, Washington, D.C.

Mitchell, Robert Cameron. 1984. "Public Opinion and Environmental Politics in the 1970s and 1980s," in Norman J. Vig and Michael E. Kraft, eds., *Environmental Policy in the 1980s: The Impact of the Reagan Administration* (Washington, D.C., Congressional Quarterly Press).

Mitchell, Robert Cameron, and Richard T. Carson. 1981. "An Experiment in Determining Willingness to Pay for National Water Quality Improvements," draft report to the U.S. Environmental Protection Agency, Washington, D.C.

Mitchell, Robert Cameron, and Richard T. Carson. 1984. *A Contingent Valuation Estimate of National Freshwater Benefits: Technical Report to the U.S. Environmental Protection Agency* (Washington, D.C., Resources for the Future).

Mitchell, Robert Cameron, and Richard T. Carson. 1985. "Comment on Option Value: Empirical Evidence from a Case Study of Recreation and Water Quality," *Quarterly Journal of Economics* vol. 100, no. 1, pp. 291–294.

Mitchell, Robert Cameron, and Richard T. Carson. 1986a. "Some Comments on the State of the Arts Report," in Ronald G. Cummings, David S. Brookshire, and William D. Schulze, eds., *Valuing Environmental Goods* (Totawa, N.J., Rowman and Allanheld).

Mitchell, Robert Cameron, and Richard T. Carson. 1986b. "Property Rights, Protest, and the Siting of Hazardous Waste Facilities," *American Economic Review* vol. 76, no. 2, pp. 285–290.

Mitchell, Robert Cameron, and Richard T. Carson. 1986c. "Valuing Drinking Water Risk Reductions Using the Contingent Valuation Method: A Methodological Study of Risks from THM and Giardia," draft report to the U.S. Environmental Protection Agency, Washington, D.C.

Mitchell, Robert Cameron, and Richard T. Carson. 1987a. "Evaluating the Validity of Contingent Valuation Studies," Discussion Paper QE87-06, Quality of the Environment Division, Resources for the Future, Washington, D.C.

Mitchell, Robert Cameron, and Richard T. Carson. 1987b. "How Far Along the Learning Curve Is the Contingent Valuation Method?" Discussion Paper QE87-07, Quality of the Environment Division, Resources for the Future, Washington, D.C.

Moeller, George H., Michael A. Mescher, Thomas A. Moore, and Elwood L.

Shafer. 1980. "The Informal Interview as a Technique for Recreation Research," *Journal of Leisure Research* vol. 12, no. 2, pp. 174–180.

Moore, William. C. 1982. "Concept Testing," *Journal of Business Research* vol. 10, pp. 279–294.

Morey, Edward R. 1981. "The Demand for Site-Specific Recreational Activities: A Characteristics Approach," *Journal of Environmental Economics and Management* vol. 8, no. 4, pp. 345–371.

Morey, Edward R. 1984. "Confuser Surplus," *American Economic Review* vol. 74, no. 1, pp. 163–173.

Morgan, James N. 1978. "Multiple Motives, Group Decisions, Uncertainty, Ignorance, and Confusion: A Realistic Economics of the Consumer Requires Some Psychology," *American Economic Review* vol. 68, no. 2, pp. 58–63.

Morgenstern, Otto. 1973. *On the Accuracy of Economics Observations* (2d ed., Princeton, Princeton University Press).

Morrison, Donald G. 1979. "Purchase Intentions and Purchase Behavior," *Journal of Marketing* vol. 43, pp. 65–74.

Moser, David A., and C. Mark Dunning. 1986. *A Guide for Using the Contingent Value Methodology in Recreation Studies, National Economic Development Procedures Manual—Recreation,* vol. 2, IWR report 86-R-5 (Fort Belvoir, Va., Institute for Water Resources).

Mosteller, Frederick, and John W. Tukey. 1977. *Data Analysis and Regression* (Reading, Mass., Addison-Wesley).

Mueller, Dennis C. 1979. *Public Choice* (New York, Cambridge University Press).

Mueller, Dennis C. 1986. "Rational Egoism Versus Adaptive Egoism as a Fundamental Postulate for a Descriptive Theory of Human Behavior," *Public Choice* vol. 51, no. 1, pp. 3–23.

Mueller, Eva. 1963. "Public Attitudes Toward Fiscal Programs," *Quarterly Journal of Economics* vol. 77, pp. 210–235.

Mulligan, Patricia J. 1978. "Willingness to Pay for Decreased Risk from Nuclear Plant Accidents," Working Paper no. 43, Center for the Study of Environmental Policy, Pennsylvania State University.

Munley, Vincent G., and V. Kerry Smith. 1976. "Learning-by-Doing and Experience: The Case of Whitewater Recreation," *Land Economics* vol. 52, no. 4, pp. 545–553.

Musgrave, Richard A. 1959. *The Theory of Public Finance* (New York, McGraw-Hill).

Myers, James H., and Edward Tauber. 1977. *Market Structure Analysis* (Chicago, American Marketing Association).

Nachtman, Steven C. 1983. "Valuation of the Maroon Valley Mass Transit System for Recreational Visitors" (M.A. thesis, School of Forestry and Natural Resources, Colorado State University).

Nash, Christopher A. 1983. "The Theory of Social Cost Measurement," in Robert H. Haveman and Julius Margolis, eds., *Public Expenditure and Policy Analysis* (3d ed., Boston, Houghton Mifflin).

National Opinion Research Center. 1983. *General Social Surveys, 1972–1983: Cumulative Codebook* (Storrs, Conn., The Roper Center).

Nisbett, Richard, and Lee Ross. 1980. *Human Inference: Strategies and Shortcomings of Social Judgement* (Englewood Cliffs, N.J., Prentice-Hall).

Norton, Bryan G., ed. 1986. *The Preservation of Species. The Value of Biological Diversity* (Princeton, Princeton University Press).

Ofiara, D. D., and J. R. Allison. 1985. "On Assessing the Benefits of Public Mosquito Control Practices," manuscript, Georgia Agricultural Experiment Station, University of Georgia.

Olson, Mancur. 1965. *The Logic of Collective Action* (Cambridge, Mass., Harvard University Press).

O'Neill, William B. n.d. "Estimating the Recreational Value of Maine Rivers: An Experiment with the Contingent Valuation Technique," manuscript, Colby College, Waterville, Maine.

Oppenheimer, Joe A. 1985. "Public Choice and Three Ethical Properties of Politics," *Public Choice* vol. 45, no. 3, pp. 241–255.

Orbell, John N., Peregrine Schwartz-Shea, and Randy T. Simmons. 1984. "Do Cooperators Exit More Readily Than Defectors?" *American Political Science Review* vol. 78, no. 1, pp. 147–162.

Orchard, T., and M. A. Woodbury. 1972. "A Missing Information Principle: Theory and Applications," in the *Sixth Berkeley Symposium on Mathematical Statistics and Probability,* vol. 1 (Berkeley, University of California Press).

Oster, Sharon. 1974. "The Incidence of Local Water Pollution Abatement Expenditures: A Case Study of the Merrimack River Basin" (Ph.D. dissertation, Harvard University).

Oster, Sharon. 1977. "Survey Results on the Benefits of Water Pollution Abatement in the Merrimack River Basin," *Water Resources Research* vol. 13, pp. 882–884.

Palfrey, T. R., and H. Rosenthal. 1984. "Participation and the Provision of Discrete Public Goods: A Strategic Analysis," *Journal of Public Economics* vol. 24, pp. 171–193.

Panel on Incomplete Data. 1983. *Incomplete Data in Sample Surveys,* vol. 3 (New York, Academic Press).

Panel on Survey Management of Subjective Phenomena, Committee on National Statistics. 1981. *Surveys of Subjective Phenomena: Summary Report,* prepared for the Assembly of Behavioral and Social Sciences, National Research Council (Washington, D.C., National Academy Press).

Parsons, Talcott. 1951. *The Social System* (Glencoe, Ill., Free Press).

Pauly, Mark, Howard Kunreuther, and James Vaupel. 1984. "Public Protection Against Misperceived Risks: Insights from Positive Political Economy," *Public Choice* vol. 43, no. 1, pp. 45–64.

Payne, Stanley L. 1951. *The Art of Asking Questions* (Princeton, Princeton University Press).

Pendse, Dilip, and J. B. Wyckoff. 1976. "Measurement of Environmental Tradeoffs and Public Policy: A Case Study," *Water Resources Bulletin* vol. 12, no. 5, pp. 919–930.

Penz, G. Peter. 1986. *Consumer Sovereignty and Human Interests* (New York, Cambridge University Press).

Peskin, Henry M., and Eugene P. Seskin, eds. 1975. *Cost-Benefit Analysis and Water Pollution Policy* (Washington, D.C., Urban Institute).

Pessemier, E. A. 1960. "An Experimental Method for Estimating Demand," *Journal of Business* vol. 33, pp. 373–383.

Peterson, George L., Beverly L. Driver, and P. J. Brown. 1986. "Benefits of a Recreation: Dollars and Sense." Paper presented at the First Annual Symposium on Social Science and Resource Management, Corvallis, Ore., May.

Peterson, George L., Beverly L. Driver, and Robin Gregory, eds. Forthcoming. *Valuation of Public Amenity Resources and Integration of Economics and Psychology* (State College, Pa., Venture Publishers).

Peterson, George L., and Alan Randall, eds. 1984. *Valuation of Wildland Resource Benefits* (Boulder, Colo., Westview Press).

Philips, Louis. 1974. *Applied Consumption Analysis* (New York, American Elsevier).

Phillips, D. L., and K. J. Clancy. 1970. "Response Bias in Field Studies of Mental Illness," *American Sociological Review* vol. 35, pp. 503–515.

Phillips, D. L., and K. J. Clancy. 1972. "Some Effects of Social Desirability in Survey Studies," *American Journal of Sociology* vol. 77, pp. 921–940.

Pierce, John C., and Douglas D. Rose. 1974. "Nonattitudes and American Public Opinion: The Examination of a Thesis," *American Political Science Review* vol. 68, no. 2, pp. 626–649.

Pigou, A. C. 1952. *The Economics of Welfare* (4th ed., London, Macmillan).

Plott, Charles R. 1982. "Industrial Organization Theory and Experimental Economics," *Journal of Economic Literature* vol. 20, pp. 1485–1527.

Plummer, Mark L., and Richard C. Hartman. 1986. "Option Value: A General Approach," *Economic Inquiry* vol. 24, pp. 455–471.

Polemarchakis, H. M. 1983. "Expectations, Demand, and Observability," *Econometrica* vol. 51, no. 3, pp. 565–574.

Polinsky, A. Mitchell. 1979. "Controlling Externalities and Protecting Entitlements: Property Right, Liability Rule, and Tax Subsidy Approaches," *Journal of Legal Studies* vol. 8, no. 1, pp. 1–48.

Portney, Paul R. 1975. "Voting, Cost-Benefit Analysis, and Water Pollution Policy," in Henry M. Peskin and Eugene P. Seskin, eds., *Cost-Benefit Analysis and Water Pollution Policy* (Washington, D.C., Urban Institute).

Portney, Paul R. 1981. "Housing Prices, Health Effects, and Valuing Reductions of Risk of Death," *Journal of Environmental Economics and Management* vol. 8, pp. 72–78.

Portney, Paul R., and Jon C. Sonstelie. 1979. "Super-Rationality and School Tax Voting," in Clifford S. Russell, ed., *Collective Decision Making: Applications from Public Choice Theory* (Baltimore, The Johns Hopkins University Press for Resources for the Future).

Posner, Richard A. 1977. *Economic Analysis of Law* (2d ed., Boston, Little, Brown).

Praet, Peter, and J. Vuchelen. 1984. "The Contribution of EC Consumer Surveys in Forecasting Consumer Expenditures for Four Major Countries," *Journal of Economic Psychology* vol. 5, pp. 101–124.

Pregibon, D. 1981. "Logistics Regression Diagnostics," *Annals of Statistics* vol. 9, pp. 704–724.

Pregibon, D. 1982. "Resistant Fits for Some Commonly Used Logistic Models with Medical Applications," *Biometrics* vol. 38, pp. 485–495.

Presser, Stanley. 1984. "The Use of Survey Data in Basic Research in the Social

Sciences," in Charles F. Turner and Elizabeth Martin, eds., *Surveying Subjective Phenomena,* vol. 2 (New York, Russell Sage Foundation).

Rae, Douglas A. 1981a. "Visibility Impairment at Mesa Verde National Park: An Analysis of Benefits and Costs of Controlling Emissions in the Four Corners Area," report to the Electric Power Research Institute by Charles River Associates, Boston.

Rae, Douglas A. 1981b. "Benefits of Improving Visibility at Great Smoky National Park," report to the Electric Power Research Institute by Charles River Associates, Boston.

Rae, Douglas A. 1982. "Benefits of Visual Air Quality in Cincinnati," report to the Electric Power Research Institute by Charles River Associates, Boston.

Rae, Douglas, A. 1983. "The Value to Visitors of Improving Visibility at Mesa Verde and Great Smoky National Parks," in Robert D. Rowe and Lauraine G. Chestnut, eds., *Managing Air Quality and Scenic Resources at National Parks and Wilderness Areas* (Boulder, Colo., Westview Press).

Rahmatian, Morteza. 1982. "Estimating the Demand for Environmental Preservation" (Ph.D. dissertation, University of Wyoming).

Randall, Alan. 1986a. "Human Preferences, Economics, and the Preservation of Species," in Bryan G. Norton, ed., *The Preservation of Species* (Princeton, Princeton University Press).

Randall, Alan. 1986b. "The Possibility of Satisfactory Benefit Estimation with Contingent Markets," in Ronald G. Cummings, David S. Brookshire, and William D. Schulze, eds., *Valuing Environmental Goods* (Totawa, N.J., Rowman and Allanheld).

Randall, Alan, Glenn C. Blomquist, John P. Hoehn, and John R. Stoll. 1985. "National Aggregate Benefits of Air and Water Pollution Control," interim report to the U.S. Environmental Protection Agency, Washington, D.C.

Randall, Alan, Orlen Grunewald, Angelos Pagoulatos, Richard Ausness, and Sue Johnson. 1978. "Reclaiming Coal Surface Mines in Central Appalachia: A Case Study of the Benefits and Costs," *Land Economics* vol. 54, no. 4, pp. 427–489.

Randall, Alan, and John P. Hoehn. Forthcoming. "Benefit Estimation for Complex Policies," in H. Folmer and E. van Ireland, eds., *Economics and Policy-Making for Environmental Quality* (Amsterdam, North-Holland).

Randall, Alan, John P. Hoehn, and David S. Brookshire. 1983. "Contingent Valuation Surveys for Evaluating Environmental Assets," *Natural Resources Journal* vol. 23, pp. 635–648.

Randall, Alan, John P. Hoehn, and George S. Tolley. 1981. "The Structure of Contingent Markets: Some Experimental Results." Paper presented at the Annual Meeting of the American Economic Association, Washington, D.C., December.

Randall, Alan, Berry C. Ives, and Clyde Eastman. 1974. "Bidding Games for Valuation of Aesthetic Environmental Improvements," *Journal of Environmental Economics and Management* vol. 1, pp. 132–149.

Randall, Alan, and George L. Peterson. 1984. "The Value of Wildlife Benefits: An Overview," in George L. Peterson and Alan Randall, eds., *Valuation of Wildland Resource Benefits* (Boulder, Colo., Westview Press).

Randall, Alan, and John R. Stoll. 1980. "Consumer's Surplus in Commodity Space," *American Economic Review* vol. 70, no. 3, pp. 449–455.

Randall, Alan, and John R. Stoll. 1982. "Existence Value in a Total Valuation Framework," in Robert D. Rowe and Lauraine G. Chestnut, eds., *Managing Air Quality and Scenic Resources at National Parks and Wilderness Areas* (Boulder, Colo., Westview Press).

Rapoport, A. 1967. "Escape from Paradox," *Scientific American* vol. 217, pp. 50–56.

Rausser, Gordon C., and Eithan Hochman. 1979. *Dynamic Agricultural Systems: Economic Prediction and Control* (New York, North-Holland).

Ray, Anandarup. 1984. *Cost-Benefit Analysis: Issues and Methodologies* (Baltimore, The Johns Hopkins University Press/World Bank).

Reizenstein, Richard C., Gerald E. Hills, and John W. Philpot. 1974. "Willingness to Pay for Control of Air Pollution: A Demographic Analysis," in Ronald C. Curhan, ed., *American Marketing Association 1974 Combined Proceedings,* series no. 36 (Chicago, American Marketing Association).

Research Triangle Institute. 1979. *Field Interviewer's General Manual* (Research Triangle Park, N.C.).

Rhoads, Steven E. 1985. *The Economist's View of the World: Government, Markets, and Public Policy* (New York, Cambridge University Press).

Rich, C. L. 1977. "Is Random Digit Dialing Really Necessary?" *Journal of Marketing Research* vol. 14 (August) pp. 300–305.

Ridker, Ronald G. 1967. *Economic Costs of Air Pollution* (New York, Praeger).

Ridker, Ronald G., and John A. Henning. 1967. "The Determinants of Residential Property Values with Special Reference to Air Pollution," *Review of Economics and Statistics* vol. 49, pp. 246–257.

Riesman, David. 1958. "Some Observations on the Interviewing in the Teacher Apprehension Study," in Paul Lazarsfeld and Wagner Thielens, Jr., eds., *The Academic Mind: Social Scientists in a Time of Crisis* (Glencoe, Ill., Free Press).

Roberts, J. 1976. "Incentives for the Correct Revelation of Preferences and the Number of Consumers," *Journal of Public Economics* vol. 6, pp. 359–374.

Roberts, Kenneth J., Mark E. Thompson, and Perry W. Pawlyk. 1985. "Contingent Valuation of Recreational Diving at Petroleum Rigs, Gulf of Mexico," *Transactions of the American Fisheries Society* vol. 114, no. 2, pp. 214–219.

Romer, T. 1975. "Individual Welfare, Majority Voting, and the Properties of a Linear Income Tax," *Journal of Public Economics* vol. 4, pp. 163–185.

Romer, T., and H. Rosenthal. 1978. "Political Resource Allocation, Controlled Agendas, and the Status Quo," *Public Choice* vol. 33, pp. 27–44.

Romer, T., and H. Rosenthal. 1979. "The Elusive Median Voter," *Journal of Public Economics* vol. 12, pp. 143–170.

Roper, Burns, 1982. "The Predictive Value of Consumer Confidence Measures," *Public Opinion Quarterly* vol. 46, no. 3, pp. 361–367.

Roper, Burns. 1983. "The Polls Malfunction in 1982," *Public Opinion* vol. 5, no. 6, pp. 41–42.

Rosen, Sherwin. 1974. "Hedonic Prices and Implicit Markets: Product Differentiation in Pure Competition," *Journal of Political Economy* vol. 82, pp. 34–55.

Rosen, Sherwin. 1986. "Comments," in Ronald G. Cummings, David S. Brookshire, and William D. Schulze, eds., *Valuing Environmental Goods* (Totawa, N.J., Rowman and Allanheld).

Rossi, Peter H., James D. Wright, and Andy B. Anderson. 1983. "Sample Surveys: History, Current Practice, and Future Prospects," in Peter H. Rossi, James D. Wright, and Andy B. Anderson, eds., *Handbook of Survey Research* (New York, Academic Press).

Rowe, Robert D., and Lauraine G. Chestnut, 1982. *The Value of Visibility: Economic Theory and Applications for Air Pollution Control* (Cambridge, Mass., Abt Books).

Rowe, Robert D., and Lauraine G. Chestnut. 1983. "Valuing Environmental Commodities Revisited," *Land Economics* vol. 59, pp. 404–410.

Rowe, Robert D., and Lauraine G. Chestnut. 1984. "Valuing Changes in Morbidity WTP versus COI Measures" (Denver, Energy and Resource Consultants).

Rowe, Robert D., Ralph C. d'Arge, and David S. Brookshire. 1980. "An Experiment on the Economic Value of Visibility," *Journal of Environmental Economics and Management* vol. 7, pp. 1–19.

Rubinfeld, Daniel. 1977. "Voting in a Local School Election: A Micro Analysis," *Review of Economics and Statistics* vol. 59, no. 1, pp. 30–42.

Rustemeyer, A. 1977. "Measuring Interviewer Performance in Mock Interviews," *Proceedings of the American Statistical Association (Social Statistics Section)* (Washington, D.C.) pp. 341–346.

Ruud, Paul A. 1986. "Review of Visibility Survey Data Using Rank-Ordered Responses," manuscript, University of California, Berkeley.

Samples, Karl C., John A. Dixon, and Marcia M. Gower. 1985. "Information Disclosure and Endangered Species Valuation." Paper presented at the Annual Meeting of the American Agricultural Economics Association, Ames, Iowa, August.

Samuelson, Paul. 1947. *Foundations of Economic Analysis* (Cambridge, Mass., Harvard University Press).

Samuelson, Paul. 1954. "The Pure Theory of Public Expenditure," *Review of Economics and Statistics* vol. 36, pp. 387–389.

Samuelson, Paul. 1955. "Diagrammatic Exposition of a Theory of Public Expenditure," *Review of Economics and Statistics* vol. 37, pp. 350–356.

Samuelson, Paul. 1958. "Aspects of Public Expenditure Theories," *Review of Economics and Statistics* vol. 40, pp. 332–338.

Samuelson, Paul. 1969. "Pure Theory of Public Expenditure and Taxation," in T. Margolis and H. Buitton, eds., *Public Economics* (London, Macmillan).

SAS Institute. 1985. *SAS User's Guide: Statistics Version 5* (Cary, N.C., SAS Institute).

Satterwaite, M. 1975. "Strategy-Proofness and Arrow Conditions: Existence and Correspondence Theorems for Voting Procedures and Welfare Functions," *Journal of Economic Theory* vol. 10, pp. 187–217.

Schall, L. D. 1972. "Interdependent Utilities and Pareto Optimality," *Quarterly Journal of Economics* vol. 86, pp. 19–24.

Scheffe, Henry. 1959. *The Analysis of Variance* (New York, Wiley).

Scheffe, Henry. 1973. "A Statistical Theory of Calibration," *Annals of Statistics* vol. 1, no. 1, pp. 1–37.

Schelling, Thomas C. 1968. "The Life You Save May Be Your Own," in Samuel B. Chase, ed., *Problems in Public Expenditure Analysis* (Washington, D.C., Brookings Institution).

Schelling, Thomas C. 1978. "Altruism, Meanness, and Other Potentially Strategic Behaviors," *American Economic Review* vol. 68, no. 2, pp. 229–230.

Scherr, B. A., and E. M. Babb. 1975. "Pricing Public Goods: An Experiment with Two Proposed Pricing Systems," *Public Choice* vol. 23, pp. 35–48.

Schmalensee, Richard. 1972. "Option Demand and Consumer Surplus: Valuing Price Changes Under Uncertainty," *American Economic Review* vol. 62, pp. 813–824.

Schneider, Friedrich, and Werner W. Pommerehne. 1981. "Free Riding and Collective Action: An Experiment in Public Microeconomics," *Quarterly Journal of Economics* vol. 97, pp. 689–702.

Schoemaker, Paul J. H. 1982. "The Expected Utility Model: Its Variants, Purposes, Evidence, and Limitations," *Journal of Economic Literature* vol. 20, pp. 529–563.

Schulze, William D., and David S. Brookshire. 1983. "The Economic Benefits of Preserving Visibility in the National Parklands of the Southwest," *Natural Resources Journal* vol. 23, pp. 149–173.

Schulze, William D., and Ralph C. d'Arge. 1978. "On the Valuation of Recreational Damages." Paper presented to the Association of Environmental and Resource Economists, New York, December.

Schulze, William D., Ralph C. d'Arge, and David S. Brookshire. 1981. "Valuing Environmental Commodities: Some Recent Experiments," *Land Economics* vol. 57, no. 2, pp. 151–169.

Schulze, William D., David S. Brookshire, E. G. Walther, K. Kelley, Mark A. Thayer, R. L. Whitworth, S. Ben-David, W. Malm, and J. Molenar. 1981. "The Benefits of Preserving Visibility in the National Parklands of the Southwest," vol. 8 of *Methods Development for Environmental Control Benefits Assessment,* final report to the Office of Exploratory Research, Office of Research and Development, U.S. Environmental Protection Agency, grant no. R805059010.

Schulze, William D., Ronald G. Cummings, David S. Brookshire, Mark A. Thayer, R. Whitworth, and M. Rahmatian. 1983. "Methods Development in Measuring Benefits of Environmental Improvements: Experimental Approaches for Valuing Environmental Commodities," vol. 2, draft manuscript of a report to the Office of Policy Analysis and Resource Management, U.S. Environmental Protection Agency, Washington, D.C.

Schulze, W[illiam D.], G. McClelland, B. Hurd, and J. Smith. 1986. "A Case Study of a Hazardous Waste Site: Perspectives from Economics and Psychology," vol. 4, *Improving Accuracy and Reducing Costs of Environmental Benefit Assessments,* draft report to U.S. Environmental Protection Agency, Washington, D.C.

Schuman, Howard, and Michael P. Johnson. 1976. "Attitudes and Behavior," in Alex Inkeles, ed., *Annual Review of Sociology,* vol. 2 (Palo Alto, Annual Reviews, Inc.) pp. 161–207.

Schuman, Howard, and Graham Kalton. 1985. "Survey Methods," in Gardner Lindzey and Elliot Aronson, eds., *Handbook of Social Psychology,* vol. 1 (New York, Random House).

Schuman, Howard, and Stanley Presser. 1981. *Questions and Answers in Attitude Surveys: Experiments on Question Form, Wording, and Context* (New York, Academic Press).

Schwarz, Norbert, Hans-J. Hippler, Brigitte Deutsch, and Fritz Strack. 1985. "Response Scales: Effects of Category Range on Reported Behavior and Comparative Judgments," *Public Opinion Quarterly* vol. 49, no. 3, pp. 388–395.

Scitovsky, Tibor. 1976. *The Joyless Economy* (New York, Oxford University Press).

Scott, Anthony. 1965. "The Valuation of Game Resources: Some Theoretical Aspects," *Canadian Fisheries Report* vol. 4, pp. 27–47.

Scott, W. D., and Company. 1980. "State Pollution Control Commission: Survey of Community Attitudes of Public Willingness to Pay for Clean Air" (Sydney, Australia).

Sellar, Christine, Jean-Paul Chavas, and John R. Stoll. 1986. "Specification of the Logit Model: The Case of Valuation of Nonmarket Goods," *Journal of Environmental Economics and Management* vol. 13, pp. 382–390.

Sellar, Christine, John R. Stoll, and Jean-Paul Chavas. 1985. "Validation of Empirical Measures of Welfare Change: A Comparison of Nonmarket Techniques," *Land Economics* vol. 61, no. 2, pp. 156–175.

Sen, Amartya K. 1977. "Rational Fools: A Critique of the Behavioral Foundations of Economic Theory," *Philosophy and Public Affairs* vol. 6, pp. 317–344.

Sen, Amartya K. 1986. "Social Choice Theory," in Kenneth J. Arrow and Michael D. Intrilligator, eds., *Handbook of Mathematical Economics,* vol. 3 (Amsterdam, North-Holland).

Seneca, Joseph T., and Michael K. Taussig. 1974. *Environmental Economics* (Englewood Cliffs, N.J., Prentice-Hall).

Shanks, J. Merrill, W. Michael Denny, J. Stephen Hendricks, and Richard A. Brody. 1981. "Citizen Reasoning About Public Issues and Policy Trade-Offs: A Progress Report on Computer Assisted Political Surveys," manuscript, Survey Research Center, University of California, Berkeley.

Shapiro, H. T. 1972. "The Index of Consumer Sentiment and Economic Forecasting: A Reappraisal," in Burkhard Strumpel, James N. Morgan, and Ernest Zahn, eds., *Human Behavior in Economic Affairs: Essays in Honor of George Katona* (San Francisco, Jossey-Bass).

Sheatsley, Paul B. 1983. "Questionnaire Construction and Item Writing," in Peter H. Rossi, James D. Wright, and Andy B. Anderson, eds., *Handbook of Survey Research* (New York, Academic Press).

Shubik, Martin. 1970. "Game Theory, Behavior, and the Paradox of the Prisoner's Dilemma: Three Solutions," *Journal of Conflict Resolution* vol. 14, pp. 181–193.

Shubik, Martin. 1981. "Game Theory Models and Methods," in Kenneth J. Arrow and Michael D. Intrilligator, eds., *Handbook of Mathematical Economics,* vol. 1 (Amsterdam, North-Holland).

Shubik, Martin. 1982. *Game Theory in the Social Sciences: Concepts and Solutions* (Cambridge, Mass., MIT Press).

Shure, Gerald H., and Robert J. Meeker. 1978. "A Minicomputer System for Multi-Person Computer-Assisted Telephone Interviewing," *Behavior Research Methods and Instrumentation* vol. 10, pp. 196–202.

Siegel, Sidney. 1956. *Nonparametric Statistics for the Behavioral Sciences* (New York, McGraw-Hill).

Silk, Alvin J., and Glenn L. Urban. 1978. "Pre-test-market Evaluation of New Packaged Goods: A Model in Measurement Methodology," *Journal of Market Research* vol. 15, pp. 171–191.

Silverberg, Eugene. 1978. *The Structure of Economics: A Mathematical Analysis* (New York, McGraw-Hill).

Silvey, S. D. 1980. *Optimal Design* (London, Chapman and Hall).

Simon, Herbert A. 1957. *Models of Man: Social and Rational* (New York, Wiley).

Simon, Herbert A. 1982. *Models of Bounded Rationality* (Cambridge, Mass., MIT Press).

Sinclair, William F. 1976. "The Economic and Social Impact of the Kemano II Hydroelectric Project on British Columbia's Fisheries Resources," report to the Fisheries and Marine Service, Department of the Environment, Vancouver.

Sinden, John A. 1974. "Valuation of Recreation and Aesthetic Experiences," *American Journal of Agricultural Economics* vol. 56, pp. 64–72.

Sinden, John A., and Albert C. Worrell. 1979. *Unpriced Values: Decisions Without Market Prices* (New York, Wiley-Interscience).

Sinden, John A., and J. Wyckoff. 1976. "Indifference Mapping: An Empirical Methodology for Economic Evaluation of the Environment," *Regional Science and Urban Economics* vol. 6, pp. 81–103.

Slovic, Paul. 1969. "Differential Effects of Real versus Hypothetical Payoffs on Choices Among Gambles," *Journal of Experimental Psychology* vol. 80, no. 3, pp. 434–437.

Slovic, Paul, Baruch Fischhoff, and Sarah Lichtenstein. 1980. "Facts versus Fears: Understanding Perceived Risk," in W. A. Albers, ed., *Societal Risk Assessment: How Safe Is Enough?* (New York, Plenum).

Slovic, Paul, Baruch Fischhoff, and Sarah Lichtenstein. 1982. "Response Mode, Framing, and Information-Processing Effects in Risk Assessment," in Robin M. Hogarth, ed., *Question Framing and Response Consistency* (San Francisco, Jossey-Bass).

Slovic, Paul, and Sarah Lichtenstein. 1971. "Comparison of Bayesian and Regression Approaches to the Study of Information Processing in Judgement," *Organizational Behavior and Human Performance* vol. 6, pp. 649–744.

Smith, N. 1980. "A Comparison of the Travel Cost and Contingent Valuation Methods of Recreation Valuation at Cullaby Lake County Park" (M.A. thesis, Oregon State University).

Smith, Tom W. 1983. "The Hidden 25 Percent: An Analysis of Nonresponse on the 1980 General Social Survey," *Public Opinion Quarterly* vol. 47, no. 3, pp. 386–404.

Smith, Tom W. 1984. "Nonattitudes: A Review and Evaluation," in Charles F. Turner and Elizabeth Martin, eds., *Surveying Subjective Phenomena,* vol. 2 (New York, Russell Sage Foundation).

Smith, V. Kerry. 1983. "Option Value: A Conceptual Overview," *Southern Economic Journal* vol. 49, pp. 654–668.

Smith, V. Kerry, ed. 1984a. *Environmental Policy Under Reagan's Executive Order: The Role of Benefit-Cost Analysis* (Chapel Hill, University of North Carolina Press).

Smith, V. Kerry. 1984b. "Some Issues in Discrete Response Contingent Valuation Studies," *Northeastern Journal of Agriculture and Resource Economics* vol. 14, pp. 1–4.

Smith, V. Kerry. 1984c. "A Bound for Option Value," *Land Economics* vol. 60, no. 3, pp. 292–296.

Smith, V. Kerry, 1986a. "Intrinsic Value in Benefit Cost Analysis," draft manuscript, Department of Economics, Vanderbilt University.

Smith, V. Kerry. 1986b. "To Keep or Toss the Contingent Valuation Method," in Ronald G. Cummings, David S. Brookshire, and William D. Schulze, eds., *Valuing Environmental Goods* (Totawa, N.J., Rowman and Allanheld).

Smith, V. Kerry. 1987a. "Nonuse Values in Benefit Cost Analysis," *Southern Economic Journal* vol. 51, pp. 19–26.

Smith, V. Kerry. 1987b. "Uncertainty, Benefit-Cost Analysis, and the Treatment of Option Value," *Journal of Environmental Economics and Management* vol. 14, pp. 283–292.

Smith, V. Kerry, and William H. Desvousges. 1985. "The Generalized Travel Cost Model and Water Quality Benefits: A Reconsideration," *Southern Journal of Economics* vol. 52, pp. 371–381.

Smith, V. Kerry, and William H. Desvousges. 1986a. "The Value of Avoiding a LULU: Hazardous Waste Disposal Sites," *Review of Economics and Statistics* vol. 78, no. 2, pp. 293–299.

Smith, V. Kerry, and William H. Desvousges. 1986b. *Measuring Water Quality Benefits* (Boston, Kluwer-Nijhoff).

Smith, V. Kerry, and William H. Desvousges. 1987. "An Empirical Analysis of the Economic Value of Risk Changes," *Journal of Political Economy* vol. 95, no. 1, pp. 89–114.

Smith, V. Kerry, William H. Desvousges, and Ann Fisher. 1983. "Estimates of the Option Values for Water Quality Improvements," *Economics Letters* vol. 13, pp. 81–86.

Smith, V. Kerry, William H. Desvousges, and Ann Fisher. 1986. "A Comparison of Direct and Indirect Methods for Estimating Environmental Benefits," *American Journal of Agricultural Economics* vol. 68, no. 2, pp. 280–290.

Smith, V. Kerry, William H. Desvousges, and A. Myrick Freeman III. 1985. "Valuing Changes in Hazardous Waste Risks: A Contingent Valuation Approach," draft report to the U.S. Environmental Protection Agency, Research Triangle Institute, N.C.

Smith, V. Kerry, and John V. Krutilla. 1982. "Toward Reformulating the Role of

Natural Resources in Economic Models," in V. Kerry Smith and John V. Krutilla, eds., *Explorations in Natural Resource Economics* (Baltimore, The Johns Hopkins University Press for Resources for the Future).

Smith, Vernon L. 1977. "The Principle of Unanimity and Voluntary Consent in Social Choice," *Journal of Political Economy* vol. 85, no. 6, pp. 1125–1139.

Smith, Vernon L. 1979. "Incentive Compatible Experimental Processes for the Provision of Public Goods," in Vernon L. Smith, ed., *Research in Experimental Economics,* vol. 1 (Greenwich, Conn., JAI Press).

Smith, Vernon L. 1980. "Experiments with a Decentralized Mechanism for Public Good Decisions," *American Economic Review* vol. 70, no. 4, pp. 584–599.

Smith, Vernon L. 1986. "Comments," in Ronald G. Cummings, David S. Brookshire, and William D. Schulze, eds., *Valuing Environmental Goods* (Totawa, N.J., Rowman and Allanheld).

Smith, Vernon L., Arlington W. Williams, W. Kenneth Bratton, and Michael G. Vannoni. 1982. "Competitive Market Institutions: Double Auction vs. Sealed Bid Auctions," *American Economic Review* vol. 72, pp. 58–77.

Sonquist, J. A., E. L. Baker, and J. N. Morgan. 1974. *Searching for Structure* (Ann Arbor, Institute for Social Research, University of Michigan).

Sonstelie, Jon C. 1982. "The Welfare Cost of Free Public Schools," *Journal of Political Economy* vol. 90, no. 4, pp. 795–808.

Sonstelie, Jon C., and Paul R. Portney. 1980. "Take the Money and Run: A Theory of Voting in Local Referenda," *Journal of Urban Economics* vol. 8, pp. 187–195.

Sorg, Cindy F. 1982. "Valuing Increments and Decrements of Wildlife Resources: Further Evidence" (M.A. thesis, University of Wyoming).

Sorg, Cindy F., and David S. Brookshire. 1984. "Valuing Increments and Decrements of Wildlife Resources—Further Evidence," report to the Rocky Mountain Forest and Range Experiment Station, U.S. Forest Service, Fort Collins, Colo.

Sorg, Cindy F., John B. Loomis, Dennis M. Donnelly, George L. Peterson, and Louis J. Nelson. 1985. "Net Economic Value of Cold and Warm Water Fishing in Idaho," Resources Bulletin RM-11, Rocky Mountain Forest and Range Experiment Station, U.S. Forest Service, Fort Collins, Colo.

Sorg, Cindy F., and Louis J. Nelson. 1986. "Net Economic Value of Elk Hunting in Idaho," Resource Bulletin RM-12, Rocky Mountain Forest and Range Experiment Station, U.S. Forest Service, Fort Collins, Colo.

Spaulding, Irving A. n.d. "Factors Influencing Willingness to Pay for Use of Marine Recreational Facilities: Sand Beach," Marine Technical Report no. 51, University of Rhode Island.

Spofford, Walter O., Jr. 1982. "The Development of Methods for Assigning Water Pollution Control Benefits to Individual Discharge Regulations," manuscript, Resources for the Future, Washington, D.C.

Starrett, D. A. 1972. "Fundamental Non-Convexities in the Theory of Externalities," *Journal of Economic Theory* vol. 4, pp. 180–199.

Stephens, Susan A., and John W. Hall. 1983. "Measuring Local Policy Options: Question Order and Question Wording Effects." Paper presented at the Annual Conference of the American Association for Public Opinion Research, Buck Hill Falls, Pa., May.

Stigler, G. J. 1974. "Free Riders and Collective Action," *Bell Journal of Economics* vol. 5, pp. 359–365.

Stigler, S. M. 1977. "Do Robust Estimators Work with Real Data?" *Annals of Statistics* vol. 5, pp. 1055–1098.

Stiglitz, Joseph E. 1977. "The Theory of Local Public Goods," in Martin S. Feldstein and Robert P. Inman, eds., *The Economics of Public Services* (New York, Macmillan).

Stiglitz, Joseph E. 1986. *Economics of the Public Sector* (New York, Norton).

Stinchcombe, Arthur, Calvin Jones, and Paul B. Sheatsley. 1981. "Nonresponse Bias for Attitude Questions," *Public Opinion Quarterly* vol. 45, pp. 359–375.

Stoker, Thomas M. 1985. "Bounds on Welfare Measures," in R. L. Basmann and George F. Rhodes, Jr., eds., *Advances in Econometrics,* vol. 4 (Greenwich, Conn., JAI Press).

Stoll, John R. 1983. "Recreational Activities and Nonmarket Valuation: The Conceptualization Issue," *Southern Journal of Agricultural Economics* vol. 15, pp. 119–125.

Stoll, John R., and Lee Ann Johnson. 1985. "Concepts of Value, Nonmarket Valuation, and the Case of the Whooping Crane," Texas Agricultural Experiment Station Article no. 19360, Department of Agricultural Economics, Texas A&M University.

Stouffer, S. A., and A. A. Lumsdaine. 1949. *The American Soldier: Combat and Its Aftermath,* vol. 2 (Princeton, Princeton University Press).

Strauss, Robert P., and G. David Hughes. 1976. "A New Approach to the Demand for Public Goods," *Journal of Public Economics* vol. 6, no. 3, pp. 191–204.

Stryker, Sheldon, and Anne Statham. 1985. "Symbolic Interaction and Role Theory," in Gardner Lindzey and Elliot Aronson, eds., *Handbook of Social Psychology,* vol. 1 (New York, Random House).

Stynes, D. J., George L. Peterson, and D. H. Rosenthal. 1986. "Log Transformation Bias in Estimating Travel Cost Models," *Land Economics* vol. 62, pp. 84–103.

Subramanian, Shankar, and Richard T. Carson. 1984. "Robust Regression: A Review and Synthesis of Developments Involving Distribution Theory and Non-Spherical Errors." Paper presented at the Econometric Society Winter Meeting, Dallas.

Sudman, Seymour. 1976. *Applied Sampling* (New York, Academic Press).

Sudman, Seymour, and Norman M. Bradburn. 1982. *Asking Questions: A Practical Guide to Questionnaire Design* (San Francisco, Jossey-Bass).

Sugden, Robert T. 1982. "On the Economics of Philanthropy," *Economic Journal* vol. 92, pp. 341–350.

Sugden, Robert T. 1986. *The Economics of Rights, Co-operation, and Welfare* (Oxford, Basil Blackwell).

Sutherland, Ronald J. 1983. "The 'Cost' of Recreational Travel Time," report no. LA-UR-83-175, Los Alamos National Laboratory.

Sutherland, Ronald J., and Richard G. Walsh. 1985. "Effect of Distance on the

Preservation Value of Water Quality," *Land Economics* vol. 61, no. 3, pp. 281–291.

Takayama, A. 1982. "On Consumer's Surplus," *Economics Letters* vol. 10, pp. 35–42.

Talhelm, Daniel R. 1983. "Unrevealed Extra Market Values: Values Outside the Normal Range of Consumer Choices," in Robert D. Rowe and Lauraine G. Chestnut, eds., *Managing Air Quality and Scenic Resources at National Parks and Wilderness Areas* (Boulder, Colo., Westview Press).

Tallis, G. M., and S. S. Y. Young. 1962. "Maximum Likelihood Estimation of the Parameters of Normal, Log-Normal, Truncated Normal, and Bivariate Normal Distributions from Grouped Data," *Australian Journal of Statistics* vol. 4, pp. 49–54.

Tauber, Edward M. 1973. "Reduce New Product Failures: Measure Needs as Well as Purchase Interest," *Journal of Marketing* vol. 37, pp. 61–64.

Taylor, Michael. 1976. *Anarchy and Cooperation* (New York, Wiley).

Thaler, Richard H. 1981. "Some Empirical Evidence on Dynamic Inconsistence," *Economics Letters* vol. 8, pp. 201–207.

Thaler, Richard H., and Sherwin Rosen. 1976. "The Value of Saving a Life," in Nestor E. Terleckyj, ed., *Household Production and Consumption* (New York, National Bureau of Economic Research).

Thayer, Mark A. 1981. "Contingent Valuation Techniques for Assessing Environmental Impacts: Further Evidence," *Journal of Environmental Economics and Management* vol. 8, pp. 27–44.

Theil, Henri. 1971. *Principles of Econometrics* (New York, Wiley).

Throsby, C. D. 1984. "The Measurement of Willingness-to-Pay for Mixed Goods," *Oxford Bulletin of Economics and Statistics* vol. 67, pp. 333–340.

Thurow, Lester C. 1978. "Psychic Income: Useful or Useless?" *American Economic Review* vol. 68, no.. 2, pp. 142–145.

Tideman, T. N. 1983. "An Experiment in the Demand-Revealing Process," *Public Choice* vol. 41, no. 3, pp. 387–401.

Tiebout, T. 1956. "A Pure Theory of Local Expenditures," *Journal of Political Economy* vol. 64, pp. 416–424.

Tietenberg, Tom. 1984. *Environmental and Natural Resource Economics* (Glenview, Ill., Scott, Foresman).

Tietenberg, Tom. 1985. *Emissions Trading: An Exercise in Reforming Pollution Policy* (Washington, D.C., Resources for the Future).

Tihansky, Dennis. 1975. "A Survey of Empirical Benefit Studies," in Henry M. Peskin and Eugene P. Seskin, eds., *Cost-Benefit Analysis and Water Pollution Policy* (Washington, D.C., Urban Institute).

Tolley, George S., and Lyndon Babcock. 1986. "Valuation of Reductions in Human Health Symptoms and Risks," University of Chicago, final report to the Office of Policy Analysis, U.S. Environmental Protection Agency.

Tolley, George S., and Alan Randall, with G. Blomquist, R. Fabian, G. Fishelson, A. Frankel, J. Hoehn, R. Krumm, and E. Mensah. 1983. "Establishing and Valuing the Effects of Improved Visibility in the Eastern United States," interim report to the U.S. Environmental Protection Agency.

Tolley, George S., and Alan Randall, with G. Blomquist, M. Brien, R. Fabian, M. Grenchik, G. Fishelson, A. Frankel, J. Hoehn, A. Kelly, R. Krumm, E. Mensah, and T. Smith. 1985. "Establishing and Valuing the Effects of Improved Visibility in the Eastern United States," final report to the U.S. Environmental Protection Agency.

Toro-Vizcarrondo, Carlos, and T. D. Wallace. 1968. "A Test of the Mean Square Error Criterion for Restrictions in Linear Regression," *Journal of the American Statistical Association* vol. 63, no. 322, pp. 558–572.

Torrance, George W., Michael H. Boyle, and Sargent P. Horwood. 1982. "Application of Multi-Attribute Utility Theory to Measure Social Preferences for Health States," *Operations Research* vol. 30, no. 6, pp. 1043–1069.

Tourangeau, Roger, Kenneth A. Rasinski, Robert P. Abelson, Roy D'Andrade, and Norman Bradburn. 1985. "Cognitive Aspects of Survey Responding: Attitudes and Explanations," preliminary report on pilot studies, National Opinion Research Center, Chicago.

Tsutakawa, Robert K. 1980. "Selection of Dose Levels for Estimating a Percentage Point of a Logistic Quantal Response Curve," *Applied Statistics* vol. 29, pp. 25–33.

Tull, Donald S., and Del I. Hawkins. 1984. *Marketing Research: Measurement and Method* (3d ed., New York, Macmillan).

Turner, A. 1972. "The San Jose Methods Test of Known Crime Victims," National Criminal Justice Information and Statistical Service, Washington, D.C.

Turner, Charles F. 1984. "Why Do Surveys Disagree? Some Preliminary Hypotheses and Some Disagreeable Examples," in Charles F. Turner and Elizabeth Martin, eds., *Surveying Subjective Phenomena,* vol. 2 (New York, Russell Sage Foundation).

Turner, Charles F., and Elizabeth Martin, eds. 1984. *Surveying Subjective Phenomena,* 2 vols. (New York, Russell Sage Foundation).

Tversky, Amos, and Daniel Kahneman. 1973. "Availability: A Heuristic for Judging Frequency and Probability," *Cognitive Psychology* vol. 5, pp. 207–232.

Tversky, Amos, and Daniel Kahneman. 1974. "Judgement Under Uncertainty: Heuristics and Biases," *Science* vol. 185, pp. 1124–1131.

Tversky, Amos, and Daniel Kahneman. 1982. "The Framing of Decisions and the Psychology of Choice," in Robin M. Hogarth, ed., *Question Framing and Response Consistency* (San Francisco, Jossey-Bass).

Tyrrell, T. J. 1982. "Estimating the Demand for Public Recreation Areas: A Combined Travel Cost-Hypothetical Valuation Approach," Working Paper no. 11, Department of Resource Economics, University of Rhode Island, Kingston.

van der Zouwen, Johannes, and Wil Dijkstra. 1982. "Conclusions," in Wil Dijkstra and Johannes van der Zouwen, eds., *Response Behavior in the Survey-Interview* (New York, Academic Press).

Varian, Hal R. 1984. *Microeconomic Analysis* (2d ed., New York, Norton).

Vartia, Yrgo. 1983. "Efficient Methods of Measuring Welfare Change and Compensated Income in Terms of Ordinary Demand Functions," *Econometrica* vol. 51, pp. 79–98.

Vaughan, William J., and Clifford S. Russell. 1982. *Freshwater Recreational Fish-*

ing: The National Benefits of Water Pollution Control (Washington, D.C., Resources for the Future).

Vaughan, William J., Clifford S. Russell, and Richard T. Carson. 1981. "A Survey of Recreation Fee Fishing Enterprises," *Farm Pond Harvest* vol. 15, no. 4.

Vaughan, William J., John Mullahy, Julie A. Hewitt, Michael Hazilla, and Clifford S. Russell. 1985. "Aggregation Problems in Benefit Estimation: A Simulation Approach," report to the U.S. Environmental Protection Agency by Resources for the Future.

Vickrey, W. S. 1961. "Counterspeculation, Auctions, and Competitive Sealed Tenders," *Journal of Finance* vol. 16, pp. 8–37.

Viladus, Joseph. 1973. *The American People and Their Environment,* vol. 1 (Washington, D.C., U.S. Environmental Protection Agency).

Vinokur-Kaplan, Diane. 1978. "To Have—or Not to Have—Another Child: Family Planning Attitudes, Intentions, and Behavior," *Journal of Applied Social Psychology* vol. 8, no. 1, pp. 29–46.

Walbert, M. S. 1984. "Valuing Policies which Reduce Environmental Risk: An Assessment of the Contingent Valuation Method" (Ph.D. dissertation, University of New Mexico, Albuquerque).

Walsh, Richard G. 1986. "Comparison of Pine Beetle Control Values in Colorado Using the Travel Cost and Contingent Valuation Methods." Paper presented at the U.S. Department of Agriculture Conference on Research Issues in Resource Decisions Involving Marketed and Nonmarketed Goods, San Diego, February.

Walsh, Richard G., and Lynde O. Gilliam. 1982. "Benefits of Wilderness Expansion with Excess Demand for Indian Peaks," *Western Journal of Agricultural Economics* vol. 7, pp. 1–12.

Walsh, Richard G., John B. Loomis, and Richard A. Gillman. 1984. "Valuing Option, Existence, and Bequest Demands for Wilderness," *Land Economics* vol. 60, no. 1, pp. 14–29.

Walsh, Richard G., Nicole P. Miller, and Lynde O. Gilliam. 1983. "Congestion and Willingness to Pay for Expansion of Skiing Capacity," *Land Economics* vol. 59, no. 2, pp. 195–210.

Walsh, Richard G., Larry D. Sanders, and John B. Loomis. 1985. *Wild and Scenic River Economics: Recreation Use and Preservation Values,* report to the American Wilderness Alliance (Department of Agriculture and Natural Resource Economics, Colorado State University).

Walsh, Richard G., R. K. Ericson, J. R. McKean, and R. A. Young. 1978. "Recreation Benefits of Water Quality, Rocky Mountain National Park, South Platte River Basin, Colorado," Technical Report no. 12, Colorado Water Resources Research Institute, Colorado State University, Fort Collins.

Water Resources Council. 1979. "Procedures for Evaluation of National Economic Development (NED): Benefits and Costs in Water Resources Planning (Level C), Final Rule," *Federal Register* vol. 44, no. 242 (December 14), pp. 72892–977.

Water Resources Council. 1983. *Principles and Guidelines for Water and Related Land Resources Implementation Studies* (Washington, D.C.).

Wegge, Thomas C., W. Michael Hanemann, and Ivar E. Strand. 1985. "An

Economic Analysis of Recreational Fishing in Southern California," report to the National Marine Fisheries Service.

Weigel, R. H., D. T. A. Vernon, and L. N. Tognacci. 1974. "The Specificity of the Attitude as a Determinant of Attitude-Behavior Congruence," *Journal of Personality and Social Psychology* vol. 30, pp. 724–728.

Weisbrod, Burton A. 1964. "Collective Consumption Services of Individual-Consumption Goods," *Quarterly Journal of Economics* vol. 78, no. 3, pp. 471–477.

Welle, Patrick G. 1985. "Potential Economic Impacts of Acid Rain in Minnesota: The Minnesota Acid Rain Survey," paper prepared for the Minnesota Pollution Control Agency.

Wellman, J. D., E. G. Hawk, J. W. Roggenbuck, and G. J. Buhyoff. 1980. "Mailed Questionnaire Surveys and the Reluctant Respondent: An Empirical Examination of Differences Between Early and Late Respondents," *Journal of Leisure Research* vol. 12, no. 2, pp. 164–172.

Welsh, Michael P. 1986. "Exploring the Accuracy of the Contingent Valuation Method: Comparisons with Simulated Markets" (Ph.D. dissertation, University of Wisconsin, Madison).

Wen, Frank, William Easter II, and Theodore Graham-Tomasi. 1987. "External Damage Cost from Soil Erosion on the Upper-Lower Mississippi," Staff Report, Department of Agricultural and Applied Economics, University of Minnesota.

Whipple, Chris, and Vincent T. Covello, eds. 1985. *Risk Analysis in the Private Sector* (New York, Plenum).

Whittemore, Alice. 1981. "Sample Size for Logistic Regression with Small Response Probability," *Journal of the American Statistical Association* vol. 76, pp. 27–32.

Whittington, Dale, John Briscoe, Mu Shinming, William Barron, and Tom Bourgeois. 1986. "Estimating the Willingness to Pay for Water Services in Developing Countries: A Case Study of the Use of Contingent Valuation Surveys in Southern Haiti." Paper presented at the Regional Science Association Annual Meeting, Columbus, Ohio.

Wicksell, K. 1967. "A New Principle of Just Taxation," in R. A. Musgrave and A. T. Peacock, eds., *Classics in the Theory of Public Finance* (New York, St. Martin's).

Wildavsky, Aaron B. 1964. *The Politics of the Budgetary Process* (Boston, Little, Brown).

Williams, Bill. 1978. *A Sampler on Sampling* (New York, John Wiley and Sons).

Willig, Robert D. 1973. "Consumer's Surplus: A Rigorous Cookbook," Technical Report no. 98, Institute for Mathematical Studies in the Social Sciences, Stanford University.

Willig, Robert D. 1976. "Consumer's Surplus Without Apology," *American Economic Review* vol. 66, no. 4, pp. 587–597.

Willis, C., and J. Foster. 1983. "The Hedonic Approach: No Panacea for Valuing Water Quality Changes," *Journal of the Northeastern Agricultural Council* vol. 12, pp. 53–56.

Wilman, Elizabeth A. 1980. "The Value of Time in Recreation Benefit Studies,"

Journal of Environmental Economics and Management vol. 7, pp. 272–286.

Wilson, James Q., and Edward Banfield. 1964. "Public Regardingness as a Value Premise in Voting Behavior," *American Political Science Review* vol. 4, 876–887.

Wilson, James Q., and Edward Banfield. 1965. "Voting Behavior on Municipal Public Expenditures: A Study in Rationality and Self-Interest," in Julius Margolis, ed., *The Public Economy of Urban Communities* (Baltimore, The Johns Hopkins University Press).

Wilson, L. A. 1981. "Citizen Preferences for Public Expenditures: A Survey of Opinion in Tempe, Arizona," draft manuscript, Center for Public Affairs, Arizona State University.

Winerip, Michael. 1986. " 'Mr. Garbage' and His Million-Dollar Offer," *New York Times,* May 6.

Winter, S. G. 1969. "A Simple Remark on the Second Optimality Theorem of Welfare Economics," *Journal of Economic Theory* vol. 1, pp. 99–103.

Wyckoff, J. B. 1971. "Measuring Intangible Benefits—Some Needed Research," *Water Resources Bulletin* vol. 7, no. 1, pp. 11–16.

Wyer, Robert S., Jr., and Thomas K. Srull. 1981. "Category Accessibility: Some Theoretical and Empirical Issues Concerning the Processing of Social Stimulus Information," in E. Tory Higgins, C. Peter Herman, and Mark P. Zanna, eds., *Social Cognition: The Ontario Symposium* (Hillsdale, N.J., L. Erlbaum Associates).

Yang, E. J., R. C. Dower, and M. Menefee. 1984. "The Use of Economic Analysis in Valuing Natural Resource Damages," report to the U.S. Department of Commerce by the Environmental Law Institute, Washington, D.C.

Yates, F. 1980. *Sampling Methods for Censuses and Surveys* (4th ed., London, Griffin).

Young, H. P., ed. 1985. *Cost Allocation: Methods, Principles, Applications* (Amsterdam, North-Holland).

Zeckhauser, Richard. 1973. "Voting Systems, Honest Preferences, and Pareto Optimality," *American Political Science Review* vol. 67, pp. 934–946.

Zeller, Richard A., and Edward G. Carmines. 1980. *Measurement in the Social Sciences: The Link Between Theory and Data* (New York, Cambridge University Press).

Ziemer, Rod F., Welsley N. Musser, Fred C. White, and R. Carter Hill. 1982. "Sample Selection Bias in Analysis of Consumer Choice: An Application to Warmwater Fishing Demand," *Water Resources Research* vol. 18, no. 2, pp. 215–219.

Name Index

Subject Index